Publishing Addiction Science: A Guide for the Perplexed

Third Edition

Edited by
Thomas F. Babor, Kerstin Stenius,
Richard Pates, Michal Miovský,
Jean O'Reilly and Paul Candon

Published by
Ubiquity Press Ltd.
6 Windmill Street
London W1T 2JB
www.ubiquitypress.com

First published 2017

Cover design by Amber MacKay, developed from 2nd Edition cover by
Matthew West of Vasco Graphics. Images used in the cover design were
sourced from Pixabay and are licensed under CC0 Public Domain.
Reproductions of journal covers and logos used with permission.

Printed in the UK by Lightning Source Ltd.
Print and digital versions typeset by Siliconchips Services Ltd.

ISBN (Paperback): 978-1-911529-08-8
ISBN (PDF): 978-1-911529-09-5
ISBN (EPUB): 978-1-911529-10-1
ISBN (Mobi): 978-1-911529-11-8

DOI: https://doi.org/10.5334/bbd

The full text of this book has been peer-reviewed to ensure high academic
standards. For full review policies, see http://www.ubiquitypress.com/

Suggested citation:
Babor, T F, Stenius, K, Pates, R, Miovský, M, O'Reilly, J and Candon, P (eds.)
2017 *Publishing Addiction Science: A Guide for the Perplexed.* London:
Ubiquity Press. DOI: https://doi.org/10.5334/bbd. License: CC-BY 4.0

To read the free, open access version of this
book online, visit https://doi.org/10.5334/bbd
or scan this QR code with your mobile device:

*We dedicate this book to two people
who changed addiction science for the better:
Lenka Čablová (1986–2016) and Griffith Edwards (1928–2012)*

Contents

Foreword to the Third Edition

The health and social burden attributable to psychoactive substance use is enormous. Alcohol, tobacco and illicit drug use taken together are by far the most important preventable risk factors to a population's health. According to the latest WHO estimates, the harmful use of alcohol alone results in around 3.3 million deaths every year. With rapid social and cultural changes taking place in many countries, alcohol and drug use are becoming increasingly embedded in social matrices, often with strong commercial forces playing a role in promoting the use of legal intoxicating and dependence-producing substances. A number of jurisdictions have undertaken major changes in the regulation of psychoactive substances controlled under international drug treaties. New Psychoactive Substances (NPS), with their health effects and distribution channels, present new challenges for public health authorities. Debates around alcohol and drugs are at the forefront of social policy processes in many countries, with significant variations in societal responses. Unfortunately, these debates are often not based on solid data or research evidence, and in many cases the relevant data simply does not exist. Significant caveats exist in the evaluation of existing policy responses and policy changes made in different jurisdictions. There is an urgent need to strengthen the evidence base for the development of adequate program and policy responses to substance use and substance use disorders at different levels.

It is difficult to overestimate the role of research and scientific data in shaping policy and program responses at all scales, from local communities to the international level. A consistent and common issue is the lack of sufficient resources for research on substance use and substance use disorders, and very often even those resources available are not utilized to their maximum potential. One of the biggest problems is when investment in research does not result in the publication and dissemination of results, preferably in peer-reviewed journals. This is a particularly prevalent issue in less-resourced countries where opportunities for publishing results of research on substance use and substance use disorders are limited, and where no specialized journals on addiction exist.

The third edition of *Publishing Addiction Science: A Guide for the Perplexed* is an important resource for researchers around the world, especially for those who work in low and middle-income countries. It is hoped that this resource will facilitate the dissemination of new data and knowledge in this area, given that research remains very much skewed towards a limited number of high-income countries with well-developed research and publishing infrastructures. The International Society of Addiction Journal Editors (ISAJE) continues to work towards increasing the publishing competence of researchers from all over the world, with this work often being implemented in consultation with our program in the World Health Organization. Such efforts make a significant and much needed contribution to capacity building in research on substance use and substance use disorders, particularly in less-resourced countries, and the WHO Department of Mental Health and Substance Abuse welcomes the third edition. We look forward to continued collaboration with ISAJE in this area.

Dr Shekhar Saxena
Director
Department of Mental Health and Substance Abuse
World Health Organization

Preface

An Idea Whose Time Has Come

The development of this book had many complex motives but a single purpose. The motives include improving scientific integrity in the field of addiction studies, sharing information with junior investigators, and strengthening addiction specialty journals. The single purpose of this volume, however, is to provide a practical guide to scientific publishing in the addiction field that is used often enough to affect personal decisions, individual careers, institutional policies, and the progress of science. The time is ripe for such an ambitious undertaking: The field of addiction research has grown tremendously in recent years and has spread to new parts of the world. With that growth has come a concomitant increase in competition among researchers, new bureaucratic regulations, and a growing interest in addiction research by health agencies, policy-makers, treatment and prevention specialists, and the alcohol industry. New professional societies, research centers, and university programs have taken root, and regulatory responsibilities such as conflict of interest declarations, human and animal subjects assurances, and the monitoring of scientific misconduct are now common.

The journal-publishing enterprise, the main organ of scientific communication in the field, has an important role to play in all of these developments, and the third edition of *Publishing Addiction Science* is designed to meet this need. The inspiration for the first edition of this volume came from the International Society of Addiction Journal Editors (ISAJE), which is not only the first society for addiction journal editors, it is also the first international organization specifically devoted to the improvement of scientific publishing in the addiction field.

From its inception, ISAJE has recognized a need for ethical guidelines for member journals. There are several reasons why ethical issues are particularly important in the addiction science field. Strong industries, such as pharmaceutical manufacturers, tobacco companies, and alcohol producers, have important financial interests to protect, and they pay special attention to the work of addiction scientists. Further, many addiction-related issues are politically loaded, a situation that could affect the objectivity of researchers. Many of the individuals who are the object of addiction research are vulnerable and in need of special protections. Finally, the field of science has become much more ethically challenging because of its growing importance and complexity. Although ISAJE offers a set of ethical guidelines, abstract policy statements and moral pronouncements are rarely read carefully or applied to the day-to-day business of conducting research and communicating ideas to the scientific community. This book aims to improve transparency in addiction publishing and, in the process, show how young investigators can negotiate the complex and sometimes bewildering ethical challenges faced on the path to a successful career in the field.

Rationale for the Third Edition

There are several reasons why a third edition of *Publishing Addiction Science* is necessary. First, rapid developments in the field of addiction publishing necessitate revisions of parts of this book, particularly the move to online and open-access publication options, the launching of many new addiction specialty journals, and the new ethical and technological challenges facing addiction publishing. For example, more than 30 new journals have been identified since the second edition of the book was published in 2008, many of them launched by for-profit enterprises with little appreciation for scientific quality or peer review.

Another reason for the third edition is related to experience from our *Publishing Addiction Science* workshops, which have been conducted during the past few years in many parts of the world, including Denmark, Finland, Greece, Jordan, Nigeria, South Korea, Uganda, the United Kingdom, and the United States. The workshops identified new areas of interest that needed attention. To make *Publishing Addiction Science* even more relevant to its target market of advanced students and young professionals, the third edition has accordingly added new material on publication issues faced by postdoctoral researchers, the ethical challenges of research funding, how to write a research paper, and procedures for peer-reviewing manuscripts,. The development of new online training material will enable the book to continue to be used as a textbook for research ethics in colleges and universities and in training workshops at scientific meetings.

E-Attachments

e-Attachments are additional supplementary materials that can be used to deepen your understanding of the concepts in *Publishing Addiction Science*. e-Attachments comprise additional information sources, readings, examples and exercises that can improve your skills and help you practice your first steps in the publishing world. You can find 6 different kinds of e-Attachments on our websites: readings, exercises, examples of good practice, simple Power-Point presentations, videos and full e-learning lessons. Some items are used for more than one chapter while others are quite specific to their chapters. For your effective use of the e-Attachments and the book, please follow the instructions on our website.

All e-Attachments are free to download from the website of the International Society of Addiction Journal Editors (ISAJE) on www.isaje.net. e-Attachments will be updated continually.

There are six kinds of e-Attachments, each with a different purpose:

- *Readings* provide additional information about a chapter or issues discussed in more than one chapter. Some of these documents provide more contextual information or are original documents to which the chapter refers.
- *Exercises* are materials for practicing and training. They are appropriate for individual or group application.
- *Examples of good practice* provide a better understanding of topics or themes discussed in the chapters.
- *Simple PowerPoint presentations* are mainly designed for use by teachers and lecturers but students and readers may find them useful as simple e-learning documents that provide well-structured information complementary to the full chapter text.
- *Videos*, like the PowerPoint presentations, provide actual presentations or workshop/training lectures given by the chapter author(s) or one of more of their colleagues from ISAJE.
- *Full e-learning lessons* provide more sophisticated e-leaning support. They combine PowerPoint slides with the full text of a presentation and finish with a knowledge test that lets you check your understanding of the lesson.

Sponsorship/Acknowledgements

The publication costs for revising and reprinting this book were covered by the book's primary sponsors: the international journal *Addiction*, the (U.K.) Society for the Study of Addiction, and the International Society of Addiction Journal Editors (ISAJE). We are also grateful to the academic institutions that enabled the authors and editors to work on this collaborative effort, including The University of Connecticut Alcohol Research Center (Farmington, Connecticut, USA, NIAAA Grant # 5P60AA003510-39), the Nordic Welfare Centre (Helsinki, Finland), the Rutgers University Center of Alcohol Studies (Piscataway, New Jersey, USA), and the Department of Addictology, Charles University in Prague (Czech Republic; Grant No. PRVOUK-P03/LF1/9). A number of individuals provided key contributions to the third edition; in particular, we thank Deborah Talamini and Melissa Feulner.

We also thank sponsors who have provided financial support for developing online supplementary materials, training materials, workshops and translations related to *Publishing Addiction Science*. These include the American Academy of Addiction Psychiatry (AAAP), Charles University in Prague's Department of Addictology, the International Order of Good Templars (IOGT International), the U.S. National Institute on Drug Abuse (NIDA), the Research Society on Alcoholism (RSA), Substance Abuse Librarians & Information Specialists (SALIS), and Wiley.

Finally, we thank an even larger number of organizations that have helped us to disseminate the book's contents and its online materials. These include all organizations mentioned in the two paragraphs above as well as the American Society of Addiction Medicine (ASAM), the College on Problems of Drug Dependence (CPDD), the Nigerian Centre for Research and Information on Substance Abuse (CRISA), the European Monitoring Centre for Drugs and Drug Addiction (EMCDDA), the International Confederation of Addiction Research Associations (ICARA), the International Society for the Study of Drug Policy (ISSDP), and the Kettil Bruun Society (KBS).

Closing

We hope that the third edition of this book will aid the training of young researchers and the continuing education of seasoned addiction scientists around the world. Given the book's continued focus on supporting young scientists who are entering the field and its goal of improving the integrity and ethicality of addiction science, we dedicate this edition of the book to Lenka Čablová (1986–2016) and Griffith Edwards (1928–2012). Lenka was the lead author of Chapter 9. She was a promising young scientist whose short professional life was nevertheless filled with creative work on the interconnections among substance use, ADHD and nutrition, and an overarching concern with addiction and risk to families.

Griffith's career as an addiction scientist, master clinician, research center director, and policy analyst served not only as an inspiration for this third edition
of *Publishing Addiction Science,* but also as a model for the kind of addiction
scientist the book's content would like to inspire.

The Editors

About the Authors

Thomas F. Babor is Professor and Chair, Department of Community Medicine and Health Care, University of Connecticut School of Medicine, USA. He is Editor-in-Chief of the *Journal of Studies on Alcohol and Drugs*.

Robert L. Balster is the Luther A. Butler Professor of Pharmacology and Toxicology and Research Professor of Psychology and Psychiatry at Virginia Commonwealth University in Richmond, Virginia, USA. He is former Editor-in-Chief of the journal *Drug and Alcohol Dependence*.

Gerhard Bühringer is Professor for Addiction Research in the Department of Clinical Psychology and Psychotherapy at the Technische Universität in Dresden, Germany.

Lenka Čablová, now deceased, was a Postdoctoral Research Fellow at the Department of Addictology, First Faculty of Medicine, Charles University in Prague, and the General University Hospital in Prague, Czech Republic.

Paul Candon is Managing Editor of the *Journal of Studies on Alcohol and Drugs*, based at the Center of Alcohol Studies at Rutgers, The State University of New Jersey, USA.

Erikson F. Furtado is a full-time tenured Assistant Professor of Child and Adolescent Psychiatry in the Department of Neurosciences and Behavior in the Faculty of Medicine at Ribeirão Preto, University of São Paulo, Brazil.

Florence Kerr-Corrêa is Professor of Psychiatry, Department of Neurology, Psychology and Psychiatry, Botucatu Medical School, São Paulo State University (UNESP), Brazil.

Roman Gabrhelík is Assistant Professor, Department of Addictology, First Medical Faculty, Charles University in Prague, Czech Republic. He is Executive Editor of the Czech journal *Adiktologie* (Addictology).

Tom Kettunen is Editor of the journal *Nordic Studies on Alcohol and Drugs* (*Nordisk alkohol- & narkotikatidskrift*).

Phil Lange has retired. He was Editor of the *Journal of Gambling Issues*.

Klaus Mäkelä, now deceased, was Research Director of the Finnish Foundation for Alcohol Studies.

Thomas McGovern is Professor Emeritus of Psychiatry and Founder Emeritus of the Center for Ethics/ Humanities/ Spirituality at the School of Medicine, Texas Tech University Health Sciences Center, Lubbock, Texas, USA. He is Editor-in-Chief of *Alcoholism Treatment Quarterly*.

Peter Miller is Professor of Violence Prevention and Addiction Studies at the School of Psychology, Deakin University, Australia. He was also the Commissioning Editor of the journal *Addiction* from 2006–2016.

Michal Miovský is Head of the Department of Addictology, 1st Faculty of Medicine at Charles University in Prague, Czech Republic. He is also a clinical psychologist, psychotherapist and supervisor. He is Deputy Editor-in-Chief of the Czech Journal *Adiktologie* (Addictology) and leads a creative team to establish academic study programs in addictions in Prague.

Andrea L. Mitchell is the Executive Director of Substance Abuse Librarians & Information Specialists (SALIS).

Dominique Morisano is a clinical psychologist and research/evaluation consultant and appointed as Assistant Professor (Status-only), Dalla Lana School of Public Health, University of Toronto; Collaborator Scientist, Institute for Mental Health Policy Research, Centre for Addiction and Mental Health;

Visiting Scholar, Rotterdam School of Management, Erasmus University (Netherlands); and Faculty, Centre for Mindfulness Studies.

Neo Morojele is Deputy Director of the Alcohol, Tobacco & Other Drug Research Unit of the South African Medical Research Council. She is an Associate Editor for the *African Journal of Drug and Alcohol Studies* and the *International Journal of Alcohol and Drug Research* and Associate Editor for Africa of the *Journal of Substance Use*.

Jonathan Noel is a doctoral candidate in the Graduate Program in Public Health, Department of Community Medicine and Health Care, University of Connecticut School of Medicine, USA.

Jean O'Reilly is Editorial Manager for the journal *Addiction* and a consulting book editor.

Isidore Obot is Professor, Department of Psychology, University of Uyo, and Director, Centre for Research and Information on Substance Abuse (CRISA), Uyo, Nigeria. He has been Editor-in-Chief of the *African Journal of Drug and Alcohol Studies* since 2000.

Richard Pates is a consultant clinical psychologist who worked in treatment of addiction problems in the NHS in the UK for 30 years. He now works at a secure children's home. He has been Editor of *The Journal of Substance Use* for the past 16 years. He holds an honorary post at the University of Worcester.

Maria Cristina Pereira Lima is an Associate Professor of Psychiatry, Department of Neurology, Psychology and Psychiatry, Botucatu Medical School, São Paulo State University (UNESP), Brazil.

Katherine Robaina is a researcher at the Department of Community Medicine and Health Care, University of Connecticut School of Medicine, USA and is a member of Delta Omega, the Honorary Society in Public Health (Beta Rho chapter).

Kerstin Stenius is guest professor at the Centre for Social Research on Alcohol and Drugs (SoRAD) at Stockholm University, Sweden. Until 2017 she was Editor-in-Chief of the journal *Nordisk alkohol- & narkotikatidskrift* (Nordic Studies on Alcohol and Drugs) at The Nordic Welfare Centre, Helsinki, Finland.

Ian Stolerman is Emeritus Professor of Behavioural Pharmacology at the Institute of Psychiatry, Psychology and Neuroscience, King's College London, UK.

He served as President of ISAJE and as co-editor of the journal *Drug and Alcohol Dependence*.

Judit H. Ward is Science Reference/Instruction Librarian at Rutgers, The State University of New Jersey. She is Field Editor of the *Journal of Studies on Alcohol and Drugs*.

Robert West is Professor of Health Psychology and Director of Tobacco Studies, Cancer Research UK Health Behaviour Research Centre, Department of Epidemiology and Public Health, University College London, UK. He is Editor-in-Chief of the journal *Addiction*.

Erin L. Winstanley is an Assistant Professor of Health Outcomes at the James L. Winkle College of Pharmacy at the University of Cincinnati and the Director of Health Services Research, Mercy Health, USA.

Supporting Institutions

INTERNATIONAL SOCIETY OF ADDICTION JOURNAL EDITORS

American Academy *of*
Addiction Psychiatry

Translating Science. Transforming Lives.

CRISA
Centre for Research and Information on
Substance Abuse

CHARLES UNIVERSITY
First Faculty of Medicine

Publication was supported from institutional programme of Charles
University No. PRVOUK-P03/LF1/9

European Monitoring Centre
for Drugs and Drug Addiction

International Confederation of ATOD
Research Associations

INTERNATIONAL

THE INTERNATIONAL SOCIETY FOR THE STUDY OF DRUG POLICY

International Program

Nordic Welfare Centre

Research Society on Alcoholism

RUTGERS
Center of Alcohol Studies

SALiS
An International Association

WILEY

SECTION I

Introduction

CHAPTER 1

A Guide for the Perplexed

Thomas F. Babor, Kerstin Stenius and Jean O'Reilly

"I do not presume to think that this treatise settles every doubt in the minds of those who understand it, but I maintain that it settles the greater part of their difficulties."

Maimonides, Guide for the Perplexed (ca. 1190)

To be perplexed is to be puzzled or even confused by the intricacy of a situation. One way to deal with perplexing situations is to find a guide who can provide advice, information, and direction. Many such guides have risen to the occasion throughout the ages, providing useful knowledge for the perplexed students of literature, religion, philosophy, and science. One of the most influential philosophical treatises, for example, was Maimonides' *Guide for the Perplexed*. In a time of religious, moral, and political change, Maimonides (1135–1204) sought to harmonize Greco-Roman, Christian, Jewish, and Arabic thought into a philosophical guide for those seeking meaning in life. In a sense, *Publishing Addiction Science* is intended to be a similar (albeit less ambitious!) guide for those of us who from time to time are perplexed about how to find our way through the complex world of addiction science. The chapters in this book constitute a virtual guide through the practical, scientific, moral, and even philosophical issues with which we must become acquainted if we are to succeed, either as temporary visitors to the field or as career scientists dedicating our lives to the study of addiction.

It is our contention—and a guiding theme of the book—that the key to successful publishing in addiction science is to understand not only how to write a scientific article and where to publish it but also how to do these things honestly and ethically. Therefore, in addition to the practical business of publishing

How to cite this book chapter:
Babor, T F, Stenius, K and O'Reilly, J. 2017. A Guide for the Perplexed. In: Babor, T F,
Stenius, K, Pates, R, Miovský, M, O'Reilly, J and Candon, P. (eds.) *Publishing
Addiction Science: A Guide for the Perplexed*, Pp. 3–8. London: Ubiquity Press.
DOI: https://doi.org/10.5334/bbd.a. License: CC-BY 4.0.

scientific articles in both multi-disciplinary and addiction specialty journals, the ultimate goal of this book is to enhance scientific integrity in the publication process, giving special consideration to the main organ of scientific communication, the scholarly journal.

What is a Journal?

According to Lafollette (1992, p. 69), "a journal is a periodical that an identifiable intellectual community regards as a primary channel for communication of knowledge in its field *and* as one of the arbitrators of the authenticity or legitimacy of that knowledge." Journals establish intellectual standards, provide a forum of communication among scientists, bring valuable information to the public, set the agenda for a field of study, provide an historical record of a particular area of knowledge, and confer implicit certification on authors for the authenticity and originality of their work (Lafollette, 1992). In addition, journals have the potential to serve the interests of career advancement and personal reward for scholarly achievement.

Journals are joint enterprises typically managed through a division of labor among owners, publishers, editors, reviewers, and authors. How this cast of characters is organized into an integrated set of players varies from one journal to another. The owners of a journal can be nonprofit organizations (such as learned societies, universities, or professional organizations), government agencies, or private publishers. The publishers of a journal range from small printers to large-scale, multi-national organizations that distribute often hundreds of journals. Journal editors tend to be appointed by the owners, society officers, or publishers. Editors of some of the larger scientific and medical journals are paid for their services and have full-time staff at their disposal. Editors of smaller journals are generally unpaid and have a small editorial staff with some volunteer assistant editors. Reviewers are usually established investigators who have specialized knowledge of the subject matter. Without remuneration and as a service to the field, reviewers provide critical and often anonymous evaluations of manuscripts written by their peers.

Without journals, addiction science—or any science—would have a limited audience and a short half-life. Therefore, scientists who wish to search for truth and to help humankind must understand the inner workings and current complexities of the journal publication process.

Purpose of the Guide

The addiction field has grown tremendously in the past 35 years, and addiction publishing has been no exception. Currently there are more than 120 journals devoted primarily to the dissemination of scholarly information about addiction and related health problems, and many more journals publish addiction science as part of their broader mission. Despite the growing amount of published material in addiction science and the increasing opportunities for publication, there exists no other guide designed to inform prospective authors about the opportunities, requirements, and challenges of publishing addiction science. Moreover, the addiction field has become perhaps one of the first areas of science in which interdisciplinary collaboration between biomedical and psychosocial researchers is essential to progress (see Edwards, 2002). At the same time, however, as Matilda Hellman (2015) argues, we appear to be moving into an age of academic compartmentalization, with increasingly narrow fields of study in which researchers are encouraged to specialize. It is therefore important that addiction science, a field that is perhaps unfashionably collaborative, has a publishing guide that looks at the field as an inter-related whole rather than as a collection of separate disciplines.

Within this context, the primary purpose of *Publishing Addiction Science* is to advise potential authors of articles in the addiction field of the opportunities for publishing their work in scholarly journals, with an emphasis on addiction specialty journals. Although all prospective authors will find such a guide useful, it should be particularly helpful to students, younger investigators, clinicians, and professional researchers.

The book's broader purpose is to improve the quality of scientific publishing in the addiction field by educating authors about the kinds of ethical and professional issues with which the International Society of Addiction Journal Editors (ISAJE) has long been concerned: scientific misconduct, ethical decision making, the publication process, and the difficulties experienced by authors whose first language is not English.

Guide to the Guide

Publishing Addiction Science is organized into five sections. The first section provides an overview of this book and a chapter ("Infrastructure and Career Opportunities in Addiction Science") describing the development and underlying structure of the field of addiction science.

The second section covers general issues of how and where to publish. The initial overview chapter (Chapter 3, "How to Choose a Journal: Scientific and Practical Considerations") deals with choosing where to submit your article, a very important decision in the publication process. The chapter describes the range of journals that publish articles related to addiction and psychoactive substances; summarizes the growth in addiction journals, including the move into open-access journals; and explains 10 steps to choosing a journal. It also provides two tables containing practical information about 45 addiction specialty journals (e.g., areas of interest, acceptance rates, author fees) to assist authors with the selection of an appropriate journal. The next chapter in this section ("Beyond the Anglo-American World: Advice for Researchers from Developing and Non-English-speaking Countries") describes the practical and professional issues addiction scientists face in countries that are less resourced or in which English is not the main language, how authors who come from these countries can improve their chances of publishing in English-language journals, the possibilities for authors to publish in both English and an additional language so they can communicate with different audiences, and how to decide whether an article may better serve the public by being published in the author's mother tongue. Chapter 5 ("Getting Started: Publication Issues for Graduate Students, Postdoctoral Fellows, and other Aspiring Addiction Scientists") describes the challenges and rewards of publishing early in one's professional career, including authorship issues, timetables, ethical dilemmas, and the pressure to publish. Lastly, Chapter 6 ("Addiction Science for Professionals Working in Clinical Settings") looks at research and publication issues specific to clinicians who work in the field of addiction. It offers advice for identifying types of clinical research that lend themselves to research articles, planning and funding such research, and avoiding common pitfalls in the journey to publication.

The third section provides a detailed guide to the practical side of addiction publishing. Chapter 7 ("How to Write a Scientific Research Article for a Peer-reviewed Journal") describes the development of a typical data-based research article from the planning stage to the completion of the final draft, emphasizing scientific writing techniques, the structure of a scientific article, common reporting guidelines for specific types of articles, effective methods of scientific communication, and resources for improving one's writing. The following chapter ("How to Write Publishable Qualitative Research") explores the differences and commonalities between qualitative and quantitative research, identifies the hallmarks of exemplary qualitative research, and offers practical advice not only for writing a qualitative article but also for getting it published. Chapter 9 ("How to Write a Systematic Review Article and Meta-analysis") provides a step-by-step process for designing, researching, and writing a comprehensive synthesis of existing research—typically a much larger undertaking than a single research article—and describes some of the best databases and guidelines available to authors. Chapter 10 ("Use and Abuse of Citations") describes appropriate and less-appropriate citation practices with recommendations for good behavior and gives a critical appraisal of citation metrics, particularly the journal impact factor, which is used to evaluate the importance attributed to different journals. Chapter 11 ("Coin of the Realm: Practical Procedures for Determining Authorship") deals with the often vexing question of how to assign authorship credits in multi-authored articles. We offer practical recommendations to provide collaborating authors with a process that is open, fair, and ethical. Chapter 12 ("Preparing Manuscripts and Responding to Reviewers' Reports: Inside the Editorial Black Box") focuses on how to negotiate the peer-review process. It describes how the process works and how journal editors make decisions about publishing an article. It also considers editors' criteria for selecting articles and explains how to revise an article when an editor asks for a response to the reviewers' comments. The final chapter in this section ("Reviewing Manuscripts for Scientific Journals") covers the peer-review process, what journal editors expect from reviewers, and how to prepare a constructive critical review.

The fourth section of *Publishing Addiction Science* is devoted to ethical issues. The first article in this section (Chapter 14, "Dante's Inferno: Seven Deadly Sins in Scientific Publishing and How to Avoid Them") reviews seven types of scientific misconduct in the context of a broader definition of scientific integrity. The seven "sins" are carelessness in citing and reviewing the literature, redundant publication, unfair authorship, failure to declare a conflict of interest, failure to conform to minimal standards of protection for animal or human subjects, plagiarism, and scientific fraud. We discuss these ethical improprieties in terms of their relative importance and possible consequences and suggest procedures for avoiding them. Chapter 15 ("The Road to Paradise: Moral Reasoning in Addiction Publishing") discusses the same issues in the context of a framework for making ethical decisions. We use case studies to illustrate the seven ethical topics, with a commentary on each case that demonstrates a practical approach to making sound decisions. Chapter 16 ("Relationships with the Alcoholic Beverage Industry, Pharmaceutical Companies, and Other Funding Agencies: Holy Grail or Poisoned Chalice?") reviews recent trends in the funding of addiction research and the ethical risks involved in accepting funding from industry as well as nonindustry sources.

The fifth and final section contains the book's concluding chapter (Chapter 17: "Addiction Publishing and the Meaning of [Scientific] Life"), in which the editors describe the pursuit of scientific integrity as a journey worth taking, as much for the joy of honest discovery as for the achievement of fame and fortune.

How to Use This Guide Effectively

The authors have collectively striven to present practical advice as well as "best practices." In most cases, such as in resolving authorship disputes or ethical problems, the solutions are not always simple or obvious but rather depend on the situation and on an open dialogue among colleagues. For these cases, we offer advice on how to use effective problem-solving techniques that will allow the reader to develop skills that can be applied to a variety of situations. The authors emphasize that no researcher, no matter how experienced in the game of science, can argue that she or he has all the right answers. This book is best seen as providing a basis for discussions about concrete problems in various research environments.

Although the book's chapters can be read in sequence, each chapter also functions as a self-contained unit and can be downloaded and read separately.

As a result, there is some repetition among chapters, more so that would occur in a book designed to be read from cover to cover, as more than one chapter may discuss similar issues in slightly different ways.

The chapters are also meant for use as background readings for lectures, workshops, and practical exercises that accompany many of the chapters. The ISAJE website (www.isaje.net) contains supplementary readings, exercises, slides, and other materials for each chapter, all free to download.

Recognizing that there are important institutional responsibilities in the ethical conduct of addiction research, we hope that this book will also inspire research institutions to develop guidelines and policies that support the ethical practices considered in these chapters. Although we have subtitled the book as *A Guide for the Perplexed,* we point out that its chapters will be helpful as well to those who believe they have all the answers, including established investigators at professional organizations and scientific institutions.

Please visit the website of the International Society of Addiction Journal Editors (ISAJE) at www.isaje.net to access supplementary materials related to this chapter. Materials include additional reading, exercises, examples, PowerPoint presentations, videos, and e-learning lessons.

References

Edwards, G. (Ed.). (2002). *Addiction: Evolution of a specialist field.* Oxford, UK: Blackwell Publishing.

Hellman, M. (2015). The compartmentalisation of social science: What are the implications? *Nordic Studies on Alcohol and Drugs, 32,* 343–346.

Lafollette, M. C. (1992). *Stealing into print: Fraud, plagiarism and misconduct in scientific publishing.* Berkeley CA: University of California Press.

Maimonides, M. (2004). *The Guide for the Perplexed.* Translated by M. Friedländer [1903] (Barnes & Noble Library of Essential Reading) Paperback.

CHAPTER 2

Infrastructure and Career Opportunities in Addiction Science: The Emergence of an Interdisciplinary Field

Thomas F. Babor, Dominique Morisano,
Jonathan Noel, Katherine Robaina,
Judit H. Ward and Andrea L. Mitchell

Introduction

During the latter part of the 20th century, there was rapid growth in the number of people employed in the societal management of social and medical problems associated with the use of alcohol, tobacco, and illicit drugs (Edwards & Babor, 2012). At the same time, similar growth occurred in the number of institutions and individuals engaged in addiction science. The current worldwide infrastructure of addiction science includes numerous research funding sources, more than 90 specialized scholarly journals, scores of professional societies, over 200 research centers, more than 80 specialty training programs, and thousands of scientists.

The purpose of this chapter is to describe the global infrastructure supporting addiction science and the career opportunities available to addiction scientists. The current global infrastructure is evaluated from two perspectives: (a) its ability to produce basic knowledge about the causes of addiction and the mechanisms by which psychoactive substances affect health and well-being and (b) its ability to address substance-related problems throughout the world at both the individual and the population levels. The first perspective speaks

How to cite this book chapter:
Babor, T F, Morisano, D, Noel, J, Robaina, K, Ward, J H and Mitchell, A L. 2017.
  Infrastructure and Career Opportunities in Addiction Science: The Emergence of
  an Interdisciplinary Field. In: Babor, T F, Stenius, K, Pates, R, Miovský, M, O'Reilly, J
  and Candon, P. (eds.) *Publishing Addiction Science: A Guide for the Perplexed*,
  Pp. 9–34. London: Ubiquity Press. DOI: https://doi.org/10.5334/bbd.b. License:
  CC-BY 4.0.

to the mission of science to produce fundamental knowledge. The second is a public health mission that is often used to justify societal investments in clinical and translational research.

This chapter begins with a discussion of the meaning of addiction science as an interdisciplinary field of study. We then consider six areas of infrastructure development: (a) specialty journals; (b) research centers; (c) professional societies; (d) specialized libraries and documentation centers; (e) training and education programs; and (f) funding agencies. We close with a discussion of the career opportunities and future directions of addiction science.

What is Addiction Science?

The multidisciplinary area of "addiction studies" (variously called addictology, narcology, alcohology) is generally devoted to the understanding, management, and prevention of health and social problems connected with the use of psychoactive substances. Within this area of addiction studies, addiction science represents a more specialized subarea of research activity applying the scientific method to the study of addiction. Over the past 150 years, addiction science has developed its own terminology, concepts, theories, methods, workforce, and infrastructure. Addiction science merges biomedical, psychological, and social perspectives within a transdisciplinary, issue-driven research framework. The goal is sometimes stated as an attempt to advance physical, mental, and population health by contributing to prevention, treatment, and harm reduction.

The field of addiction science, like other interdisciplinary areas of research, often requires expertise and collaborations across traditional disciplinary boundaries as well as transdisciplinary research efforts (Choi & Pak, 2006) that involve scientists trained in the basic sciences, medicine, and public health, as well as the social, biological, and behavioral sciences. It also encourages integration of nonacademic participants, such as policymakers, service providers, public interest groups, and persons in recovery from substance use disorders. The basic underlying framework, or infrastructure, of current addiction science consists of research centers, scholarly journals, professional societies, education programs, specialized services, specialized libraries, funding agencies, and the people to populate these institutions and services.

Box 2.1 provides an abbreviated chronology of major events in the development of addiction science in North America, Europe, and other parts of the world.

The first wave of activity consisted of establishing organizational and communication structures such as the American Association for the Study and Cure of Inebriety in 1870, and its British counterpart, the Society for the Study and Cure of Inebriety in 1884. The emergence of addiction science was driven primarily by societal concerns about the problems of alcohol and, later, about cocaine and opiates. Addiction science initially flowered and then nearly expired in concert

*William White
has broader
dates*

- **First Wave: Organizational and Communication Structures**
  - 1870 – American Association for the Study and Cure of Inebriety
  - 1884 – Society for the Study and Cure of Inebriety (United Kingdom)
  - 1907 – International Bureau Against Alcoholism

- **Second Wave: Institutional Support for Research**
  - Early 1940s – Yale Center of Alcohol Studies, New Haven, Connecticut, United States
  - 1949 – Addiction Research Foundation, Toronto, Canada
  - 1950 – Finnish Foundation for Alcohol Studies, Helsinki, Finland
  - 1960 – National Institute for Alcohol Research, Oslo, Norway
  - 1967 – Addiction Research Unit, London, United Kingdom
  - 1971 – U.S. National Institute on Alcohol Abuse and Alcoholism
  - 1973 – U.S. National Institute on Drug Abuse

- **Third Wave: The Modern Era**
  - Addiction research centers
  - Addiction specialty journals
  - Addiction-focused professional societies
  - Addiction-focused education and training programs
  - Addiction-focused libraries

Box 2.1: Major milestones in the history of addiction science.

with the rise and fall of the temperance movement in America and Europe. During a 40-year period (1875–1915), an international cadre of addiction specialists emerged from various areas of medicine and science to advance knowledge about addiction problems. This was done by means of professional societies, international meetings, scientific journals, scholarly books, and expert committee reports (Babor, 1993a,b; 2000; Billings et al., 1905; Bühringer & Watzl, 2003; Sournia, 1996). Although the research produced by these organizations was unsophisticated by current standards, there were some notable advances in toxicology, clinical diagnosis, epidemiology, and policy research during this time (Babor, 1993a, 2000; Billings et al., 1905; Sournia, 1996), especially in the United Kingdom, the United States, France, Germany, and Sweden. The demise of addiction studies followed the imposition of prohibition legislation in the United States, Scandinavia, the United Kingdom, and many other countries in the aftermath of the First World War. It was not until the 1940s that addiction research regained a sense of identity and purpose and not until the 1970s when it gained enough scientific respectability to be considered a legitimate part of society's public health response to alcohol and other drug problems.

The second wave of addiction science is characterized by the growth of institutional support for research, beginning with the establishment of the Yale Center of Alcohol Studies in New Haven, Connecticut, in the United States in the early 1940s; the Addiction Research Foundation, Toronto, Canada, in 1949; and similar organizations in Finland, Germany, Norway, and other countries. With the creation of government funding agencies at the federal level in the United States in the early 1970s, the stage was set for the modern era.

As part of the developing biomedical establishment in the United States, addiction science experienced phenomenal growth, which was paralleled by similar developments in Europe. That growth—the third wave—can be characterized by at least four megatrends (Babor, 1993b): (a) the rapid expansion of scientific publishing of addiction research, (b) the development of addiction research centers and related organizational structures, (c) international collaboration in research, and (d) the development of significant scientific breakthroughs in addiction science and medicine. We now consider these trends in the context of the seven types of infrastructure that have emerged in the modern era described above.

Addiction Specialty Journals

One indication that addiction science has emerged as a separate discipline is the appearance of specialty academic journals that serve as a medium of communication among clinicians and scientists. The first journals specifically publishing addiction science were the (quarterly) *Journal of Inebriety* (1876–1914), the *British Journal of Inebriety* (1884–present; now *Addiction*) and the *International Monthly Journal for the Fight against Drinking Practices* (1890-present with two World War interuptions; now *SUCHT*). After a relative lapse of interest in addiction science, the *Quarterly Journal of Studies on Alcohol* (now the *Journal of Studies on Alcohol and Drugs*) was established in 1940 and revived scientific interest in alcoholism, a development that began the modern era of addiction research.

Figure 2.1 traces the cumulative growth of addiction specialty journals since 1884. The journals are characterized in terms of their language of publication (English and non-English), but there are other important distinctions that are discussed in more detail in Chapter 3. The dominance of English as the international language of science has facilitated communication far beyond national boundaries. With the development of online publishing and the "open access" trend to make scientific research freely available to the scientific community and the general public, there has been a proliferation of online open-access English-language journals that have transformed the way that scientific information is published and distributed. However, as discussed in Chapter 3, many of the new online open access journals that have been established in the last decade are produced by "predatory publishers," organizations that engage in

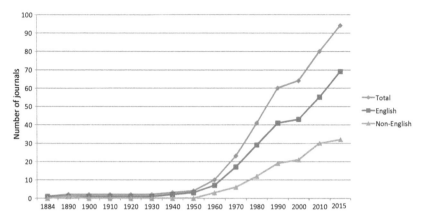

Fig. 2.1: Growth of addiction specialty journals.

questionable practices with regard to journal management, marketing activities, peer review, and page fees (Beall, 2012).

Addiction specialty journals provide a communication forum for scientists and clinicians. They deliver valuable information to practitioners, scientists, and the general public. They set the agenda for a field of study and maintain ethical and quality standards. Another function is to archive the historical record for an area, allowing permanent access to articles for future use by scientists, clinicians, administrators, policymakers, and historians. Finally, by means of the peer-review process, journals certify the authenticity and originality of an author's work (LaFollette, 1992). For these reasons, scientific journals are the institutional memory of a field.

In addition to the growth in specialty journals, addiction science is also published by discipline-oriented journals dealing with medicine, pharmacology, biochemistry, neurobiology, psychology, sociology, and epidemiology. When the addiction articles of these journals are combined with the publications in addiction specialty journals, it becomes possible to estimate trends in the volume of research in addiction science by means of historical records and bibliometric analyses. Between 1900 and 1950, for example, approximately 500 scientific articles were published per year on alcohol (Keller, 1966). Between 1950 and 1970, the number of publications doubled each decade. By the late 1980s, more than 3,000 scholarly publications on alcohol were appearing per year, and the trend has continued unabated until the present.

To estimate the current output of scientific publications, we used bibliometric procedures to extract journal publications in SCOPUS from 2000 through 2014 that dealt with addiction research (e.g., "alcohol use disorder" and "tobacco use disorder"). We then categorized the publications by area of focus across four areas of research: alcohol, tobacco, other drugs, and gambling. The SCOPUS database was selected for its inclusion of all MEDLINE journals. It should be noted that there is no single database that covers the entire output of scholarly publications in addiction science, after the major databases that previously collected, indexed, and abstracted addiction literature ceased operations over the past 15 years (ETOH in 2003, Rutgers Alcohol Studies Database in 2007, CORK in 2015). In the absence of a comprehensive database, it is difficult to estimate the number of articles published in the field, and it is not possible to give an accurate account of other addiction-related publications (e.g., books, reports). The estimates provided in this chapter should therefore be considered conservative and better suited to the identification of relative growth trends than to the estimation of the absolute number of publications.

The four searches yielded 233,970 results published since the year 2000. We identified 212,891 unduplicated journal publications for all four areas of research, of which 79,585 were published between 2010 and 2014. Figures 2.2 and 2.3 show the trends in document production. The trend is generally positive for all areas until 2009 when a decline begins for tobacco and nicotine research, followed by lesser declines in 2013 for alcohol and other drugs. The decline in publications may be attributed to reductions in public research funding in the major research-producing countries as well as the global economic recession that began in 2008. This interpretation is supported by the absence of a decline in gambling research, which is mainly supported by the gambling industry or by tax revenues from state lotteries.

The geographical dispersion of the research publications was also examined. The country of origin of each article was determined from the address of the first or corresponding author. Publication contributions between 2010 and 2014 from the most research-prolific countries are shown in Table 2.1.

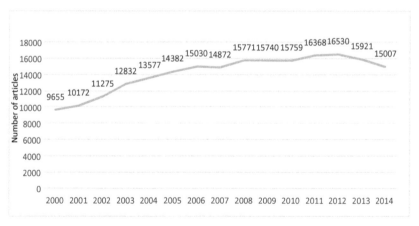

 Fig. 2.2: Total number of addiction articles per year (2000–2014).

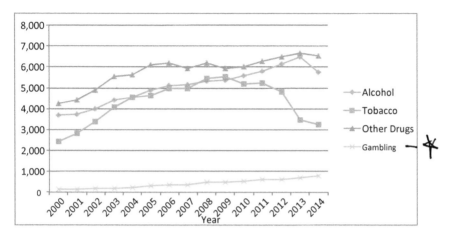

Fig. 2.3: Total number of addiction articles, by year and category (2000–2014).

	Alcohol	Tobacco	Other drugs	Gambling	Population-adjusted publication rate*
United States	12,479	9,115	14,201	1,067	10.45
United Kingdom	2,421	2,236	2,601	382	10.99
Australia	1,674	1,027	1,723	345	18.71
Germany	1,430	879	1,280	206	4.35
Canada	1,297	1,252	1,738	399	12.04
Italy	996	780	1,233	159	4.90
France	995	686	1,137	134	4.03
Spain	978	661	1,322	108	5.99
The Netherlands	902	707	817	105	13.83
Brazil	838	303	786	64	0.90
China	791	649	1,010	148	0.18
India	755	614	553	18	0.14
Switzerland	568	367	693	59	19.06

Table 2.1: Publications by country and research category.
*Rates based on unduplicated totals from total population estimates from 2013;
Source: World Bank (2013).

When looking at the number of publications across all four categories combined (totals not shown), the top five producing countries are the United States, United Kingdom, Australia, Canada, and Germany. The United States accounts for approximately 42% of the total production, but on a population-adjusted basis several other countries (Australia, the United Kingdom, Canada, the Netherlands, and Switzerland) make even greater contributions. In the emerging economies of the world, China, India, and Brazil are beginning to produce significant amounts of the research published in the English-language literature as well. An important consideration regarding the geographic concentration of research in the United States and Europe is that the findings may not generalize to other parts of the world—especially nations in Africa, Asia, and Latin America—facing epidemics of alcohol abuse, nicotine dependence, other drug dependence, or pathological gambling. In general, these analyses indicate that the steady growth of addiction science during the latter part of the 20th century has continued unabated into the first part of the 21st century.

Addiction Research Centers

Although addiction research in many countries is conducted by independent scientists whose primary affiliation is to an academic department in a university or by clinicians who work in treatment facilities, in recent years there has been an expansion of specialized centers whose primary purpose is to support alcohol, tobacco, and other drug research. As such, they provide a good indicator of growth trends in research infrastructure.

Centers provide dedicated facilities to groups of scientists and supporting staff so that long-term programmatic research can be carried out. Centers constitute an optimal environment for researchers, one that is relatively free of administrative, clinical, and teaching responsibilities. Not only are the positions dedicated exclusively to research, but the centers also provide the prospect of long-term support and career advancement. Training of junior investigators is another important function of research centers.

Building on earlier estimates of the annual growth in research centers (Babor, 1993b), we conducted an Internet search to identify the location and other characteristics of addiction research centers, including the dates they were established. We estimate that the number of research centers devoted to addiction research now number approximately 275 worldwide. The largest number of centers is located in the United States, the Nordic countries, the United Kingdom, Russia, Brazil, Canada, and Japan.

The growth of research centers is indicative of a more general trend in addiction science and clinical services. Over the last 45 years, the number of research centers has increased exponentially, from fewer than 20 before 1970 to more than 150 at the end of the century. By the year 2000, the multi-disciplinary research center had become the dominant setting for basic, clinical, and

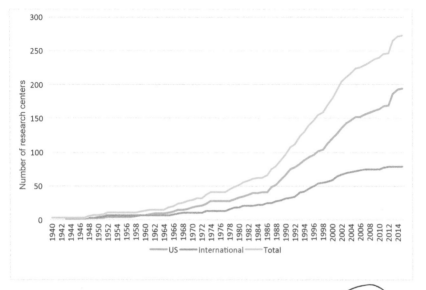

Fig. 2.4: Cumulative growth of addiction research centers (1940–2015).

psychosocial research on addictive substances. Figure 2.4 shows the exponential growth in addiction research centers in both the United States and globally over a 75-year period. The scope of these centers varies, with 70.5% focusing on drugs and alcohol, 57.4% on alcohol alone, 36.0% on tobacco, and 2.9% on other addictions (e.g., problem gambling).

The type of addiction research varies across centers, with 55.6% conducting studies on addiction treatment, 54.2% on the psychosocial factors involved in addiction, 51.3% on policy or prevention programs, and 33.1% on the biological underpinnings of addiction. Approximately 8% of research centers are known to have more than 50 affiliated research scientists; 50% house fewer than 25 investigators; and 21% have fewer than 10.

As the number of centers has grown, collaborative networks have been formed to better leverage existing resources, conduct cross-national projects, train doctoral and postdoctoral candidates, write scientific publications, provide policy consultations, and increase the media coverage of addiction science. In Germany, the Federal Ministry of Education and Research implemented a long-term research funding program (1994–2008 with nearly 35 million euros) to enhance drug research and collaborations, disseminate findings, improve addiction-science information exchange across professionals, and advise the public and policymakers on addiction-related topics. The program supported 18 single projects and, from 2001 onwards, four consortia among 12 research centers (composed of MDs and psychologists) engaged in behavioral, clinical, neurobiological and genetic research (Mann, 2010). In that context, the first chair in addiction research was created in 1999 at the Central Institute of

Mental Health Mannheim (University of Heidelberg) and the second in 2005 at the University of Dresden. In part because of the success of these networks, Germany is now investing substantially more in addiction research.

In the United States, the National Institute on Alcohol Abuse and Alcoholism (NIAAA) and the National Institute on Drug Abuse (NIDA) support research centers and research networks through several funding mechanisms. NIAAA supports 20 research centers through its National Alcohol Research Centers Program and also funds large-scale cooperative agreements among researchers collaborating on high-priority projects such as Project MATCH (Matching Alcoholism Treatments to Client Heterogeneity; Babor & DelBoca, 2003), the multisite trial of Combined Pharmacotherapies and Behavioral Interventions for Alcohol Dependence (COMBINE; Anton et al., 2006), and the Collaborative Study on Genetics of Alcoholism (COGA) project (Agrawal & Bierut, 2012). NIDA also supports a Clinical Trials Network (Wells et al., 2010) devoted to treatment research. These kinds of large-scale, cross-site collaborations facilitate rapid, standardized data-collection projects that would not be possible at a single small site, and they permit more generalizable conclusions and data applications.

Addiction research centers provide core facilities and laboratories, training opportunities for new scientists, and resources to sustain career investigators. In addition, research centers facilitate links between scientists, policymakers, and the general public. During the 75-year period depicted in Figure 2.4, there was parallel growth in governmental institutes and private funding agencies devoted to the sponsorship of addiction research. The combination of categorical support for addiction research and academic freedom to engage in addiction science as a career contributed substantially to the information and productivity explosion in the addiction field discussed in subsequent sections of this chapter (Babor, 1993a,b; Babor et al., 2008).

Professional Societies

In the addiction field, professional societies have been operating for almost 150 years, with the oldest continuing society being the Society for the Study of Addiction, established in 1884 in the United Kingdom. These societies include national and international organizations and sections of larger organizations that are devoted to addiction treatment, prevention, policy, and research. Membership comprises clinical, prevention, and research professionals, including psychologists, physicians, psychiatrists, social workers, addiction counselors, and other professional groups. Figure 2.5 documents the growth of professional societies, based on an earlier compilation of alcohol-related associations (NIAAA, 1985) and a review of Internet sources. The number of professional societies grew dramatically between 1970 and 2005, particularly in the United States. A more recent trend has been the growth of international organizations and confederations of societies.

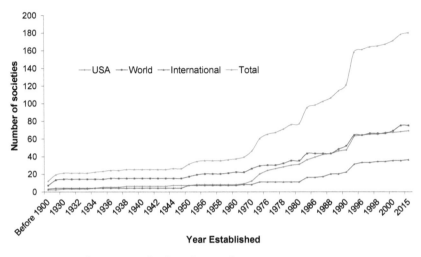

Fig. 2.5: Cumulative growth of professional societies.

A minority of these societies, perhaps no more than 40 in number, can be classified as addiction research organizations because their mission statements suggest primary involvement in issues related to research on alcohol, tobacco, other drugs, and behavioral addictions. Twelve countries have national-level research societies, and there are 14 international organizations. Only a few societies are located in developing countries. These organizations can be classified into three broad categories: multi-disciplinary, professional specialty, and research societies.

Multi-disciplinary societies are open to professionals of all disciplines who work in the addiction area, including treatment, prevention, research, policy, and education. The Brazilian Association for the Study of Alcohol and Other Drugs (ABEAD) is a good example of a multi-disciplinary national society, as is the British Society for the Study of Addiction. Professional specialty societies are typically special-interest groups organized within larger disciplinary societies, such as the Alcohol, Tobacco and Other Drugs section of the American Public Health Association. Several of these specialty societies are international in scope, such as the International Society of Addiction Journal Editors. Research societies provide a forum for new scientific developments and networking for potential investigative collaborations, usually within the context of an annual meeting. The Research Society on Alcoholism, College on Problems of Drug Dependence, and International Society for Biomedical Research on Alcoholism are examples of this type of organization.

Table 2.2 shows professional societies that sponsor scientific journals in terms of their year of foundation, membership numbers, and journal (adapted from Edwards & Babor, 2008). These are among the largest societies devoted to research, representing more than 7,000 members, even taking into account multiple memberships by the same individuals across societies.

Name of organization	Year established	Number of members	Society journal(s)
Society for the Study of Addiction (United Kingdom)	1884	478	*Addiction, Addiction Biology*
SOCIDROGALCOHOL: Spanish Scientific Society for the Study of Alcohol, Alcoholism and other Drug Dependencies	1969	816	*Adicciones*
Association for Medical Education and Research in Substance Abuse (United States)	1976	300	*Substance Abuse*
Research Society on Alcoholism (United States)	1977	1,500	*Alcoholism: Clinical and Experimental Research*
ABEAD, Brazilian Association for the Study of Alcohol and Other Drugs	1978	840	*Society Bulletin and the Brazilian Journal on Chemical Dependence (Jornal Brasileiro de Dependências Químicas)*
German Society for Addiction Research and Addiction Treatment	1978	400	*SUCHT*
Société Française d'Alcoologie et Addictologie(French Society of Alcoholism and Addiction)	1978	807	*Alcoologie et Addictologie*
Japanese Society of Alcohol-Related Problems	1979	543	*Journal of the Japanese Society of Alcohol-Related Problems*
Australasian Professional Society on Alcohol & Other Drugs	1981	382	*Drug and Alcohol Review*
Kettil Bruun Society for Social and Epidemiological Research on Alcohol	1987	197	*International Journal of Alcohol and Drug Research*
Society for Research on Nicotine and Tobacco	1994	1,000	*Nicotine & Tobacco Research*

Table 2.2: Selected addiction societies according to year of foundation, membership, and journal sponsorship.

Although the activities of professional societies are diverse, first and foremost they run meetings, ranging from large annual events to small topic-based workshops and thematic conferences. Networking—encouraging professionals to communicate and work with each other—is a major function, if not primary purpose, of these organizations. As noted in Table 2.2, many sponsor scientific journals. Some organizations influence national policy. ABEAD (Dias da Silva et al., 2002), for example, is close to the Brazilian government. Others stay clear of political involvement and focus on "science as science"; the German Society for Addiction Research and Addiction Treatment (Mann & Batra, 2008) has supported the renaissance of the national addiction science base. Publications are another significant product of many societies, highlighting relevant research and achievements in the form of journals, yearbooks, bulletins, guidelines, and educational materials. Some societies provide continuing education to interested parties, with several offering professional certifications in addiction medicine or other relevant topics. Most societies share a common concern for enhancing the addiction field's status as an important area of research and clinical practice, with the aim of overcoming patient stigma and government neglect.

Some countries have just one major body dealing with alcohol and other drugs, whereas others have a plethora. Japan, for instance, has the Japanese Society of Alcohol-Related Problems, the Japanese Medical Society on Alcohol and Drug Studies, the National Society of Biomedical Research on Alcohol, the Society of Psychiatric Research on Alcohol, and a society focused on addiction behavior (Maruyama & Higuchi, 2004).

Rather than being the products of government intention, many addiction societies were formed spontaneously by small groups of professionals who identified an emerging need and resolved to work together to address it. The British Society for the Study of Addiction, for example, was formed by an alliance of physicians in 1884 (Tober, 2004) to mobilize parliamentary support for the compulsory treatment of "inebriates." The impetus to the foundation in 1977 of the Research Society on Alcoholism was the expansion in research funding following the initiation of NIAAA (Israel & Lieber, 2002). The Italian Association on Addiction Psychiatry (SIPDip) (Nizzoli & Foschini, 2002) was established in 1989 to create a role for psychiatry in the face of political chaos and the neglect of addiction-related problems. Each of these societies was shaped by national trends in substance use, assumptions about the proper role of voluntary action, and the role of professional disciplines in the national response to addiction problems.

In the late 19th and early 20th centuries, when the world temperance movement and specialized asylums for addiction treatment had reached a high level of maturity, large umbrella organizations or confederations were formed to facilitate communication among diverse addiction-related entities around the world. The first example of such a coalition of individuals and organizations was the The International Bureau Against Alcoholism, founded in 1907, which became, in 1964, the International Council on Alcohol and Alcoholism. More recently, confederations of research organizations have again begun to take shape in the addiction field with the creation of the European Federation of Addiction Societies (EUFAS) and the International Confederation of Addiction Research Associations (ICARA) (Stenius, 2012). The aim of ICARA is to provide a forum for the discussion of issues such as governance, organizational management, relationships with governments, advocacy for addiction science, and the promotion of treatment services. Another sign of the consolidation of infrastructure is the formation in 2001 of the International Society of Addiction Journal Editors (Edwards & Babor, 2001).

According to Krimsky (2003), professional societies, along with a network of academic journals, define "acceptable scholarship and certifiable knowledge" (p. 107). Professional organizations, especially research societies, are a major resource for scientists working in biomedical and psychosocial research. They distribute news and scientific information to their members, publish journals and newsletters, engage in advocacy for research, coordinate scientific meetings, and at times facilitate collaborative research. These organizations, in turn, provide a means of networking and communication for their members. They confer prestige and often serve as advocates for professional issues such as research funding, the training of scientists, and evidence-based policy.

Specialized Libraries and Databases

Information services—including libraries, resource centers, and clearinghouses—are an integral part of any research program. A specialized library in the addiction field provides information resources, such as books and journals on addiction, as well as reports, pamphlets, and historical documents. Addiction libraries are usually managed by universities, government agencies, and nongovernmental organizations. With the growth of digital databases, addiction libraries have provided easy access to the international addiction literature.

Substance Abuse Librarians & Information Specialists (SALIS) is a professional organization established in 1978 with assistance from NIDA and NIAAA. As an international association of individuals and organizations interested in the exchange of information on alcohol, tobacco, and other drugs (ATOD), SALIS provides a good example of the growth of specialization in addiction science. A major aim of SALIS is to promote the dissemination of accurate knowledge about the use and consequences of ATOD.

Figure 2.6 shows the cumulative growth and decline of specialized addiction libraries over the past 85 years in the United States and other parts of the world. The figure is based in part on an inventory compiled by SALIS (Mitchell, 1991) to document ATOD libraries, clearinghouses, and resource centers. From it, specialized libraries and collections that primarily serve an academic or research purpose were identified, although some documentation centers were also included. Libraries and other collections reporting fewer than 500 books were not included, nor were mental health libraries, those with no identifiable start date, resource centers, clearinghouses, or trade/industry libraries, unless they served an academic purpose. The figure plots the cumulative number of functioning libraries by year established, subtracting any documented closures, based on a 2015 review that identified closures over the past 25 years.

The first specialized libraries were established in Europe (1907) and the United States (1940) during the early part of the 20th century. Starting in the 1940s, more ATOD libraries were added, a trend that accelerated in the 1960s. The global network of specialized libraries that SALIS now represents has followed a growth curve similar to other parts of the addiction science infrastructure, but there have also been signs of decline. The decline in the number of libraries after 1995 could be because of budget cuts that have affected libraries and databases in both North America and Europe, resulting in downsizing, service reduction, and closures. Another explanation is a change in information-seeking habits, with more professionals using the Internet to access information through their computers and smartphones (McTernan, 2016).

Regardless of the reason, specialized addiction libraries are declining in number, as are the number of specialized librarians. For example, in 2006, NIDA closed its library—which contained a collection dating from 1935. The U.S. Substance Abuse and Mental Health Services Administration (SAMHSA) also closed its prevention library and cut support to Regional Alcohol and Drug Awareness Resource (RADAR) centers, which were created to disseminate

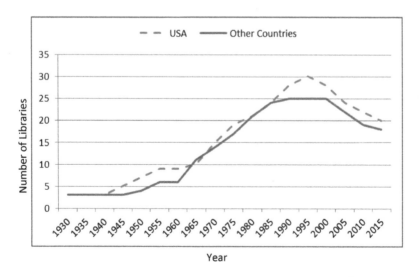

Fig. 2.6: Cumulative growth of specialized addiction libraries.

government agency publications related to alcohol and other drugs. Europe joined the culling effort with library closures or downsizing at the Trimbos Institute (the Netherlands); Alcohol Concern, Drug Scope, and the Temperance Alliance (United Kingdom); Toxibase (France); and Gruppo Abele (Italy). Some of these organizations maintain online information portals, but collections have been packed up, databases and catalogues terminated, and staff positions eliminated. More than 25 libraries or databases have closed in the past decade (Mitchell et al., 2012). Not only have these closures resulted in a reduction in the ATOD information base, but they also have reduced the pool of librarians who have expertise and knowledge of valuable historical material. Print collections have been de-funded and neglected without ensuring archival preservation (Mitchell et al., 2012).

Budget reductions have been justified by the assumption that online access is "free," but the majority of scholarly literature cannot be accessed readily through search engines or websites because of copyright and the proprietary nature of information. Excluding PubMed, most research databases are available only through paid subscription. Furthermore, most do not provide full-text articles without a fee.

In addition to specialized libraries, more than 100 companies and institutions currently offer abstracting and indexing services that provide digital access to abstracts and titles pertaining to the world literature on alcohol, other drugs, tobacco, and the behavioral addictions (e.g., problem gambling). There are approximately 20 main electronic databases that index the published literature by author, topic, and bibliographic reference and provide abstracts of articles for potential readers in search of particular types of information (see Chapter 3). Abstracting and indexing services provide detailed information about the content of scientific journal articles, including abstracts, which are invaluable for those without immediate access to the full text of the article. Some of the more specialized databases were established before the digital revolution in the 1990s, and, as their functions have been taken over by more generic databases, they have fallen into decline and neglect. For example, the Alcohol and Alcohol Problems Science Database, informally known as ETOH, was a comprehensive online resource covering all aspects of alcohol abuse and alcoholism, including journal articles, books, conference papers and proceedings, reports and studies, dissertation abstracts, and chapters in edited works. Unfortunately, it ceased operations in 2003. Two other specialized databases, Project CORK and DrugScope, were closed in 2015, leaving the addiction field without a comprehensive digital repository of the world's addiction literature.

To the extent that library closures and downsizing of other information sources could be a bellwether of the future of addiction science, they are perhaps an indication that the exponential growth of the field has begun to slow or even decline.

Education and Training Programs in Addiction Studies

Without career professionals to populate its infrastructure and develop its products, the addiction field would not exist. To fill the need for a growing professional workforce in treatment, prevention, and research, specialized education and training programs have been created throughout the world. Most of them focus on the training of clinicians, but several are devoted to addiction science.

In general, the concept of addiction studies can be used as a framework to describe the emerging education programs that focus on the interactions between science, clinical practice, and social policy and across a range of addiction topics (e.g., opiate addiction, nicotine dependence, gambling behavior, alcoholism). Figure 2.7 shows the cumulative growth in university-based degree programs in addiction studies. Some of these programs offer undergraduate- or graduate-level degrees, and they are often interdisciplinary, involving training in genetics, neuroscience, psychology, epidemiology, and public health.

Other programs, not included in the figure, offer postbaccalaureate, postdoctoral, or even single-workshop–based training options geared toward a variety of individuals interested in improving their clinical skills, research methods, and professional qualifications for positions in research, clinical services, prevention, and policy. The aim of addiction studies programs is not to replace other professions but to work with them to promote the integration of research findings, prevention activities, and clinical approaches. Table 2.3 describes some of the training programs in addiction studies.

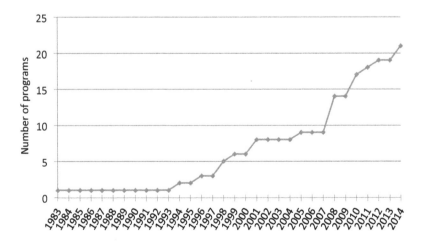

Fig. 2.7: Cumulative growth in degree programs in addiction studies.

University	Country	Degree	Program
Middlesex University, Aarhus University and University del Piemonte Orientale "A Avogadro"	England, Denmark and Italy	Master's degree	European Masters in Drug and Alcohol Studies
National Addiction Centre at the Institute of Psychiatry, Maudsley Hospital, King's College London	England	Master of science degree	Clinical and Public Health Aspects of Addiction
King's College London, Virginia Commonwealth University and University of Adelaide	England, United States and Australia	Joint master's-level degree	International Programme in Addiction Studies
Department of Addictology, First Faculty of Medicine, Charles University	Czech Republic	Bachelor's, master's, and doctoral degrees	Academic Study Programs in "Addictology" (Addiction Science
University of Auckland, School of Populations Sciences	New Zealand	Postbaccalaureate certificate, postbaccalaureate diploma, full master's degree	Postbaccalaureate specialization in addiction science: Alcohol and Other Drugs Program
Center for Addiction Science Specialties, Sahmyook University	South Korea	Connective major for bachelor degree in Substance Addiction and Behavioral Addiction Prevention	Departments of Nursing, Health Management, Counselling and Physical Therapy
University of Dresden (TUD) and Dresden International University (DIU)	Germany, open for PhD/MD students from Europe	Certificate as basis for the MD/PhD degree at the home university	European Graduate School in Addiction Research (ESADD)

Table 2.3: Examples of specialized addiction-studies programs.

An Internet search conducted by Charles University (Pavlovska et al., 2015) identified 79 university study programs at 24 different universities. The programs were distributed across all education levels, that is, bachelor's, master's, and doctorate, with 35 programs located in Europe, 34 in the United States and Canada, 7 in Australia and New Zealand, and 3 in Asia.

The ultimate goal of this new academic area is to advance research-based knowledge, practice, and policies to further improve prevention and treatment of disorders and problems related to substance use. Despite the growth of programs for the training of addiction psychiatrists, narcologists, psychologists, social workers, psychiatric nurses, and addiction counselors, there has been little attention to the development of specialized training programs for addiction scientists. The value of having specially trained addiction scientists is to maintain, if not expand the global infrastructure for social, behavioral, biological, epidemiological and health services research.

The size of the addiction science workforce needed in a country will depend on the extent of addiction-related problems, the delegation of professional responsibilities, and the funding provided by governments to manage the problems of addiction. Globally, there is now a network of perhaps 10,000 people worldwide who identify addiction science as part of their career identity (Babor, 2012). Membership in the 10 professional societies listed in Table 2.2, which includes both basic and clinical scientists, is comparable to this number. Without more systematic attention to workforce monitoring, it is impossible to say whether the current number of addiction scientists is sufficient to meet the needs and the demands for scientific information about addiction.

Funding Sources and Patronage

How society allocates its resources to support the infrastructure of addiction science is not only testimony to its values, but it also is an indication of current priorities in relation to the management of society's addiction-related problems. As in other areas of science, the addiction field relies on patronage. In some cases, the support and sponsorship comes from private sources, such as when a philanthropist creates an endowment for a research center or an academic chair. More often, however, the patronage comes from public sources. During the past 50 years, a variety of funding mechanisms across the globe have provided support for addiction research and research infrastructure, which in turn has made possible much of the growth in professional careers (Babor, 2012). National research institutes, for example, have been created in many high- and middle-income countries to plan, support, and conduct scientific research on addiction (Babor, 1993b). Examples of such organizations include the Norwegian National Institute for Alcohol and Drug Research, the Indian National Drug and Alcohol Institute, the National Institute of Public Policy for Alcohol and Other Drugs (INPAD) in Brazil, and the National Research Centre on Addictions (Russian Federation). Many of these organizations have been established to support the development of scientific expertise with a clinical and sometimes a public health orientation, via the direct funding of research scientists, research training, public education, and the coordination of international activities.

Another source of support for addiction research comes from the private sector, especially pharmaceutical companies. There has also been an increase in funding opportunities from the alcohol and gambling industries, both through direct support for research projects and programs and indirect support from organizations funded by these industries. As described in Chapter 16, there are some important ethical considerations involved in the acceptance of industry funding, not the least of which is financial conflict of interest.

Another issue is the role of funding agencies in the determination of the research agenda. Increasingly, the dollars dictate the science. Alcohol industry funding has been questioned because the agenda is often set by commercial objectives rather than by public health priorities. But even in the public sector, governments can shape the research agenda toward topics that may not address the most effective solutions for addiction problems.

Midanik (2006), for example, identified a bias in U.S. research-funding agencies' priorities toward biomedical (vs. psychosocial) approaches to alcohol-related problems. This has led to the majority of U.S. publications on drugs and alcohol being devoted to basic science and clinical interventions, which conflicts with the interests of policymakers on research related to supply control and demand reduction. In the European Union as well, there is a relative disconnect between research published on illicit drugs and the priorities advanced by policymakers who are responsible for funding research and using its results to lessen the

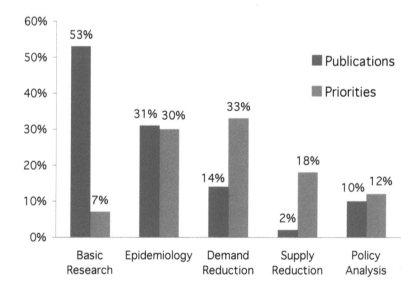

Fig. 2.8: Percentage distributions of research publications (N = 3,028) and research priority ratings (N = 57) across five research areas, based on data from European Union Member states (N = 27). (Source: Bühringer et al., 2009).

suffering of those who experience addiction-related problems (Bühringer et al., 2009). This disconnect between research and policy is reflected in the data presented in Figure 2.8, which contrasts the distribution of research publications in Europe with research priority ratings obtained from 57 policymakers from 27 European Union Members States. The figure shows an inverse relationship between the types of scientific evidence being published and the priorities of policymakers who fund the research behind the publications.

Addiction Science as a Career Option

As described in the infrastructure areas reviewed in this chapter, the field is built around institutions that help to define its roles and responsibilities. Professional societies, research centers, national institutes, addiction journals, specialized libraries, and specialized treatment programs constitute the major ingredients of the addiction field's infrastructure, but, as previously suggested (Edwards & Babor, 2012), addiction careers constitute its building blocks and its human capital.

Today, the field of addiction science is populated by a variety of creative people: basic scientists in pursuit of knowledge for its own sake, clinical investigators searching for new or better treatments, and applied researchers trying to solve difficult social problems (Edwards & Babor, 2012). How do people select a career in an emerging field that for most of its existence had no name or identity? As suggested by personal accounts derived from a long series of interviews published in the journal *Addiction* (Edwards & Babor, 2012), the answer is as varied as the field itself. Personal experience with substance misuse, the influence of a mentor, the need to make a living, and the love of science are all mentioned. Some researchers and addiction professionals developed their interest in the field from personal, even tragic, experience. Others describe serendipity or "opportunity knocking."

With an identity defined by the work of a diverse group of career scientists and the prominence of mentors from a wide variety of disciplines, the career of an addiction scientist is no longer a risk or a mystery. Addiction science as such can now be perceived as an independent, professional career (Babor, 2012; Edwards, 2002).

Conclusion

In the past 50 years, there has been dramatic growth in the demand for and production of addiction science, both globally and in specific countries. Addiction science has evolved to become part of a specialized academic field, with its own training programs, professional organizations, research centers, funding mechanisms, and communication channels. It is devoted both to the pursuit of basic knowledge about addiction and the application of that knowledge to treatment and prevention activities.

By integrating itself with the postwar biomedical establishment (particularly psychiatry), the addiction field experienced phenomenal growth. As suggested by the information presented in this chapter and elsewhere, that growth has been characterized by a number of "megatrends" (Babor, 1993b, 2000), as depicted in Figure 2.9. These trends include the following: (a) the emergence of public and private financing mechanisms to support treatment, prevention, and research programs; (b) development of an institutional base consisting of research centers, specialized clinical facilities, and related organizational structures; (c) the growth of professional societies to give the field a sense of identity and purpose; and (d) the rapid expansion of scientific communication outlets and publication opportunities to facilitate information exchange and dissemination. The final ingredient of the addiction field depicted in the figure is the result of all this effort—that is, basic and applied knowledge about addiction.

Although opinions will differ as to what constitutes the collective "products" of professional careers in academia and the health sector, from a societal perspective, the tangible products of the addiction field can be measured in terms of scientific knowledge, evidence-based clinical and prevention services, and policy interventions designed to address the consequences of psychoactive substance use. Ultimately, the cumulative and collective impact of these efforts should be the reduction of substance-related harm, suffering, and mortality.

The growth of addiction science has fostered increasing communication and collaboration on an international level. Part of this has been the result of the explosion of communications technology and the ease of international travel,

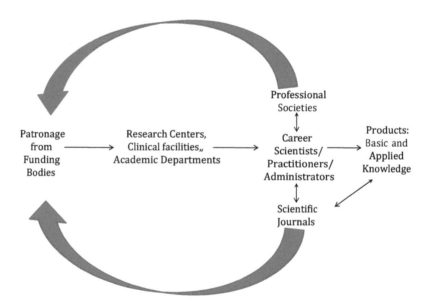

Fig. 2.9: "Megatrends" in addiction science.

but it may also be the result of the globalization of alcohol and other drug distribution networks, which are bringing addictive substances to locations and populations that were previously unexposed. Examples include the marketing by transnational alcohol producers of new alcohol products to women and young adults and the growth of illicit drug use in the major population areas of Africa, Latin America, and Asia.

Perhaps most importantly, what impact does the modern addiction research infrastructure have on the health of the populations it is intended to serve? Countries invest in research on alcohol and other drugs for a reason. Typically, the purpose is to reduce human suffering caused by psychoactive substance use and to prevent further problems.

In most low- and middle-income countries, however, in which addiction presents the same harms as in more developed countries, addiction-research infrastructure is weak or absent. That a journal series on addiction societies and addiction research centers (Edwards & Babor, 2008) could locate in the developing world only a few societies, centers, and journals devoted to the addictions suggests the need to support addiction science in less-resourced countries that have substantial addiction problems. Established groups could aid the development of such societies in large parts of the world that do not at present have this kind of resource. Any such initiatives would need to be culturally sensitive. Even in countries in which resources might not easily allow development of specialist treatment services, specialist research centers, or the publication of national journals, international collaboration combined with voluntary action catalyzed by local associations may constitute entirely feasible kinds of initiatives capable of considerable impact.

If research were the main vehicle for the development of a cure for addiction-related problems, however, by now there should have been breakthroughs in translating research findings into effective prevention policy. As previously mentioned, there is a gap between the bulk of scientific research currently conducted and the interests of policymakers who set the agenda for prevention and treatment funds. Despite the field's apparent growth in many areas, the question of whether the modern infrastructure (surveillance, treatment, prevention, research) has a population-level impact remains unanswered. Until policymakers and addiction experts achieve a greater sense of mission and purpose, nation states will continue to struggle with the question of how best to configure a rational response to the problems of substance abuse.

Please visit the website of the International Society of Addiction Journal Editors (ISAJE) at www.isaje.net to access supplementary materials related to this chapter. Materials include additional reading, exercises, examples, PowerPoint presentations, videos, and e-learning lessons.

References

Agrawal, A., & Bierut, L. J. (2012). Identifying genetic variation for alcohol dependence. *Alcohol Research and Health, 34*(3), 274–281.

Anton, R. F., O'Malley, S. S., Ciraulo, D. A., Cisler, R. A., Couper, D., Donovan, D. M., . . . , Zweben, A. (2006). Combined pharmacotherapies and behavioral interventions for alcohol dependence: The COMBINE study: A randomized controlled trial. *JAMA, 295,* 2003–2017. DOI: https://doi.org/10.1001/jama.295.17.2003

Babor, T. F. (2000). Past as prologue: The future of addiction studies [Editorial]. *Addiction, 95,* 7–10. DOI: https://doi.org/10.1046/j.1360-0443.2000.95172.x

Babor, T. F. (2012). Addictionology as biography: One hundred ways to have a successful career in addiction science. *Addiction, 107,* 464–466. DOI: https://doi.org/10.1111/j.1360-0443.2012.03788.x

Babor, T. F., & Del Boca, F. K. (Eds.). (2003). *Treatment Matching in Alcoholism.* Cambridge University Press: Cambridge, UK.

Babor, T. F., Morisano, D., Stenius, K., Winstanley, E. L., & O'Reilly, J. (2008). How to choose a journal: Scientific and practical considerations. In T. F. Babor, K. Stenius, S. Savva, & J. O'Reilly (Eds.), *Publishing addiction science: A guide for the perplexed* (2nd ed., pp. 15–32). Brentwood, UK: Multi-Science Publishing.

Babor, T. F., Stenius, K., & Romelsjö, A. (2008). Alcohol and drug treatment systems in public health perspective: Mediators and moderators of population effects. *International Journal of Methods in Psychiatric Research, 17*(Supplement 1), S50–S59. DOI: https://doi.org/10.1002/mpr.249

Beall, J. (2012). Predatory publishers are corrupting open access. *Nature, 489,* 179. DOI: https://doi.org/10.1038/489179a

Billings, J. S., Eliot, C. W., Farnam, H. W., Greene, J. L., & Peabody, F. G. (1905). *The liquor problem: A summary of investigations conducted by the committee of fifty, 1893–1903.* Boston, MA: Houghton Mifflin.

Bühringer, G., & Watzl, H. (2003). On the history and new layout of the journal *SUCHT*: Old wine in new bottles? [Editorial] *SUCHT, 49,* 12–15. DOI: https://doi.org/10.1024/suc.2003.49.1.4

Bühringer, G., Farrell, M., Kraus, L., Marsden, J., Pfeiffer-Gerschel, T., Piontek, D., & Stillwell, G. (2009). *Comparative analysis of research into illicit drugs in the European Union. Executive summary.* Brussels, Belgium: European Commission.

Choi, B. C., & Pak, A. W. (2006). Multidisciplinarity, interdisciplinarity and transdisciplinarity in health research, services, education and policy: 1. Definitions, objectives, and evidence of effectiveness. *Clinical & Investigative Medicine, 29,* 351–364.

Dias da Silva, J. C., Ramos, S. d. P., Luz, E. L. Jr., de Oliveira, E. M., Requião, D. H., & Campana, A. A. M. (2002). The Brazilian Association for the Study of Alcohol and Other Drugs. *Addiction, 97,* 9–13. DOI: https://doi.org/10.1046/j.1360-0443.2002.00013.x

Edwards, G. (Ed.). (2002). *Addiction: Evolution of a specialist field.* Oxford, England: Blackwell.

Edwards, G., & Babor, T. F. (2001). The International Society of Addiction Journal Editors (ISAJE) has become established. *Addiction, 96,* 541–542. DOI: https://doi.org/10.1046/j.1360-0443.2001.9645411.x

Edwards, G. E., & Babor, T. F. (2008). Closing remarks: Addiction societies as valuable assets. *Addiction, 103,* 9–12. DOI: https://doi.org/10.1111/j.1360-0443.2007.02095.x

Edwards, G., & Babor, T. F. (Eds.). (2012). *Addiction and the making of professional careers.* Piscataway, NJ: Transaction Publishers.

Israel, Y., & Lieber, C. S. (2002). The Research Society on Alcoholism. *Addiction, 97,* 483–486. DOI: https://doi.org/10.1046/j.1360-0443.2002.00012.x

Keller, M. (Ed.). (1966). *International bibliography of studies on alcohol.* New Brunswick, NJ: Rutgers Center of Alcohol Studies.

Krimsky, S. (2003). *Science in the private interest: Has the lure of profits corrupted biomedical research?* Lanham, MD: Rowman & Littlefield.

Lafollette, M. C. (1992). *Stealing into print. Fraud, plagiarism, and misconduct in scientific publishing.* Berkeley, CA: University of California Press.

Mann, K., & Batra, A. (2008). The German Society for Addiction Research and Addiction Treatment. *Addiction, 103,* 6–8. DOI: https://doi.org/10.1111/j.1360-0443.2007.02028.x

Mann, K. (2010). Addiction Research Centres and the Nurturing of Creativity: Department of Addictive Behaviour and Addiction Medicine, Central Institute of Mental Health, Mannheim, University of Heidelberg. *Addiction,* 105, 2057–2061. DOI: https://doi.org/10.1111/j.1360-0443.2010.02974.x

Maruyama, K., & Higuchi, S. (2004). The Japanese society of alcohol-related problems. *Addiction, 99,* 419–424. DOI: https://doi.org/10.1111/j.1360-0443.2004.00712.x

McTernan, J. (2016, March 29). Don't mourn the loss of libraries – the internet has made them obsolete. *The Telegraph.* Retrieved from http://www.telegraph.co.uk/news/12206475/Dont-mourn-the-loss-of-libraries-the-internet-has-made-them-obsolete.html.

Midanik, L. T. (2006). *Biomedicalization of alcohol studies: Ideological shifts and institutional challenges.* New Brunswick, NJ: Aldine Transaction.

Mitchell, A. (1991). *SALIS directory 1991: An international directory of alcohol, tobacco, and other drug libraries, clearinghouses, resource and information centers: Includes state and specialty regional alcohol and drug awareness resource (RADAR) centers.* San Francisco, CA: Alcohol Research Group, Medical Research Institute of San Francisco.

Mitchell, A. L., Lacroix, S., Weiner, B. S., Imholtz, C., & Goodair, C. (2012). Collective amnesia: Reversing the global epidemic of addiction library closures. *Addiction, 107,* 1367–1368. DOI: https://doi.org/10.1111/j.1360-0443.2012.03813.x

National Institute on Alcohol Abuse and Alcoholism. (1985). *Resource directory of national alcohol-related associations, agencies and organizations.* Rockville, MD: Author.

Nizzoli, U., & Foschini, V. (2002). ERIT-Italia, the Italian federation for professionals working in the field of drug abuse. *Addiction, 97,* 1365–1367. DOI: https://doi.org/10.1046/j.1360-0443.2002.00137.x

Pavlovská, A., Miovský, M., Gabrhelík, R., & Babor, T. (2015). European University-based study programmes in the addictions field. Unpublished manuscript, Department of Addictology, 1st Faculty of Medicine, Charles University, Prague, Czech Republic.

Sournia, J. C. (1996). *A history of alcoholism.* Basil, Blackwell.

Stenius, K. (2012). ICARA: A global forum for research associations. *Nordic Studies on Alcohol and Drugs, 29,* 439–440. DOI: https://doi.org/10.2478/v10199-012-0038-3

Taylor, P., & Keeter, S. (2010). Millennials: A portrait of generation next. *Pew Research Center Report.* Retrieved from http://www.pewsocialtrends.org/files/2010/10/millennials-confident-connected-open-to-change.pdf.

Tober, G. (2004). The Society for the Study of Addiction (SSA). *Addiction, 99,* 677–685. DOI: https://doi.org/10.1111/j.1360-0443.2004.00729.x

Wells, E. A., Saxon, A. J., Calsyn, D. A., Jackson, T. R., & Donovan, D. M. (2010). Study results from the Clinical Trials Network's first ten years: Where do they lead? *Journal of Substance Abuse Treatment, 38*(Supplement 1), S14–S30. DOI: https://doi.org/10.1016/j.jsat.2009.12.009

World Bank. (2013). Retrieved from http://data.worldbank.org/indicator/SP.POP.TOTL.

Appendix A. Search Terms Used in SCOPUS Search of Addiction Publications (2000–2014)

Search Terms	# of results	
	2010–2014	2000–2014
Alcohol (*"alcohol drinking"* **OR** *"alcohol-related problems"* **OR** *"alcohol intoxication"* **OR** *"alcohol abuse"* **OR** *"alcohol-induced disorder$"* **OR** *"alcohol use"* **OR** *"alcohol-related harm$"* **OR** *"alcoholism"* **OR** *"alcohol use disorder$"*)	28,667	74,921
Tobacco (*"tobacco smoking"* **OR** *"smoking cessation"* **OR** *"cigarette smoking"* **OR** *"tobacco use disorder$"*)	21,528	64,346
Drugs (*"street drug$"* **OR** *"illicit drug$"* **OR** *"illegal drug$"* **OR** *"drug dependence"* **OR** *"drug use disorder$"* **OR** *"drug abuse"* **OR** *"marijuana"* **OR** *"heroin"* **OR** *"hallucinogens"* **OR** *"cocaine"* **OR** *"cannabis"*)	31,425	86,402
Gambling (*"gambling"* **OR** *"pathological gambling"*)	3,192	6,115

SECTION 2

How and Where to Publish

CHAPTER 3

How to Choose a Journal: Scientific and Practical Considerations

Thomas F. Babor, Dominique Morisano,
Kerstin Stenius and Judit H. Ward

Introduction

One of the most important and least understood decisions made in the course of publishing a scientific article is the choice of a journal. The decision influences the audience reached, the context in which work is presented, and the time it takes to achieve formal publication. At best, the right choice of a journal results in the rapid publication of an article that achieves the exposure it deserves. At worst, the wrong choice results in rejection, delay, and even loss of an author's motivation to persist in seeking publication for a potentially valuable scientific contribution. And in some cases the choice of a journal operated by a predatory publisher can embarrass an author when it is learned that the journal does not conduct peer review and will publish anything for a fee.

Journal choice is little understood even by those who have spent decades in the field of addiction research. One reason for this state of affairs is that the field is rapidly changing, with new publication opportunities and formats constantly being added (e.g., electronic journals; e-pub ahead of print; open access; or interactive, supplemented with media options such as audio and video) and more traditional organs of communication (e.g., print journals) adapting to new technology. Another reason for the difficulty in choosing a publication outlet is that, until recently, there was little communication between journal editors and their potential authors. As indicated in Chapter 12, the process by which a journal decides to accept or reject a given article has been mysterious.

How to cite this book chapter:
Babor, T F, Morisano, D, Stenius, K and Ward, J H. 2017. How to Choose a Journal: Scientific and Practical Considerations. In: Babor, T F, Stenius, K, Pates, R, Miovský, M, O'Reilly, J and Candon, P. (eds.) *Publishing Addiction Science: A Guide for the Perplexed*, Pp. 37–70. London: Ubiquity Press. DOI: https://doi.org/10.5334/bbd.c. License: CC-BY 4.0.

Most journals have carefully preserved the mystery within the "black box" of editorial decision making. With virtually no formal training programs on how to write for and publish in scholarly journals, novices often find that the learning process for them has been left to chance and to the luck of finding an experienced mentor.

This chapter provides guidance on how to choose a journal for a scholarly publication on the subject of addiction, broadly defined as any topic dealing with psychoactive substances as well as behavioral addictions, such as gambling. A basic assumption of this chapter is that the primary purpose of publishing is to communicate findings and ideas to a broader audience than one's immediate circle. Our focus is on scholarly journals, which have become the primary organ (in addition to conference presentations, posters, books, and abstracts) of the scientific communication system that has evolved over the past century. Our main interest is in the addiction specialty journals, which limit their subject matter to research on psychoactive substances and related addictive behaviors. To the extent that many articles on addiction topics are also published in disciplinary journals devoted to psychology, biology, sociology, medicine, and other relevant professional disciplines, we will also consider how to choose among these journals as well.

Growth of Addiction Specialty Journals and Other Publication Sources

A scientific journal has multiple roles and functions. Journals provide a forum for scientific communication and should certify the scientific value of an individual author's work. They provide access to reliable knowledge and, at the same time, confer scholarly prestige and facilitate career advancement (see Lafollette, 1992). The number of journals focusing on addiction-related articles since the late 19th century, when addiction publishing first began, accelerated during the 1970s and 1980s, and has continued to grow dramatically since 2007. By the year 2016, there were more than 120 addiction specialty journals operating throughout the world.

A majority of the peer-reviewed addiction journals are published in English, which has emerged as the main language for international scientific communication (Babor, 1993). Details about the member journals of the International Society of Addiction Journal Editors (ISAJE) are provided in Tables 3.1 and 3.2. The data in these tables are based on the results of a 2015 survey of ISAJE journal editors. The survey results were supplemented by a review of public information sources, such as the journal's webpage (if available), print copies of the journal, and its instructions to authors.

Table 3.1 lists the titles of the English-language journals along with information about the substances or addictive behaviors they are concerned with (e.g., alcohol, tobacco, licit and illicit drugs, pathological gambling, other behavioral

Journal Name	Substances (1)	Areas of Interest (2)	Issues per Year	Acceptance Rate	Impact Factor (3)	Print/ Online	Fees	Open Access (4)	Abstracting & Indexing Services (5)
Addicta: The Turkish Journal on Addictions	A,D,T,O	T,P,Po,CE, H,PM,S	2	25%	–	Both	Free	Full	2
Addiction	A,D,T,O	HR,T,P,G,Po, SE,PE,H,S	12	15%	4.97	Both	Free	Hybrid 1	PsycINFO; Web of Science; Scopus; MEDLINE 76
Addiction Science & Clinical Practice	A,D,T	HR,T,Po	N/A	50%	–	Online Only	Processing fee	Full	Scopus; MEDLINE 9
Addictive Behaviors	A,D,T,O	HR,T,P,N,Po, SE,PE,PM,S	12	40%	2.80	Both	Free	Hybrid 1	PsycINFO; Web of Science; Scopus; MEDLINE 20
Advances in Dual Diagnosis	A,D	HR,T,P,Po, PE,B,PM,S	4	–	–	Both	Free	Hybrid 1	Scopus 5
African Journal of Drug and Alcohol Studies	A,D,T	HR,T,P,Po,SE, PE,PM,S	2	60%	–	Online Only	Free	Full	PsycINFO; Scopus 9
Alcoholism Treatment Quarterly	A	HR,T,P,Po, SE,PM,S	4	80%	–	Both	Free	Hybrid 1	PsycINFO; Scopus 11
Alcoholism: Clinical & Experimental Research	A,T,O	A,HR,T,P,N, G,Po,SE,PE, CE,B,PM,S	12 + 1–2 supplements	51%	2.83	Online Only	Page Fee	Hybrid 1	PsycINFO; Web of Science; Scopus; MEDLINE 47
American Journal of Drug and Alcohol Abuse	A,D	A,HR,T,P,N, G,Po,PE,B	6	–	1.83	Both	Free	Hybrid 1	PsycINFO; Web of Science; Scopus; MEDLINE 21

Journal Name	Substances (1)	Areas of Interest (2)	Issues per Year	Acceptance Rate	Impact Factor (3)	Print/ Online	Fees	Open Access (4)	Abstracting & Indexing Services (5)
Canadian Journal of Addiction/Le Journal Canadien d'Addiction	A,D,T,O	HR,T,P, Po,PM,S	3	80%	–	Both	Free	Full	Scopus (in process of applying to PsycINFO, MEDLINE) 9 (3 more in process)
Drug and Alcohol Dependence	A,T,D	A,HR,T, P,N,GPo, SE,PE,H, B,PM,S	12	–	3.35	Both	Free	Hybrid 1	PsycINFO; Web of Science; Scopus; MEDLINE 20
Drug and Alcohol Review	A,T,D	HR,T,P,Po, SE,PE, H,B, PM,S	6	–	2.41	Both	Free	Hybrid 1	PsycINFO; Web of Science; Scopus; MEDLINE 34
Drugs: Education, Prevention & Policy	A,T,D	HR,T,P, Po,H, S	6	–	0.76	Both	Free	Hybrid 1	PsycINFO; Web of Science; Scopus 25
Experimental and Clinical Psychopharmacology	A,D,T,O	A,HR,T,P, N,G,SE, PE,CE,B, PM	6	56%	2.14	Both	Free	Hybrid 1	PsycINFO; Scopus; MEDLINE 15
International Gambling Studies	O	A,HR,T,P,N, Po,SE,PE,H, B,PM,S	3	49%	1.23	Both	Free	Hybrid 1	PsycINFO; Web of Science; Scopus 8

Journal									
International Journal of Alcohol and Drug Research	A,D,T,O	HR,T,P,Po, SE,PM,S	2–4	89%	–	Online Only	Page Fee	Full	PsycINFO 6
Journal of Addiction Medicine	A,D,T,O	HR,T,P, Po, SE,PE,B, PM,S	6	48%	2.07	Both	Free	Hybrid 1	PsycINFO; Web of Science; Scopus; MEDLINE 39
Journal of Addictions Nursing	A,O	HR,T,P, Po,PM,S	4	–	0.48	Both	Free	Hybrid 1	PsycINFO; Web of Science; Scopus; MEDLINE 18
Journal of Addictive Diseases	A,D,T,O	HR,T,P,N, G,Po,SE, PE,CE,B, PM,S	4	–	1.78	Both	Free	Hybrid 1	PsycINFO; MEDLINE 16
Journal of Behavioral Addictions	O	HR,T,P,N, G,SE,PE, B,PM,S	4	–	2.49	Both	Free	Hybrid 1	PsycINFO; Web of Science; Scopus; MEDLINE 11
Journal of Drug and Alcohol Research	A,T,D	A,HR,T, N,G,B,PM	1	90%	–	Online Only	Processing Fee	Full	4
Journal of Gambling Issues	O	HR,T,P, Po,SE,PE, H,B,PM,S	2 (+ 1 supplement)	60%	–	Online Only	Free	Full	PsycINFO; Scopus 10
Journal of Groups in Addiction and Recovery	A,T,D,O	HR,T, PM,S	4	–	–	Both	Free	Hybrid 1	PsycINFO; Scopus 18

Journal Name	Substances (1)	Areas of Interest (2)	Issues per Year	Acceptance Rate	Impact Factor (3)	Print/ Online	Fees	Open Access (4)	Abstracting & Indexing Services (5)
Journal of Psychoactive Drugs	A,D,T,O	HR,T,N, Po,SE,PE, H,B,PM,S	5	–	1.78	Both	Free	Hybrid 1	PsycINFO; Scopus; MEDLINE 14
Journal of Studies on Alcohol and Drugs	A,D,T	HR,T,P, N,G,Po, SE,PE, B,PM,S	6	–	2.20	Both	Free	Hybrid 1	PsycINFO; Web of Science; Scopus; MEDLINE 27
Journal of Substance Abuse Treatment	A,D,T,O	HR,T,Po, PE,PM,S	10	–	2.47	Both	Free	Hybrid 1	PsycINFO; Web of Science; Scopus; MEDLINE 14
Journal of Substance Use	A,D,T,O	HR,T,P, Po,PM	6	–	0.89	Both	Free	Hybrid 1	PsycINFO; Scopus 11
Nicotine & Tobacco Research	T	A,HR,T, P,N,G, Po,SE,PE, CE,B,PM,S	6	38%	3.81	Both	Free	Hybrid 1	PsycINFO; Web of Science; Scopus; MEDLINE 16
Nordic Studies on Alcohol and Drugs (Nordisk Alkohol & Narkotikatidskrift)	A,D,T,O	T,P,Po,SE, H,PM,S	6	67%	0.77	Both	Free	Full	PsycINFO; Web of Science; Scopus 36

| Psychopharmacology | A,D,T,O | A,HR, T,P,N,G, PE,B | 24 | – | 3.54 | Both | Free | Hybrid 1 | PsycINFO; Web of Science; Scopus; MEDLINE 35 |
| Substance Abuse | A,D,T | A,HR,T, P,Po,SE, PE,CE,H, B,PM,S | 4 | – | 2.58 | Both | Free | Hybrid 1 | PsycINFO; Web of Science; Scopus; MEDLINE 13 |

Table 3.1: ISAJE-member journals published in the English language.

(1) Substances: A = alcohol; D = licit and illicit psychoactive drugs other than alcohol; T = tobacco and other nicotine products; O = other substances and addictive behaviors, including gambling and eating disorders.

(2) Areas of Interest: A = animal research; HR = human research; T = treatment; P = prevention; N = neuroscience; G = genetics; Po = policy; SE = social epidemiology; PE = psychiatric epidemiology; CE = chronic disease epidemiology; H = history; B= biological mechanisms; PM = psychological mechanisms; S = social factors.

(3) Thomson Reuters 2015 Impact Factor. From 2016 Release of Journal Citation Reports. Source: 2015 Web of Science Data.

(4) This column indicates the journal's open access policy. Full = all articles of the journal are made available to the reader for free online; Hybrid 1 = open access only for those individual articles for which the authors or the author's institution or funder pay an open access publishing fee; Hybrid 2 = all articles are open access after an embargo period, usually one year.

(5) This column lists up to four of the largest abstracting and indexing services, if the journal is included in them (PsycINFO; Web of Science; Scopus; MEDLINE), plus the total number of abstracting and indexing services that are claimed by the journal, including those listed.

Journal Name	Language	Substances (1)	Areas of Interest (2)	Issues per year	Acceptance Rate	Impact Factor (3)	Print/ Online	Fees	Open Access	Abstracting/ Indexing Services (4)
Addicta: The Turkish Journal on Addictions	Turkish & English	A,D,T,O	T,P,Po,CE, H,PM,S	2	25%		Both	Free	Full	2
Adiktologie	Articles submission in Czech, Slovak or English	A,D,T,O	A,HR,T,P,N, Po,SE,PE,CE, H,PM,S	4 + 1–2 supplements	85%		Both	Free	Hybrid 2	Scopus 4
Alkoholizm i Narkomania (Alcoholism and Drug Addiction)	Polish & English	A,D,T	A,HR,T,P, G,Po,SE,PE, H,B,PM,S	4	85%		Both	Free	Full	3
Drogues, santé et société	French (English Abstracts)	A,D,T	H,T,P,N,Po, SE,PE,H, PM,S	2	68%		Online Only	Free	Full	PsycINFO 5
Exartisis	Greek (English Abstracts)	A,D,T,O	T,P,PM,S	2	85%		Print Only	Free	No	N/A
Narcologia	Russian	A,D,T,O	HR,T,P,Po, SE,PE,H,B, PM,S	12			Print Only	N/A	N/A	N/A

Journal	Languages	Substances	Areas of Interest							Indexing
Nordisk Alkohol & Narkotikatidskrift (Nordic Studies on Alcohol and Drugs)	English, Danish, Norwegian, & Swedish	A,D,T,O	T,P,Po,SE,H,PM,S	6	67%	0.77	Both	Free	Full	PsycINFO; Web of Science; Scopus 36
SUCHT	German (with English abstracts and titles), but some articles in English	A,D,T,O	HR,T,P,Po,SE,PE,H,PM,S	12	80%		Both	Free	Hybrid 1	PsycINFO; Scopus 10
Suchtmedizin (Addiction Medicine)	German (English Summaries)	A,D,T,O	T,Po,SE,PE,B,S	6	–		Both	Free	Full	Scopus 2

Table 3.2: ISAJE-member journals published in languages other than English.

(1) Substances: A = alcohol; D = licit and illicit psychoactive drugs other than alcohol; T = tobacco and other nicotine products; O = other substances and addictive behaviors, including gambling and eating disorders.

(2) Areas of Interest: A = animal research; HR = human research; T = treatment; P = prevention; N = neuroscience; G = genetics; Po = policy; SE = social epidemiology; PE = psychiatric epidemiology; CE = chronic disease epidemiology; H = history; B = biological mechanisms; PM = psychological mechanisms; S = social factors.

(3) Thomson Reuters 2015 Impact Factor. From 2016 Release of Journal Citation Reports. Source: 2015 Web of Science Data.

(4) This column lists up to four of the largest abstracting and indexing services, if the journal is included in them (PsycINFO; Web of Science; Scopus; MEDLINE), plus the total number of abstracting and indexing services that are claimed by the journal, including those listed.

addictions); general topical areas covered (e.g., treatment, prevention, epidemiology, biological mechanisms, history); and details about the journals' frequency of publication, acceptance rate, impact factor, and dissemination channels (i.e., abstracting or indexing services). Table 3.2 provides similar information for journals published in languages other than English. These tables were last updated in January 2017; the most current list of ISAJE member journals is available on www.isaje.net.

The number of specialized addiction journals is only part of the story of how the addiction field has grown in size and complexity. A significant portion of the addiction literature is also published in scholarly journals that have a more general orientation toward disciplines such as medicine, psychology, biochemistry, sociology, economics, and public health. In an earlier version of this chapter published in the first edition of Publishing Addiction Science, we reported that 58% of the alcohol-related articles prior to 2003 were published in general or disciplinary journals and that 42% were published in addiction specialty journals. When the articles were subclassified as either "biomedical" (i.e., dealing with biological or medical topics) or "psychosocial" (i.e., dealing with topics such as treatment, prevention, epidemiology, psychology, or social policy), the addiction specialty journals published a higher percentage of articles on psychosocial topics, whereas disciplinary journals published a greater share of the biomedical articles.

Alcoholism: Clinical and Experimental Research
Drug and Alcohol Dependence
Addictive Behaviors
PLOS One
Neuropsychopharmacology
Addiction
Journal of Studies on Alcohol and Drugs
Substance use and misuse
Psychopharmacology
Journal of Substance Abuse Treatment
Psychology of Addictive Behaviors
Biological Psychiatry
Neuropharmacology
BMC Public Health
Drug and Alcohol Review

Table 3.3: Journals publishing the highest annual numbers of articles on alcohol and drug research.

Table 3.3 provides a list of the top 15 of journals publishing articles on alcohol and drug research as identified through a search in Web of Science for articles published in 2014, indexed with any of the following terms: *alcohol, alcoholism, addiction, drug abuse, drug addiction, substance use,* or *substance abuse.* The table suggests that many disciplinary journals (e. g., *Biological Psychiatry*) also publish significant amounts of addiction research.

In addition to the expanding array of journals that addiction authors have to choose from, many publishers have increased the standard number of issues released per year, added supplements or special issues, and created new electronic formats for submitting articles. With the increased number and breadth of scholarly journals covering addiction-related research, there has probably never been a greater opportunity to publish on the subject. Nevertheless, the plethora of journals has created new challenges for prospective authors, not the least of which is the proliferation of online, open-access journals, some of which have questionable publishing credentials. Other questions that arise in the rapidly changing publishing environment are the following: What are the relative merits of publishing in disciplinary versus addiction specialty journals? How does an author find the most appropriate journal for a particular article? What are the chances that an article will be accepted by a given journal? Which journals have the greatest impact on the field? How does an author know whether a journal will reach the intended audience for a specific article? What are the costs of publishing in pay-per-page journals?

To assist prospective authors in finding answers to these questions, Box 3.1 describes the kinds of decisions that must be made during the search for an

---

1. Decide first whether the article is primarily of interest to a national or an international audience.
2. Consider the language of publication.
3. Consider whether to publish in a generic, disciplinary, or addiction specialty journal.
4. Review the journal's content range (type of drug, clinical/basic science, etc.) and general culture.
5–6. Evaluate the journal's quality and integrity.
7. Gauge your article's potential exposure by reviewing the journal's indexing and abstracting services, as well as its open-access policy.
8. Evaluate your chances of acceptance.
9. Take into account time to publication and other practical matters.
10. Consider, but don't be fooled by, impact factors.

Box 3.1: Ten steps in choosing a journal.

appropriate journal. The following sections expand on this outline, discussing each step in the process. It should be noted that although our review focuses primarily on how to publish a standard article based on original research, the publication of other types of articles (e.g., review articles, theoretical articles, case reports) can also be informed by following these steps.

Ten Steps in Choosing a Journal

1. Decide First Whether the Article is Primarily of Interest to a National or an International Audience

This is partly a matter of the article's information content and partly a matter of presentation or appeal. If the topic is primarily of local or national interest (e.g., prevalence of substance abuse among Brazilian secondary-school students or an evaluation of a local treatment program) and the presentation is oriented toward professionals in a particular country, then the article should be submitted to a journal capable of reaching that audience, such as one sponsored by a national professional society. If the topic is likely to appeal to scientists or professionals in many countries and the presentation speaks to this broader audience, then an international journal should be considered. Country- or region-specific case studies of international significance and new advancements or findings with potential international follow-up or applications would also suggest the choice of an international journal. In general, the best way to determine the scope and audience of a journal is to visit the journal's website and review its mission statement.

2. Consider the Language of Publication

English has become the main language of scientific communication throughout the world. Nevertheless, significant numbers of scientific articles are published in German, Russian, Japanese, French, Spanish, Italian, Chinese, and the Scandinavian languages, as indicated by the journals listed in Table 3.2. For most researchers, choosing what language to publish in depends largely on the author's native tongue, the country in which the study was conducted, and the potential audience. Another limiting factor is the availability of an addiction journal that publishes in that language and accepts articles on the author's topic. If one is writing for an international audience, it is wise to choose an English-language journal that can be read by scholars in most countries. Under many circumstances, an article in English will have greater exposure, especially when the journal's articles are included in major abstracting and indexing services (e.g., MEDLINE, Web of Science), most of which operate in the English language. Some journals demonstrate an intentional internationalism that is expressed in a readiness to publish articles and review books submitted from many different countries.

English-language authors can choose between national, more specialized journals or the bigger international journals, depending on the quality of the article, the importance of the findings, and the audience one wishes to reach (see Step 1). If the article is likely to be of interest to an international audience but is not written in English, the author can consider publishing it in English in addition to his or her native language. Multiple publications in different languages, however, require permission from both of the journal editors involved. In some cases, reporting research findings in more than one language will result in very different publications, because the target audience will require different perspectives and background information. The rules of academic integrity and plagiarism still apply, as described in Chapter 14.

Alternatively, researchers writing in languages other than English should consider publishing in journals that provide English-language abstracts (see Table 3.2), thereby gaining entré into some of the world's major abstracting services (see Appendix A to this chapter).

In general, journals published in languages other than English provide a valuable service to national and regional audiences that have a special interest in addiction studies. For example, if an article has special relevance to French-speaking populations, the journal *Alcoologie et Addictologie* (*Alcohol and Addiction Studies*) provides immediate access to that audience not only because of the language it is written in but also because of the network through which the journal is distributed (i.e., the Société Française d'Alcoologie et Addictologie [French Society of Alcohol and Addiction Studies]). Articles written in languages other than English also fulfill an important function by maintaining language use and terminology current and relevant to the addiction field in all of these languages. With the fast-paced changes in addiction science, shifts in use of language and terminology inadvertently mirror the trends in research, society, and scholarly communication. Authors and editors play a significant role in shaping the language of addiction science and promoting use of preferred terms such as nonstigmatizing words and phrases in every language.

Overall, non–English-language journals serve as a necessary medium for communication among clinicians, scientists, and policymakers within major linguistic areas of the world. They increase the range of cultural and scientific diversity in the addiction field and, in this way, provide new opportunities for authors and readers. In some countries, for instance Finland and Norway, journals in national languages have been upgraded as publication channels, even if they do not have an impact factor. Authors whose first language is English should not ignore the advantages of publishing in these journals, which often have a higher acceptance rate and, in some cases, are open to submissions written in English. Depending on the topic and scope of the article, some journals are willing to either translate into the language of publication or publish the article directly in English.

3. Consider Whether to Publish in a Generic, Disciplinary, or Addiction Specialty Journal

The third step involves examining whether the results of a study are mainly of interest to other addiction researchers or to a more general readership. It is probably easier to get an addiction article accepted in an addiction specialty journal. Publishing in a non–addiction journal may require authors to write the article in a way that is understandable to those who do not speak the "addiction dialect."

Some journals, such as *Nature* and *Science,* are multidisciplinary and are oriented toward the general scientific community. Other journals, such as *The Lancet* and the *Bulletin of the World Health Organization,* publish articles dealing with a specific discipline, such as medicine or public health, respectively. In countries without addiction specialty journals, a journal in psychiatry can, for instance, be an important channel for addiction research.

There are several reasons for considering more broadly oriented generic and disciplinary journals. As noted above, disciplinary journals publish a considerable amount of the scientific literature on substance-related research. These journals are generally published by and oriented toward professional groups associated with the major disciplines contributing to addiction studies (i.e., biology, neuroscience, genetics, psychology, medicine, psychiatry, public health, sociology, and anthropology).

Disciplinary journals are sometimes favored by addiction researchers because they are thought to have greater prestige value within a given discipline than addiction specialty journals. Professional advancement for academic researchers is often based on such subtle considerations. Moreover, some of the most popular disciplinary journals (e.g., *The Lancet, The New England Journal of Medicine*) have higher impact factors (discussed below) than addiction specialty journals, which adds to their prestige value.

Nevertheless, the chances of publishing an article on an addiction-related subject are sometimes reduced if a journal does not have reviewers or editors familiar with the topic. If a particular disciplinary journal rarely publishes articles on addiction, it is advisable to contact the editor before submitting an article. In addition, if a disciplinary journal has a large circulation and a high impact factor, authors should make sure that the article is likely to be seen as important before submitting it for review. In the remainder of this chapter, we discuss the merits of publishing in addiction specialty journals, which offer a range of opportunities to prospective authors that are comparable to those available in the disciplinary journals.

4. Review the Journal's Content Range and General Culture

Every journal has a culture of its own, sometimes developed over many years of serving a particular professional society or through the influence of editors who sometimes place their own particular imprint on the journal. The best way to understand that culture is to review several issues of the journal in their entirety, including editorials, letters to the editor, and scientific articles. A visit to the journal's homepage will accomplish the same purpose. Prospective contributors should also read the journal's mission statement, which often describes the focus of the journal, its goals, its preferences, and its audience. Although these statements are sometimes dated and written in general terms, they often provide a broad outline of the journal's traditions, image, priorities, and aspirations.

In Tables 3.1 and 3.2, the first column describes the major substances (and addictive behaviors) that each journal considers part of its purview. Some journals (e.g., *Nicotine and Tobacco Research*) are interested in one particular substance, whereas others are quite generic (e.g., *Drug and Alcohol Dependence*). The topical areas covered by a journal are also an important consideration. Some specialize in treatment research, others in biological effects or mechanisms, and still others in prevention or policy. The less a particular article meets a journal's content areas, the more likely it is to be rejected. Even when an article is considered to be scientifically sound and relevant to the addiction field, it may be dismissed by a journal editor because it does not meet with the journal's current priorities and stated mission. It is therefore important for authors to narrow their choice of journals to those whose history and current contents have demonstrated an interest in (or at least an openness to) the topic, substance, and scope of the article being submitted. When in doubt, it is always advisable for authors to talk with colleagues and communicate with journal editors. By asking someone with experience in publishing for advice, younger or less experienced authors can obtain firsthand information about the priorities and preferences of particular journal editors.

5–6. Evaluate the Journal's Quality and Integrity

Until recently, scientific journals were usually managed by publishing companies and professional societies, which vouched for the quality and integrity of the journal. The most important criterion for quality and integrity is the peer-review process, as overseen by a qualified journal editor and the journal's editorial board. A troubling development in scientific publishing is the proliferation of publishing companies that operate online journals of questionable quality and integrity. Twenty-nine journals in the addiction field operate in open-access formats. One third of them are members of ISAJE, which evaluates their quality and integrity as a condition of membership. Of the remaining journals, several have been evaluated by Thomson Reuters and are listed in the Web of Science. Others are listed in Scopus, PsycINFO, and MEDLINE, which are indexing and abstracting services that have standards that must be met before a journal's articles are listed. And then there are a few online open-access journals that fail to fulfill the minimal criteria for a responsible scientific journal.

Conventional non–open-access journals cover publishing costs through subscriptions and single-article purchases. Some non–open-access journals provide open access after an embargo period of 6–12 months or longer. Some allow authors to post their manuscripts, before final copyediting, on their own or their institution's website. Some allow no open access (see Tables 3.1 and 3.2 for information about the journals that are members of ISAJE).

Open-access journals use a funding model that does not charge readers or their institutions for access. They allow users to read, download, copy, distribute, print, search, or link to the full texts of their articles at no cost to the user. Open-access thus provides unrestricted online access to peer-reviewed scholarly research without the need for a journal subscription or use of a university library. A few open-access journals have financial resources, for instance state support, that make it possible to provide open access without any costs for the author (platinum open access). Most open-access journals operate using a business model in which they charge authors fees to publish their articles (gold open access), but some journals waive these charges. Some unscrupulous entrepreneurs have discovered that this financial base offers an opportunity to make money by providing a publication channel without any quality control. It seems that the majority of new open-access journals levy page charges or processing fees as part of a business model in which a publishing company manages scores, sometimes hundreds, of online journals.

The term *predatory publisher* was coined by Jeffrey Beall, a University of Colorado librarian (Beall, 2012). The term refers to some of the open-access publishing companies that engage in questionable practices with regard to journal management, marketing activities, peer review, and page fees. Efforts to test the quality of the review process conducted by these journals have not been encouraging. One researcher (Davis, 2009) submitted an article with nonsense text and fictitious authors, who were listed as being affiliated with the nonexistent "Center for Research in Applied Phrenology." The author received a letter from the editor of *The Open Information Science Journal* stating that the article had been "accepted for publication after peer-reviewing process" (quoted in Davis, 2009). A publication fee of $800 was requested, to be sent to an address in the United Arab Emirates.

In another case (Bohannon, 2013), a science writer submitted a faked and fabricated cancer article to 305 online journals. The fictitious authors of the article received 157 acceptance letters and 98 rejections. Would the results have been the same had these fake articles been submitted to traditional, subscription-based journals? One would hope that fraud and mediocrity would not be rewarded as easily, but there is some evidence to suggest that the scientific enterprise is not being protected by the traditional academic publishers either.

Until it stopped operating in 2016, Beall's list of predatory publishers was the main resource for authors to verify the quality of publishers and individual journals. It has also started an entire movement of "journal watching," a grassroots movement to maintain the integrity of scholarly communication. Information on new and potentially questionable journals is voluntarily submitted as "hat tips" to the website by scholars, authors, and librarians, who report incidents of inappropriate or unethical practices.

Box 3.2 summarizes the characteristics of journals associated with predatory publishers. As indicated in the box, several tactics are used by these journals to take advantage of the situation in which publication in peer-reviewed journals is considered one of the highest distinctions for peer recognition, academic advancement, and personal accomplishment. These include flattering authors with invitations to contribute articles to be included in special issues and the promise of rapid publication in a peer-reviewed journal. Many scientists have received email invitations to serve on editorial boards by these publishers.

---

1. Rapid acceptance of articles with little or no peer review or quality control
2. Journal names or website styles that resemble those of more established journals
3. Use of poor English grammar and syntax in the journal's website and email communications
4. No issue or only single issue has been published before
5. Aggressive email marketing that urges academics to submit articles or serve on editorial board, sent to you "because of your eminence in the field"
6. Journal editors who have no academic standing or minimal scientific credentials in the topical area of the journal, or the editor cannot be identified at all
7. Article fees not apparent at the time of submission
8. Listing academics as members of editorial boards without their permission
9. No ethical guidelines, or guidelines that apply to the entire range of journals the publisher operates
10. Misleading information about the location of the publishing operation

---

Box 3.2: Characteristics of journals associated with predatory publishers (adapted from Beall, 2012, 2013).

Prospective board members, regardless of their experience or qualifications, are told that if they decide to publish in the journal, they will receive a discounted fee based on the manuscripts they secure from other authors for the journal. In addition to publishing journals, some predatory publishers host conferences, including the publication of the proceedings, for a fee. The latest development is the appearance of the predatory impact factor, an arbitrary number computed by for-profit publishers for a fee (see later section on impact factor).

In the preparation of this chapter, the authors identified several problems with the approximately 20 journals having addiction-related names that are affiliated with predatory publishers and other non-ISAJE, open-access, online, for-profit publishers. Most did not respond to an editors' survey we conducted. Almost half ($n = 9$) had no identifiable editor. Some were found to falsely list indexing/abstracting services. Many listed Google as one of their indexing/abstracting services.

What are the risks of publishing in these journals? The first risk is that your article may not reach its intended audience because these journals are poorly indexed and may not be permanently stored or archived. Many of them simply cease to exist after a few issues. A second risk is that your contribution to the publisher's profit margin may help to perpetuate journals that engage in questionable publishing practices, including the publication of fabricated articles accepted with minimal or nonexistent peer review. A third risk is that when an article published in a questionable journal is listed in a person's curriculum vitae, it may ultimately cause embarrassment to that individual, or there could be worse consequences. As these problems become more apparent in the future, the quality of journals will be evaluated more rigorously by those charged with protecting scientific integrity, as well as university committees charged with hiring, appointments, promotions, and tenure decisions. Publishing in or being listed on the editorial boards of low-quality or unverifiable journals may be a disadvantage in that these publications could count against hiring, promotion, or tenure because they represent such poor scientific quality.

What can be done to protect authors from being exploited and embarrassed by publishing in a journal that does not operate competently, ethically, and scientifically? In the addiction field, quality control is provided by ISAJE, an organization that insists that its 33 member journals subscribe to a set of core principles covering appropriate peer review, conflict of interest policies, editorial management, and transparency (Farmington Consensus, 1997). As shown in Tables 3.1 and 3.2, ISAJE has several online open access journals that are fully compliant with the Farmington Consensus. Another precaution is to find out whether the journal receives any significant citations by checking Web of Science or the Journal Citation Reports before submitting to an open-access journal. Ulrich's Periodicals Directory, now available online in most academic libraries (on a subscription basis), has been a trusted resource to find information about scholarly journals, magazines, and newsletters since 1932. The most effective precaution is not to submit an article to an open-access journal published by an organization that meets the criteria listed in Box 3.2, sponsorship by a learned society, email or telephone access to an editor who is qualified to manage manuscripts, evidence that there is a rigorous peer-review process, and the existence of a verifiable Thomson Reuters impact factor are other ways to determine whether a journal is reputable.

Finally, the reputation and scientific standing of a journal can be checked by verifying that the journal is indexed in one or more of the key indexing and abstracting services that disseminate information only about journals that meet minimal criteria for quality and integrity (e.g., MEDLINE, PsycINFO, Scopus, Web of Science). This is discussed in the following section.

7. Gauge Your Article's Potential Exposure by Reviewing the Journal's Indexing and Abstracting Services, as Well as its Open-Access Policy

One of the most important goals of scientific publication is to reach one or more specific audiences, such as the scientific community, clinical practitioners, or policymakers. A journal's ability to provide exposure to these audiences is determined by its circulation (print and electronic) and its dissemination capabilities, determined by access to abstracting and indexing services.

Print circulation refers to the number of copies printed for the journal's subscribers as well as those who receive free copies. Scholarly journals have two major types of subscribers: members of professional organizations and academic libraries. In addition, there are smaller numbers of personal and nonacademic institutional subscribers. Before the advent of the Internet, the number of journal copies in circulation was a good indicator of a journal's exposure. Today, figures describing the number of visits to homepages or the number of downloads may be better measures of how extensively and frequently a journal is read.

In addition to traditional circulation data, article-level alternative metrics beyond page visits and download counts provide evidence of the immediate impact of the article (i.e., readership, as reflected in scholarly social media; Weller, 2015). A new field of measuring scholarly performance, called *altmetrics,* compiles data on the publication's appearance in the various social media outlets, such as shares or mentions on Facebook and Twitter, or in blogs, as well as in mainstream media (Piwowar, 2013; Priem et al., 2012). As evidence of immediate exposure, the citation analysis computed by altmetrics is said to be a good indicator of an article's success when compared with traditional bibliometrics, such as citation counts (Ortega, 2015). A few journals, such as *Addiction,* already indicate these metrics on their sites at the article level, often symbolized with the so-called altmetric donut, with each color representing a different type of alternative metric.

If an article is relevant to the members of a particular learned society (e.g., the British Society for the Study of Addiction to Alcohol or Other Drugs), professional group (e.g., the Canadian Medical Association), or scientific organization (e.g., the Research Society on Alcoholism), then it may make sense to submit the manuscript to a journal that is sponsored by that organization. Many of the journals listed in Tables 3.1 and 3.2 are sponsored by professional organizations or learned societies that provide free subscriptions or reduced rates to their members. For example, *Alcoologie et Addictologie* (*Alcohol and Addiction Studies*) is sponsored by the Société Française d'Alcoologie [French Society of Alcohol Studies], which distributes free copies of the journal to its 1,400 members. *Psychology of Addictive Behaviors* is published by the American Psychological Association, which makes the journal available to members of the Society of Addiction Psychology (American Psychological Association, Division 50) at a reduced subscription rate. See Chapter 2, Table 2.2 for a complete list.

In addition to targeting organizational subscribers, exposure is also affected by the number of library subscriptions. Libraries, especially university libraries, guarantee exposure to students and scholars, thereby providing direct access to perhaps the most important audience for any scientific communication. Currently, a single subscription from a large university library might mean exposure to as many as several thousand potential readers, because journal subscriptions are based on full-time equivalents (a calculation of faculty, staff, and students). The world of library and information science has changed rapidly in the past decade, with electronic subscriptions replacing or supplementing print copies available on the library shelf. Library subscriptions remain an important conduit for a publication to reach a broader audience than the members of a particular learned society, but now subscriptions are mainly electronic and discoverability and access have become key components of exposure. University libraries and other large information sources have begun to pool resources to increase electronic availability of full-text journals. This also means that the same journal can be available from various content providers on various platforms in various subscription packages. Tools provided by modern technology to describe, organize, and access information include versions of the library catalog, the database of the content provider, and the full text of the article from the journal optimized for mobile devices. The result is an increased discoverability, more exposure, and potentially larger impact. Access to individual articles is no longer limited to content subscribed to by the library. As a result of consortia and interlibrary-loan agreements, those affiliated with an institution of higher education can have full-text articles delivered on their desktop, free of charge, as fast as the next day after placing a request. Unaffiliated readers can also benefit from the better discoverability provided by proprietary and subscription databases, as well as from the new access options for a fee, such as previewing or renting an article instead of purchasing it.

Beyond the journal's print circulation and subscriber base, an article's exposure is now determined primarily by the electronic databases that index the published literature by author, topic, and bibliographic reference and provide abstracts of articles for potential readers in search of particular types of information. Abstracting and indexing services provide detailed information about the content of scientific journal articles and eBooks by adding metadata and abstracts, which are invaluable for those without immediate access to the full text of the article. Proprietary databases use a controlled vocabulary by establishing preferred terms for each word or concept, such as Medical Subject Headings (MeSH) in MEDLINE. Added to the individual articles by a trained indexer as a subject heading or descriptor, these keywords ensure that the main ideas of the articles will become transparent and are appropriately conveyed. As a result, the article will be discoverable and retrievable via a search conducted in the database. Users can locate relevant articles, chapters, or books in the databases enhanced with abstracting and indexing services at a higher rate of precision by searching the metadata, keywords, and the abstract than they can by using a free search engine, such as Google Scholar, which searches the full text of articles, resulting in higher recall and lower precision. Those affiliated with an institution that has access to the service as well as a subscription to the particular journal can download the full text with the help of an article-linker application. If it is an open-access publication, they can immediately use the full text. Otherwise, they are usually shown information from the publisher on how to access the text. More than 100 companies and institutions currently offer abstracting and indexing services, but many may not cover subject areas related to addiction. At present, no single service or database is available to cover the entire addiction literature, as is the case, for example, for the psychology literature (PsycINFO). To the extent that most of the information summarized in this paragraph applies to the English-language literature, the reader is referred to Chapter 4 ("Beyond the Anglo-American World") for information and advice related to publishing in other languages.

Appendix A lists some of the main abstracting and indexing services used by the addiction specialty journals listed in Tables 3.1 and 3.2. These organizations provide a variety of important services that dramatically increase the potential exposure of a scholarly communication. Although some of these databases used to be available in both print and electronic versions, electronic databases have now become the information source of choice for those who are searching for topical information via the Internet. They are comprehensive and rapid, and at least some of the information is often inexpensive or free. These services differ widely in their subject matter, coverage of the literature, document types included, service features, and content provider. The major databases (e.g., MEDLINE, Scopus, Web of Science, PsycINFO) are available through libraries that pay a subscription fee and are highly selective in choosing the journals that they list in their index. They permit searches of the current and past literature according to author, title, and keywords, often providing the author's abstract for review. Other abstracting and indexing services (e.g., Sociological Abstracts) are selective and scholarly but tend to reach a smaller distribution network. Still other services (e.g., Google Scholar) are more general in nature and may not provide the best access to the audience an author is trying to reach.

Many of the journals operated by predatory publishers are indexed or listed only in services databases such as the Directory of Open Access Journals or are only crawled by Google and Google Scholar. DOAJ is a reference tool, based on the fact that a journal is available free on the web. Google and Google Scholar are search engines that aggregate information from the internet and are not abstracting and indexing services.

From the author's perspective, a journal's ability to provide a listing of its journal articles and abstracts to these secondary information sources greatly increases an article's exposure to scholars and students throughout the world. The greater the number of quality indexing and abstracting services a journal belongs to (as indicated in Tables 3.1 and 3.2), the more likely it is that an article will reach its intended audience. Although many of the non–English-language journals indicate minimal coverage in abstracting and indexing databases, this situation is changing rapidly, and most of these journals now provide English abstracts and keywords, an important first step in reaching an international audience.

8. Evaluate Your Chances of Acceptance

A major consideration in the choice of a journal is the likelihood of accept-ance. Journals vary tremendously in the criteria they use to select articles for publication and in the competition a given article will encounter in relation to other authors seeking to claim the same journal space. Some journals have high acceptance rates and are often looking for articles to publish. Other journals have a surfeit of submissions, making it necessary for editors to reject articles that would nevertheless be worthy of publication in less competitive journals. A journal's acceptance rate provides a rough estimate of an author's chances of eventual acceptance, but the rates listed in Tables 3.1 and 3.2 are subject to a number of limitations. First, some journals do not know or choose not to reveal their acceptance rate. In the ISAJE member journal survey conducted for the preparation of this chapter, we asked journal editors to tell us the proportion of articles accepted that were eligible for peer review (regardless of whether the articles were sent out for review or were returned un-reviewed).

It should be noted that several journals (e.g., *Alcohol Research: Current Reviews, Addiction Science & Clinical Practice*) operate primarily by commis-sioning authors to write articles on a topic or theme, which accounts for their high acceptance rates. Beyond a journal's acceptance rate, an author's chances of acceptance depend on many other considerations, some of them scientific, some stylistic, others administrative.

Stylistic factors include the quality of the writing and the way in which the data are presented. If the article is poorly written or not well organized, reviewers may see this as a limitation, and editors may be reluctant to take the time to work with authors to bring the article up to the journal's standards. Administrative factors include the length of the article, the amount of revision required, and the appropriateness of the topic to the journal's mission. If an article is too long, it reduces the amount of space available for equally worthy articles that are written more concisely by competing authors. If the article is not appropriate to the journal's current priorities or mission statement, it might be rejected even before it is sent out for peer review. Finally, the number of articles published by a journal could affect chances of acceptance. Journals that are published monthly or weekly need to accept more articles than journals that publish less frequently. But journals that publish more frequently also tend to be more competitive. See Chapter 12 for further discussion of factors influencing the acceptance or rejection of manuscripts.

9. Take into Account Time to Publication and Other Practical Matters

There are several other factors that should be taken into account in selecting a journal. One is the lag time to publication. Some journals take longer than others to process their manuscripts. However, most journals do not reveal how long it takes to arrive at a decision; and even when this information is available, it should be noted that the average time is affected by the number of manuscripts that are rejected before being sent out for peer review. Another factor is the time between the acceptance of a revised manuscript and its final publication. This will depend in part on the number of issues published by the journal per year, the number of accepted manuscripts, and the efficiency of the publisher. In general, journals that publish more frequently are likely to have a shorter lag time to publication. The best way to obtain information about the review process is to consult the journal's instructions to authors or the journal's website. It is best not to rely on hearsay, anecdote, or the journal's reputation.

10. Consider, but Don't Be Fooled by, Impact Factors

The Journal Impact Factor is an attempt to provide an objective measure of how often a scientific journal's published work is cited. Such a measure has also been used to judge the quality of an author's work, to the extent that publishing in a high-impact journal may reflect the quality of a particular article. The impact of a journal on a field of study is thus based on the assumption that the more a journal's articles are cited, the more influence it has on the field. In 1964, the Institute of Scientific Information began publishing the Science Citation Index. By the early 1990s, 3,200 journals belonged to the core or citation journals of Science Citation Index (Seglen, 1998).

Impact factor was originally developed to objectively compare the quality of journals listed in a particular database (i.e., Journal Citation Reports [now of Thomson Reuters], which provides tools for ranking, evaluating, categorizing, and comparing journals; Garfield, 1994). Devised by Eugene Garfield, the founder of the Institute for Scientific Information, impact factors are calculated annually based on the data of the previous years. The impact factor of a journal is the average number of citations received per article published in that journal during the two preceding years. Impact factor is widely used to compare journals in a particular field, and, as such, its use has generated a lot of debate concerning its validity as a measure of a journal's importance as well as a reflection of the quality of an author's work (for more on the history and use of impact factor, see Garfield, 1994).

Increasingly, the data used to calculate impact factors have been used as a shortcut to compare and rank individual articles, researchers, and research groups. Impact factor has been criticized almost from its inception (Seglen, 1998; Stenius, 2003), partly because its databank covers only a small share of the world's scientific journals. Different research fields have different coverage in the database. The database has a clear preference for English-language journals (particularly those based in the United States). National or regional journals in other languages are not well represented (Seglen, 1998), as indicated by only one of the journals listed in Table 3.2 having an impact factor. All journals from a field that is underrepresented will receive lower impact factors. In addition, citation frequencies and patterns vary among different research fields. Thus, it is not acceptable to compare impact factors for journals from different fields. A journal representing a field that typically favors large numbers of references will automatically get a higher impact factor, especially if the field is quickly developing. Research fields that get references from related disciplines get higher impact factors. This explains why journals focusing on basic science have higher values. The humanities are in a particularly unfavorable position. Disciplines in which national or regional research, or publications in local languages, are important also tend to get low impact factors (Rousseau, 2002).

As a measure of impact, with its two-year time frame, impact factor is more appropriate for quickly developing research fields, such as molecular medicine. Applied, clinical, or social sciences do not fare as well with the two-year window (Andersen, 1998; Luukkonen, 1994). Non-English-language journals or bilingual journals (for instance Japanese-English), even if included in Science Citation Index, will on average receive a lower impact factor. The recently introduced five-year impact factor is supposed to provide a more balanced picture of the performance of journals.

Finally, it is important to note that a citation is not necessarily an indication of research quality. Every researcher knows that there are numerous reasons (apart from its quality) for citing a scientific publication. Authors may cite or quote for polemical reasons, to flatter their readers, or to promote their own research (or that of their friends, colleagues, or patrons). West and McIlwaine

(2002) studied 79 articles published in *Addiction* between 1995 and 1998 and found no correlation between citation frequency (up to the year 2000) and an independent quality rating. Interestingly, West and McIlwaine also found that articles from the developing world received fewer citations than the quality ranking would have led them to expect. (See Chapter 10 for further discussion of citation procedures).

As a response to the narrowly defined yet widely influential impact factor, a new metric called Eigenfactor was created in 2007 to rank journals in a more comprehensive way. Eigenfactor (eigenfactor.org), also available from Thomson Reuters along with impact factor, expands the timeframe to five years and takes into account the influence level of the citing journals in its algorithm.

Similarly, a nonproprietary application, SCImago Journal Rank (SJR; scimagojr.com) attempts to rank journals with its SJR indicator based on the popular Google PageRank algorithm, which also takes into account the quality of journal citations in addition to quantity (Moed, 2006). Another metric, Source Normalized Impact per Paper (SNIP), factors the amount of potential citing sources based on the size of the field in order to normalize the numbers for direct comparison (Moed, 2010). These recent statistics indicate that no single statistic can definitively rank journals in a comprehensive way.

Although altmetrics and scholarly social media are promising alternatives to measure scientific impact both at the level of an article or the author (Ward et al., 2015), they are not treated as equivalents in most fields of science, in which impact factor predominates. However, there is a widely accepted indicator of an individual's scholarly performance, the *h*-index, which is also based on citations. Introduced by Hirsch (2005), this performance indicator computes a scholar's top-cited articles rather than considering the total citation count. The main problem with this metric is, as with impact factor, the number is computed within a particular database only and, as such, will be only as accurate as the data input. As an example, an author with 250 total publications will be underrepresented in Scopus if only 75 of these articles are listed under the author's name due to the coverage of the database. On the other hand, with its duplicates and erroneous author attributions, Google Scholar Citations can display an inflated number closer to 400. The discrepancies will lead to an embarrassing *h*-index in the first case and a falsely high one in the second, with a difference of as many as 20 points. Either way, it is beyond the author's reach to correct them.

In conclusion, impact factors should be treated with caution. Until the deficiencies in the system have been corrected and its limitations are better understood, however, impact factor remains a relatively crude index of the value of a particular journal. According to Jones (1999), authors should not be preoccupied with the impact factor of a journal. Rather, they should give more consideration to the speed and efficiency of the editorial handling of their manuscripts, the selectiveness of its abstracting and indexing services, and the quality and timeliness of the peer review.

Conclusion

Journals differ in the quality of articles they publish, the exposure they provide to an author's work, and their subject matter. Once an author or a group of authors has a clear idea of the results of a particular study or project, it is often valuable to conduct a preliminary review of the journals most likely to publish an article on that subject. As indicated in Tables 3.1 and 3.2, there are many peer-reviewed addiction specialty journals to choose from, as well as hundreds of disciplinary and multidisciplinary journals. The careful selection of a journal, when one takes into account both scientific and practical considerations, is clearly worth the effort. Not only is the process likely to save valuable time for authors, peer reviewers, and journal editors, but it also will increase the likelihood that an article will contribute as much to science as it does to the author's curriculum vitae.

> Please visit the website of the International Society of Addiction Journal Editors (ISAJE) at www.isaje.net to access supplementary materials related to this chapter. Materials include additional reading, exercises, examples, PowerPoint presentations, videos, and e-learning lessons.

References

Andersen, H. (1998). Acta Sociologica på den internationale arena: Hvad kan SSCI fortelle? [Acta Sociologica on the international arena: What can SSCI tell us?]. *Dansk Sociologi*, VII, 72–78.

Babor, T. F. (1993). Beyond the invisible college: A science policy analysis of alcohol and drug research. In G. Edwards, J. Strang, & J. H. Jaffe (Eds.), *Drugs, alcohol and tobacco: Making the science and policy connections* (pp. 48–69). Oxford, England: Oxford University Press

Babor, T. F. (1993) Megatrends and dead ends: Alcohol research in global perspective. *Alcohol Health and Research World, 17,* 177–186.

Beall, J. (2012). Predatory publishers are corrupting open access. *Nature, 489,* 179. DOI: https://doi.org/10.1038/489179a

Beall, J. (2013). Predatory publishing is just one of the consequences of gold open access. *Learned Publishing, 26,* 79–84. DOI: https://doi.org/10.1087/20130203

Bohannon, J. (2014). Who's Afraid of Peer Review? *Science, 342,* 60–65. DOI: https://doi.org/10.1126/science.342.6154.60

Cohen, J. F., Korevaar, D. A., Wang, J., Spijker, R., & Bossuyt, P. M. (2015). Should we search Chinese biomedical databases when performing systematic reviews? *Systematic Reviews,* 4, 23. DOI: http://doi.org/10.1186/s13643-015-0017-3; http://www.ncbi.nlm.nih.gov/pmc/articles/PMC4374381/

Davis, P. (2009, June 10). Open access publisher accepts nonsense manuscript for dollars. *The Scholarly Kitchen*. Retrieved from http://scholarlykitchen. sspnet.org/2009/06/10/nonsense-for-dollars/

Farmington Consensus. (1997). *Addiction, 92,* 1617–1618.

Garfield, E. (1994, June 20). *The Thomson Reuters Impact Factor.* Thomson Reuters. Retrieved from http://wokinfo.com/essays/impact-factor/.

Hirsch, J. E. (2005). An index to quantify an individual's scientific research output. *Proceedings of the National Academy of Sciences of the United States of America, 102,* 16569–16572. DOI: https://doi.org/10.1073/pnas. 0507655102

Jones, A. W. (1999). The impact of *Alcohol and Alcoholism* among substance abuse journals. *Alcohol and Alcoholism, 34,* 25–34.

Lafollette, M. C. (1992). *Stealing into print.* Fraud, plagiarism, and misconduct in scientific publishing. Berkeley, CA: University of California Press.

Luukkonen, T. (1994). Viiteanalyysi ja tutkimuksen arviointi [Reference analysis and evaluation of research]. *Signum, 27,* 130–132.

Moed, H. F. (2006). *Citation analysis in research evaluation.* Dordrecht, The Netherlands: Springer.

Moed, H. F. (2010). *The Source-Normalized Impact per Paper (SNIP) is a valid and sophisticated indicator of journal citation impact.* Retrieved from http:// arxiv.org/ftp/arxiv/papers/1005/1005.4906.pdf.

Ortega, J. L. (2015). Relationship between altmetric and bibliometric indicators across academic social sites: The case of CSIC's members. *Journal of Informetrics, 9,* 39–49. DOI: https://doi.org/10.1016/j.joi.2014.11.004

Piwowar, H. (2013). Altmetrics: Value all research products. *Nature, 493,* 159. DOI: https://doi.org/10.1038/493159a

Priem, J., Piwowar, H. A., & Hemminger, B. M. (2012). *Altmetrics in the wild: Using social media to explore scholarly impact.* Retrieved from http://arxiv. org/html/1203.4745.

Rousseau, R. (2002). Journal evaluation: Technical and practical issues. *Library Trends, 50,* 418–439.

Seglen, P. (1998). Citation rates and journal impact factors are not suitable for evaluation of research. *Acta Orthopaedica Scandinavica, 69,* 224–229.

Stenius, K. (2003). Journal impact factor - mittari joka vahvistaa tutkimusmaailman hierarkiaa [The journal impact factor—an indicator that strengthens the hierarchy of the research world]. *Tieteessä Tapahtuu, 7,* 35–39. Retrieved from http://www.tieteessatapahtuu.fi/037/stenius.pdf.

Ward, J., Bejarano, W., & Dudás, A. (2015). Scholarly social media profiles and libraries: A review. *Liber Quarterly, 24,* 174–204. DOI: https://doi. org/10.18352/lq.9958/

Weller, K. (2015). Social media and altmetrics: An overview of current alternative approaches to measuring scholarly impact. In I. M. Welpe, J. Wollersheim, S. Ringelhan, & M. Osterloh (Eds.), *Incentives and performance* (pp. 261–276). Switzerland: Springer International Publishing.

West, R., & McIlwaine, A. (2002). What do citation counts count for in the field of addiction? An empirical evaluation of citation counts and their link with peer ratings of quality. *Addiction*, *97*(5), 501–504.

Xia, J., Wright, J., & Adams, C. E. (2008). Five large Chinese biomedical bibliographic databases: accessibility and coverage. *Health Info Libr J., 25*, 55–61. DOI: https://doi.org/10.1111/j.1471-1842.2007.00734.x [PubMed] [Cross Ref]

Appendix A: An Inventory of Abstracting and Indexing Services and Databases Relevant to the Scientific Literature on Addiction

Subscription databases are available through the author's institutional subscription. Please contact your local library for information on how to access them.

Chinese Databases

According to the China National Knowledge Infrastructure, large numbers of publications may be missed when not searching Chinese databases. These databases index 2,500 journals largely not familiar to MEDLINE users. Free access, search features, record selection, ease of downloading, and cost of subscription varies considerably between databases. At a minimum, Chinese biomedical databases should be searched when performing systematic reviews. (See Xia et al. (2008); Cohen et al. (2015))

CSA Sociological Abstracts (Subscription)

CSA Sociological Abstracts provides an index and abstracts of journal articles from the international literature in sociology and related disciplines in the social and behavioral sciences. Major subject areas include evaluation research, family and social welfare, health law, substance abuse, and addiction. Its database is drawn from more than 2,000 serials publications, including a variety of sources such as journal articles, conference papers, books, dissertations, and conference papers, plus citations to important book reviews related to the social sciences. A backfile that begins in 1952 adds to the coverage with records published by the then print version of Sociological Abstracts. Because 40% of the provided content is published outside of North America, the database also provides a global perspective. The database is updated monthly with approximately 30,000 records added per year.

Current Contents (subscription)

Current Contents, a current awareness database developed at the Institute for Scientific Information, now part of Thomson Reuters, provides access to bibliographic research information from articles, editorials, meeting abstracts, and other sources from more than 8,000 scholarly journals, with separate editions for clinical medicine, life sciences, and social and behavioral sciences. Internet access is provided through Current Contents Connect. Updated daily, it provides access to complete tables of contents, abstracts, and bibliographic information from the most recently published journals and books. CC Connect offers cover-to-cover indexing that provides access to all the valuable information available in journals — not just articles.

Directory of Open Access Journals (DOAJ) (Open Access): https://doaj.org/

DOAJ is an online directory that indexes and provides access to high quality, open-access, peer-reviewed journals. Launched at Lund University, Sweden, in 2003, DOAJ is a membership organization, with membership intended to prove a commitment to quality, peer-reviewed open access. The aim of the DOAJ is to promote increased usage and impact of open-access scientific and scholarly journals by increasing their visibility and ease of use. Including more than 10,000 journals from 134 countries, it covers over 2 million articles, of which more than 6,000 are searchable at the article level. Subjects listed include broad areas such as medicine, health sciences, psychiatry, public health, and social sciences, indexing them with top-level Library of Congress Subject categories only. Keyword search is available for the full text of the article, with high recall and low precision. The directory claims to be comprehensive and cover all open-access academic journals that use an appropriate quality-control system. DOAJ is independent and is not connected to, or owned by, any other organization or business. To be included, a journal must exercise peer review with an editor and an editorial board or editorial review carried out by at least two reviewers. The DOAJ Seal of Approval for Open Access Journals is a mark of certification awarded by DOAJ to journals that achieve a high level of openness, adhere to best practices, and have high publishing standards.

DrugWise: http://www.drugwise.org.uk/

Launched in 2016 as a continuation of DrugScope, DrugWise is a new drug information service located in the United Kingdom. The full range of DrugScope archival materials is complemented with updates and new reports on drugs, alcohol, and tobacco (including e-cigarettes). In addition to drug information, such as the DrugSearch Encyclopedia and Drug-Wise reports, a new function, called I-Know, serves as an international knowledge hub. I-Know brings together international and internationally-relevant national reports and reviews covering the range of substances. The plan is to build up a library of information, policy and practice material over time.

Fetal Alcohol Spectrum Disorders (FASD) Database: http://fasdcenter.samhsa.gov/search/basic/index.aspx

The FASD Database collects information on thousands of FASD-related resources, including audiotapes, books, CD-ROMs, newsletter, magazine, newspaper, and journal articles, pamphlets and booklets, posters, videos, slide shows, and Web-based materials. It includes a quick-search function activated by typing in a keyword and selecting a media type and an advanced search option for more specific searches.

EMBASE (Elsevier) (Subscription)

Embase is a comprehensive index of the world's literature on human medicine and related disciplines. Each record is classified and indexed using terms and synonyms that assist the process of searching for specific subjects. Subject coverage includes AIDS, drug dependence, psychiatry, and public health. EMBASE provides access to articles from more than 2,900 journals from 110 countries.

CORK Database (Open Access): www.projectcork.org

Project Cork was founded at Dartmouth Medical School in 1977 through a grant from Operation Cork. The project also resulted in CORK, a searchable bibliographic database of the substance abuse literature and the emerging area of behavioral addictions. Its goal is to provide immediate access to authoritative information and materials on substance abuse and to assist health and human service professionals, educators and their students as well as those in public policy. The CORK database contains 120,500 items including journal articles, books, book chapters, conference proceedings, and special reports on substance abuse, indexed by more than 400 terms. The database was updated quarterly until 2015.

Google Scholar (Open Access): scholar.google.com

Google Scholar provides access to the scholarly literature across many disciplines and sources, including articles, theses, books, abstracts, and court opinions. Crawling millions of pages of the public and invisible web, it indexes full text of the scholarly literature, gathering information from academic publishers, professional societies, online repositories, universities, and other websites. Individual authors can be listed in Google Scholar by simply uploading their articles to a website. The main advantage of Google Scholar is the convenience of searching all scholarly publications on one platform. It allows the user to find related articles. It is currently the fastest way to locate a known item and to retrieve the full text(for either open-access items or titles that one's library subscribes to). Authors can create a publicly accessible author profile in Google Scholar Citation to showcase their work and track citations to their publications (scholar.google.com/citations). The main disadvantage of this service derives from the lack of a controlled vocabulary (i.e., instead of index terms describing the articles, as it is customary in the proprietary databases such as MEDLINE or PsycINFO, the search is performed in the full text of the publication, resulting in many irrelevant hits.)

International Alcohol Information Database (IAID): www.icap.org

Launched in 2014, the International Alcohol Information Database is a publicly accessible bibliographic resource created to provide an easily searchable database of published research on alcohol. It covers multiple disciplines, including biomedical, sociobehavioral, prevention, treatment, policy, and regulatory research fields. Citations are compiled from more than 3,550 peer-reviewed journals from around the world, and the included research is available in 30 languages and from more than 150 countries. The continually updated database has approximately 50,000 citations from peer-reviewed research journals dating back to 2003, accessible through simple or advanced search options. The advanced search allows users to refine their results through title, author, journal, or publication date, as well as through an extensive list of keywords, the countries covered in the research, or the original publication language. There is no cost to search, register, or access the database's content. The database is supported by funding from the International Alliance for Responsible Drinking, a consortium of beer, wine, and spirits producers.

MEDLINE (Subscription), PubMed, PubMed Central

MEDLINE (Medical Literature, Analysis, and Retrieval System Online) is the U.S. National Library of Medicine's journal citation database. MEDLINE is widely known as the major source for bibliographic and abstract coverage of biomedical literature, covering the topics of medicine, nursing, dentistry, as well as other areas, such as allied health, biological and physical sciences, humanities, and information science as they relate to medicine and health care, communication disorders, population biology, and reproductive biology. Started in the 1960s, MEDLINE now provides more than 22 million references and includes citations from more than 5,600 scholarly journals published in the United States and other countries. The Literature Selection Technical Review Committee reviews and recommends journals for MEDLINE considering the quality of the scientific content, including originality and the importance of the content for the MEDLINE global audience, using the guidelines found on the National Library of Medicine Fact Sheet MEDLINE Journal Selection (http://www.nlm.nih.gov/pubs/factsheets/jsel.html). Although MEDLINE is restricted to institutional subscribers, such as libraries, the content of the database can be searched free of charge via PubMed (http://www.ncbi.nlm.nih.gov/pubmed) and PubMed Central (http://www.ncbi.nlm.nih.gov/pmc). MEDLINE is the largest subset of PubMed, with the added value of using the National Library of Medicine controlled vocabulary—Medical Subject Headings (MeSH)—to index the citations in MEDLINE. MEDLINE is updated daily.

PubMed (http://www.ncbi.nlm.nih.gov/pubmed) has been available since 1996. It offers more than 25 million references, including the MEDLINE database and additional types of records, such as in-process citations, citations to articles that are out of scope, epub ahead-of-print citations, citations to author manuscripts of articles published by National Institutes of Health–funded researchers, and citations for the majority of books available on the National Center for Biotechnology Information Bookshelf. Both MEDLINE and other PubMed records may include links to full-text articles, depending on open-access and subscription-based availability.

PubMed Central (http://www.ncbi.nlm.nih.gov/pmc) was launched in 2000 as a free repository of full-text biomedical and life-sciences journal articles. It serves as a collection for scholarly literature deposited by either participating publishers or authors who submitted their manuscripts in compliance with the National Institutes of Health Public Access Policy and similar policies. Some PubMed Central journals are also indexed in MEDLINE. There are reciprocal links between the full text in PubMed Central and corresponding citations in PubMed.

PsycINFO (Subscription)

PsycINFO is the electronic version of Psychological Abstracts, which was published by the American Psychological Association monthly for 80 years and ceased in 2006. With nearly 4 million bibliographic records focusing on the scholarly literature in the behavioral sciences and mental health, the PsycINFO database provides a unique resource for locating scholarly literature for addiction researchers. It contains bibliographic records and abstracts of English-language articles from journals originating in more than 50 countries, all professionally indexed by American Psychological Association experts. PsycINFO is available through library subscriptions and to individual members of the American Psychological Association. Nearly 2,500 journal titles (99% of which are peer reviewed) are covered in the database. Articles are selected based on their relevance in psychology and related fields, such as psychiatry, management, business, education, social science, neuroscience, law, medicine, and social work. The database also covers books and book chapters, 3% and 8% of PsycINFO records, respectively. Its global perspective is proven by indexing publications from more than 50 countries, journals from 29 languages, and non–English-language titles in Roman alphabets since 1978. Easy discoverability and high precision during literature searches are ensured by 22 major categories and 135 subcategories in the classification system and by the controlled vocabulary describing the articles, with more than 8,400 terms and cross-references. Online access to a Thesaurus of Psychological Index Terms is included. PsycINFO is updated weekly.

Scopus (Elsevier) (Subscription)

Scopus is the largest abstract and citation database of peer-reviewed literature, covering scientific journals, books, and conference proceedings. It provides a comprehensive overview of the world's research output in the fields of science, technology, medicine, social sciences, arts, and the humanities. Scopus features tools to track, analyze, and visualize research. It is oriented toward researchers, teachers, and students. Scopus claims that it has twice as many titles and more than 50% more publishers listed than any other abstracting and indexing database. It contains more than 50 million records with coverage strongest in the physical sciences (7,200+ titles) and health sciences (6,800+ titles), followed by the life sciences (4,300+ titles), and finally the social sciences and humanities (5,300+ titles). More than 25,000 titles (including open-access journals) from around the world are covered in Scopus. Quick searches by document, author, or affiliation are available, but there is also an advanced search option. Scopus offers several methods of analysis, such as the Journal Analyzer, which compares the citation metrics of different journals using SCimago Journal Rank and other metrics. Authors can benefit from the citation overview function, which includes the h-index, computed from the author's publications listed in Scopus. It is updated daily.

Web of Science (Thomson Reuters) (Subscription)

Thomson Reuters provides paid subscribers with comprehensive coverage of the world's most important journals. Web of Science covers more than 12,000 international and regional journals in the natural sciences, social sciences, the arts, and humanities. Three citation indexes contain the references cited by the authors of the articles: Arts & Humanities Citation Index (from 1975 to the present), the Science Citation Index Expanded (from 1900 to the present), and the Social Sciences Citation Index (from 1900 to the present). The database provides bibliographic records, searchable abstracts, and cited references. Many factors are taken into account when evaluating journals for coverage in Web of Science, ranging from the qualitative to the quantitative. The journal's basic publishing standards, its editorial content, the international diversity of its authorship, and the citation data associated with it are all considered. Thomson Reuters also determines if an electronic journal follows international editorial conventions, which are intended to optimize retrievability of source articles. These conventions include informative journal titles, fully descriptive article titles, and author abstracts, complete bibliographic information for all cited references, and full address information for every author. Thomson Reuters editors look for international diversity among the journal's contributing authors, editors, and editorial advisory board members. For more information, see: http://wokinfo.com/essays/journal-selection-process/. Coverage is strongest in the sciences (8,000+ journals), followed by social sciences (almost 3,000 journals), and arts and humanities (approximately 1,600 journals). For impact factor information about specific journals, users are directed to the index Journal Citation Reports.

For more Information

The Substance Abuse Librarians and Information Specialists website offers a more comprehensive collection of abstracting and indexing services and other related databases (salis.org/resources). Maintained by Barbara Weiner of Hazelden–Betty Ford, the lists are monitored by the group and updated frequently both for U.S. and international services.

United States: http://www.hazelden.org/web/public/usdatabaselibrary.page

International: http://www.hazelden.org/web/public/lib_cdandaddictions.page

CHAPTER 4

Beyond the Anglo-American World: Advice for Researchers from Developing and Non–English-Speaking Countries

Kerstin Stenius, Florence Kerr-Corrêa, Isidore Obot, Erikson F. Furtado, Maria Cristina Pereira Lima and Thomas F. Babor

Introduction

Today, more than 81% of the world's population lives in nations categorized as low- and middle-income countries (LMICs) (World Bank, 2014). However, there are still few addiction journals published outside Europe, the United States, and Australia (see Table 3.2, Chapter 3), despite the growing need for specialized knowledge in many countries where addiction problems are prevalent.

Presently, between 5% and 9% of the world's population grows up with English as their first language. The dominance of English within scientific communication is, however, overwhelming. It is estimated that 80% of the world's scientific articles are published in English-language journals (Montgomery, 2004; Van Weijen, 2012). The dominance is particularly strong in the physical and life sciences, whereas local languages may still have important roles in social sciences, law, and humanities. In the addiction field, we estimate that at least three fourths of the known addiction journals communicate in English.

This chapter deals with the challenges encountered by addiction scientists who work in countries with few resources as well as those whose first language is not English. The aims of the chapter are to discuss (a) the practical and

How to cite this book chapter:
Stenius, K, Kerr-Corrêa, F, Obot, I, Furtado, E F, Lima, M C P and Babor, T F. 2017. Beyond the Anglo-American World: Advice for Researchers from Developing and Non–English-Speaking Countries. In: Babor, T F, Stenius, K, Pates, R, Miovský, M, O'Reilly, J and Candon, P. (eds.) *Publishing Addiction Science: A Guide for the Perplexed*, Pp. 71–88. London: Ubiquity Press. DOI: https://doi.org/10.5334/bbd.d. License: CC-BY 4.0.

professional issues that are faced by these scientists, (b) how authors who come from these countries can improve their chances of publishing in English-language journals, (c) the possibilities for authors to publish in both English and an additional language so they can communicate to different audiences, and (d) how to decide whether an article serves the public best by being published in the author's mother tongue and/or a local or regional journal.

The Structural Barriers

The Skewed Distribution of Scholarly Communications

There is a fundamental imbalance between available resources and resource needs in the addiction field. On the one hand, there is as noted above a disproportionate concentration of addiction science and addiction publishing in the richer and English-speaking areas (North America, Europe, and Australia). On the other hand, the majority of the world's population and an increasing share of the addiction problems can be found in LMICs and countries where the native language is not English (Room et al., 2002). For example, Russia, Mexico, and many South American countries have high rates of alcohol-related disease and disability (World Health Organization [WHO], 2011), but few addiction journals can be found in these countries. This imbalance between prevalence of problems on the one hand and scientific and publishing possibilities on the other presents a serious challenge to those interested in the most effective and efficient use of resources in the interests of public health on an international level.

In November 2003, the WHO arranged a meeting called "Mental Health Research in Developing Countries: Role of Scientific Journals." The joint statement by participating journal editors and the WHO (2004) describes the barriers to scientific publishing experienced by researchers from LMICs in the mental health research field.

The document states that the accumulation of scientific knowledge is dependent on free and accessible communication across the world. The promotion of good research increasingly requires not only the ability to access research from other parts of the world, which in many LMICs still is a problem, but also the opportunity to communicate research results. Researchers from LMICs often have difficulties in publishing their findings in scientific journals. The reasons include limited access to information, lack of advice on research design and statistics, and the difficulty of writing in a foreign language as well as material, financial, policy, and infrastructural constraints. Limited global appreciation of the research needs of LMICs and the comparative anonymity of their researchers may constitute additional barriers. According to the WHO (2004) report, many researchers from LMICs "are daunted by the seemingly insurmountable chasm between their research effort and its publication in international journals" (p. 226).

In a subsequent WHO mapping of research capacity for mental health in 114 LMICs (WHO, 2007), 66 countries had produced fewer than five articles between 1992 and 2003 that were indexed in MEDLINE or PsycINFO. On the other hand, a number of countries—Argentina, Brazil, China, India, the Republic of Korea, and South Africa—at this time all had substantive and increasing scientific production. More than half of the journals that published most of the indexed mental health research articles from LMICs were also edited in these countries.

Most of the problems in research production and indexing could be applied to the addiction field. Many countries with few resources are striving to develop scientific research capabilities in general. Efforts to strengthen addiction research do not always have sufficient political support. Politicians and decision makers in these countries—as in many others—are not necessarily interested in whether certain alcohol or other drug treatment and prevention measures are evidence based or not. Public support may be more important. Also, research results can be difficult to translate into policy. For these reasons, research and scientific publishing on addiction-specific questions may not be high on the list of political priorities. Turci et al. (2010) analyzed for instance the trends of epidemiological production in Brazil from 2001 to 2006. The authors observed that the main themes were public health nutrition, maternal and infant health, and infectious diseases; in short, there was a lack of epidemiological research on alcohol in Brazil.

Career scientists and professionally trained clinicians are needed, but except in the instance of government-sponsored university programs, there is little support for clinical, epidemiological, and policy research. Few LMIC countries have specialist addiction societies in which locally relevant and topical problems can be discussed and solutions developed. Training opportunities are lacking. In some countries, the number of master's and doctoral students has grown, as have specialization courses at the universities (see Chapter 3). But many addiction professionals entering the work force are clinicians in private practice who may do academic work voluntarily or for a small salary. Under the circumstances, the development of addiction research will be slow.

Further, communication with researchers in other countries is often restricted by lack of resources. Many libraries have run out of journal subscription funds, and addiction journals are seldom a priority. In some countries, influential research-funding agencies are now supporting programs that give most universities free access to online periodicals. These programs have improved the availability of international research. For example, the HINARI project was launched in 2002 by the WHO in collaboration with scientific publishers to make health research available in LMICs. Today it covers 13,000 journals and 30,000 e-books in many different languages (see www.who.int/hinari).

The formal communication of locally relevant addiction research is encountering other challenges. Local journals are necessary to deal with sociocultural peculiarities and the priorities of different societies. Presently there is a strong movement in several countries to publish good-quality articles, preferably in English. Because competition in the scientific field is intensifying, publication in indexed journals is a priority for researchers who need scientific credit for their work. Alcohol and other drug science is, however, a young and relatively small field. Local and non–English-language addiction journals have difficulties meeting the criteria for inclusion in U.S. and international indexing systems, such as Web of Science and MEDLINE.

A sign of how problematic the situation still can be is that no addiction journal from the Latin American region has been able to establish itself. As a consequence, many addiction scientists publish in indexed public health or mental health journals when writing for the local or regional audience in this part of the world. Only a small number of these articles are published in English. Publishing in these journals is, of course, in itself not a bad thing. But for the development of the addiction field in a particular country or region, a specialized journal can play an important role. In India, addiction researchers have since 2010 had the possibility to publish addiction research in the *Indian Journal of Psychiatry* (Murthy et al., 2010), but also the *Journal of Mental Health and Human Behavior* has articles on addiction. Researchers in African countries have the option of publishing in the *African Journal of Drug and Alcohol Studies.* In relation to the population and problems, the local publishing availability is anyhow extremely restricted. In many other countries the only option if you want to publish in an indexed addiction journal is to seek for one from outside your own country.

However important national or local journals are, it sometimes can be hard for a researcher from a country with few resources to rely on them. These journals often have limited funds, may be published irregularly, or may have long delays between submission and publication of an article. Not infrequently, these journals will find themselves in a vicious circle: They are not regarded as prestigious enough, which means that they will not get enough good articles, which in turn means that they will not get enough resources and not enough good articles.

Even if there are still relatively few addiction specialty journals outside of North America and Europe, and even fewer that are well indexed, there are some signs that the inequality in access to scientific publication, and in journals' relative status, may be leveling out. For instance, the indexing of non–English-language journals, including addiction journals, with English-language abstracts in Scopus has increased. Open-access developments and the possibilities to have online-only publications have improved the possibilities to publish without printing costs and also to add non–English-language versions of English-language articles as online-only supporting material (Meneghini & Packer, 2007). This is not yet an established practice in addiction journals but may be a model for the future. *World Psychiatry*, the journal of the World Psychiatric Association, is for instance now published not only in English but also in Arabic, Spanish, Chinese, Russian, French, and Turkish, with the aim to improve dissemination of research to clinical psychiatrists in different parts of the world (Maj, 2010).

Marginalisation of LMIC Research in the International Discourse

In academia, faculty are often evaluated by the number of their publications and the impact of the journals in which their articles are published. Publishing in high-impact journals has become the principal aim for many because grants, positions, and funding go to scientists, faculty, and departments that succeed in this respect (e.g., see Linardi et al., 1996). When research funds are in short supply, resources are concentrated in the hands of a few investigators, and the dominance of impact factors contributes to this concentration.

Thomson Reuters, which publishes the most commonly used impact factors, does not provide complete coverage of the world's scientific journals. English-language journals and especially U.S. journals are better represented. This means that, in general, research conducted in LMICs and reported in languages other than English is under-represented. However, the situation is improving in several regions. SciELO is a bibliographic database and electronic library focusing on the developing world. In 2014, it covered more than 1,000 selected journals from South America, Spain, Portugal, the Caribbean, and South Africa. The topics include health sciences and social sciences, and every article can be downloaded free. In 2013 SciELO reached an agreement with Thomson Reuters Web of Knowledge that will increase the visibility of Latin American and Portuguese language research. This development was possibly facilitated by strong efforts to increase the English language publication of Brazilian research. In Brazil, English language scientific articles now are more common than Portuguese, and there are systematic attempts to improve the quality of the published texts (Science for Brazil, 2013). The African Journals Online (AJOL), a database with nearly 500 journals, has been launched to promote access to African research. About 160 of the journals are devoted to health fields, but only one addiction journal (see above) is listed among them. The European Reference Index for the Humanities and Social Sciences (ERIH PLUS) (which expanded in 2014 to include both humanities and social sciences) is established with the aim to "enhance global visibility of high quality research in the humanities published in academic journals in various European languages all over Europe" (NSD, 2014). In Iran, several electronic databases for scientific publishing were established in 2004. Amin-Esmaili and colleagues (2009) showed that the international databases have a low coverage of Iranian addiction research but argue that, by combining the bilingual (Iranian and English) Iranian databases with big international ones such as MEDLINE, PsycINFO, and Embase, it was possible to cover as much as 80% of the Iranian addiction research publications. Similar efforts are seen in Turkey.

The problems for LMIC researchers who seek to publish internationally may be compounded by structural factors associated with the management of the English-language scientific journals. Around 2000, a survey of the editorial and advisory boards of leading international journals in the field of mental health (e.g., *Archives of General Psychiatry, American Journal of Psychiatry, Schizophrenia Bulletin, British Journal of Psychiatry, Adolescent Psychiatry*) found only 4 representatives from LMICs among 530 board members (Saxena et al., 2003). The absence of LMIC representation on the editorial boards of the major journals may explain why authors from developing countries often feel that their articles do not receive sympathetic treatment. Thus, research from LMICs is likely to be regarded as less relevant in the international discourse. This is supported by a study of articles published in *Addiction* (West & McIlwaine, 2002), which found that articles from LMICs were cited significantly less often than those ranked by independent peer reviewers to be of the same quality as those from the developed world. Other studies have shown that an increase in the number of articles published from LMICs is not paralleled by a similar increase in citation of these articles (Holmgren & Schnitzer, 2004; Volpato & Freitas, 2003).

Additional factors that may account for the relatively limited number of publications from these countries include poor research methods, inadequate sample sizes, less-sophisticated statistical analyses, lack of national or regional journals, and limited English-language competence (see for instance Gosden, 1992)

The Language and Culture Trap

English is the lingua franca of scientific research today and will be in the foreseeable future. However, as Montgomery (2004) points out, to call it "the universal language of science" is ahistorical and possibly inattentive to the complex linguistic developments taking place in the world. In the future, more and more people will be bilingual, and languages other than English will grow in importance. For the present, however, the English language has a dominant position in addiction science.

The scientific world today is dominated by a small group of rich countries. The United States is in the lead, followed by the United Kingdom, Canada, Australia, and the European nations, which are oriented toward a similar scientific tradition and in which English-language training is well developed. The disproportionate influence of research from these countries extends to basic science, prevention, epidemiology, and treatment research. American researchers tend to cite American researchers (see further discussion in Chapter 7 and in Babor, 1993). The same applies to other countries, but with the dominance of journals from the United States and other English-language countries (and English-oriented countries such as Sweden), there is a citation bias across the research field as a whole. Research that is performed in the United States may represent a priori for many Anglo-American readers and some uncritical readers as well—that is, such results may appear to represent a more universal truth than results from a study conducted in a country such as India. Researchers in some Western nations (e.g., the Nordic countries) have adapted to the dominant research paradigms and seem to manage quite well, in terms of citation measurement (Ingwersen, 2002). The under-representation of non–English-speaking nations in indexed journals and in cited research extends to several developed countries, such as Spain, Germany, and France (Maisonneuve et al., 2003), suggesting that general linguistic and cultural influences may be at work. The present dominance of a few countries' science on an international level may imply a serious bias in the selection of research topics, questions asked, methods used, and types of research conducted, and a relative neglect of problems in the developing world. There are other problems inherent in this hierarchy within addiction research. Addiction science has at least two subdivisions—basic and applied research. The former is more or less universal in its nature, and scientific knowledge from basic research can be applied everywhere in the world. The latter is contextual. Public health research, for instance, belongs to this category. Today, public health research in LMICs suffers from a double disadvantage: (a) the difficulty in getting published and quoted in the influential journals and (b) unfair competition at the national and international level with the much better funded neurobiological research (see Midanik, 2004). In short, this means that the world literature on substance misuse is rarely determined by the research priorities of the developing countries.

Commerce plays a role as well and may not favor the public health interests of the poorer parts of the world. Randomized clinical trials of new medicines, with potential markets in richer countries, have a greater probability of being published than brief interventions to treat alcohol and other drug users. Not all policymakers realize that alcohol and tobacco are more important issues than heroin and cocaine in the developing countries (Ezzati et al., 2002).

Again, we can see signs of an improvement in the situation. Warner et al. (2014) analyzed published contributions in the international journal *Tobacco Control* between 1992 and 2011. The proportion of original-article authors from LMICs during 2007–2011 compared with all the earlier years increased from 7.2% to 22.7% and LMIC lead authors increased from 4.0% to 13.7%. There was also a significant increase in articles covering LMIC issues. In another study (Zyoud et al., 2014), a considerable increase of tobacco articles with authors from Middle Eastern Arab countries was reported between 2003 and 2012.

For researchers from LMICs, some of the problems in getting published come from not being familiar with the codes of international scientific communication. In the above-mentioned survey of physics, chemistry, and biology journals (Gosden, 1992), the editors summarized the problems encountered by researchers who were not native English speakers. The most often mentioned problem was that research results and discussion were <u>not well written</u>: that is, <u>an inability</u> to communicate the <u>importance and relevance of the research.</u> Another important problem was that authors did not know the written and unwritten "<u>rules of the publishing game</u>" (pp. 132–133). For instance, they <u>failed to cite sufficient references</u> to earlier research and were <u>not familiar with</u> <u>the argumentation style</u> or <u>scientific level</u> of the journal (Gosden, 1992). Writing a good scientific article for an international audience demands not only technical skill, such as being able to <u>carefully follow the instructions to authors,</u> but also an <u>acquired competence in social communication.</u> The best way to gain this is by reading some of the journals mentioned in Chapter 3 and getting feedback on your writing from more experienced researchers. This is not always easy in an LMIC.

What Do We Know about Addiction Journals' Language and Cultural Policies?

Unfortunately, we have almost no research to show how addiction journals in general deal with articles from LMICs and only a small, and partly old, amount of information about their language policies. In two surveys conducted by the International Society of Addiction Journal Editors (ISAJE), Edwards and Savva (2002a, 2002b) mapped the language policies of 14 English-language journals and nine non–English-language journals. Half the editors of the English-language journals who responded had not mastered any language besides English. This is a handicap in a multilingual scientific world. Based on this ISAJE questionnaire, it seems that the English-language addiction journals outside the United States have greater international representation on their editorial boards. The composition of an editorial board can give an indication of the internationalism of a journal. We have no exact knowledge of how the LMICs are represented on the editorial boards, but representation is likely to be low.

Among the responding English-language journals in the 2002 survey, the share of research articles from non–English-language countries varied from 0% to 57% at this point of time. In this sample, about one third of the journals had a policy to give special support to authors with mother tongues other than English. Only three of the 14 journals declared that <u>they could not give any</u> <u>language-editing support.</u> Of the <u>non–English-language journals</u> responding to the questionnaire, the majority published only in the language of the country of publication. Several published articles that had already been published in English. Several journals were regional or had international ambitions. All the editors knew English, and several were competent in more than one foreign language. All journals had <u>English summaries.</u> The editorial boards often had representatives from other countries.

In general, because ISAJE is an international organization with particular sensitivity to the language issue, it is possible that addiction journal editors are more conscious than editors in general of the importance of supporting research from non–English-language cultures.

What Can an Author Do?

In this section we turn to some practical suggestions that may help to correct the imbalance, level the playing field, and improve the diversity of addiction science.

Crossing the Cultural Border to the English-language Publications

As noted above, it may be particularly difficult for authors from LMICs and non–English-speaking countries to get an article accepted in an English-language journal. It is thus especially important for LMIC authors to show that they have mastered the rules of the game: to carefully follow the instructions to authors, checking that the structure, the language, and the presentation of the study and its results are clear and logical and that the references are correct. If the formalities are not followed, even a study containing strong and original findings might immediately be turned down. Cultural bias may put higher demands on research from countries where resources are few. The famous Chilean pharmacologist Jorge Mardones concluded in an interview (Edwards, 1991, p. 392) after a long career:

> I do not know why there is a generalized attitude of doubt concerning results reported in papers coming from Latin American laboratories. In order to overcome this situation, we need to be extremely certain about the accuracy and high significance of our results, before submitting a paper for publication. I feel that this is an advantage, because the worst thing a scientist can do is to pollute the scientific environment with data of poor value.

Before submitting a manuscript, an author would be wise to find a mentor or an experienced investigator who could read through the article and give advice on the presentation of the results. This may however be difficult in many countries where the addiction research milieu is very small. ISAJE is able in some cases to provide support to unexperienced authors through its mentoring program, in which experienced editors and researchers will help authors to produce publishable manuscripts (see ISAJE's website, www.isaje.net).

Collaborative studies should be encouraged. A survey of Nigerian articles published in a psychology journal showed that more than 75% of the articles were published by single authors, a figure that was much higher than that found in American journals at the time (I. Obot, personal communication, 2004). One suggestion is to try to work in a team that includes people with expertise in different areas, such as statistics and social science. This may help to improve the quality of the study and enhance its appeal to a greater number of readers. Another possibility is to work within a joint project with researchers from non-LMICs or within a large, international network. This is in most cases only possible if you have already published in an international English-language journal or work with other researchers who have international contacts and reputation. International conferences can provide possibilities for networking, but to attend them you need financial resources. In Brazil, it has been possible to document publishing success with this kind of cooperation and international exchange (Barata, 2010).

Technical requirements are relatively easy to identify and follow. A more difficult challenge is that conventions about how to write an article differ among countries. Burrough-Boenisch (2013), in a text on editing problems, gives some examples that show how culturally embedded our scientific writing endeavors are. For an Anglo-American, the author states, the German tradition of writing may seem both pretentious and less well organized. The traditional writing style of some Asian cultures, such as China, Japan, Korea, and Thailand, may give an incoherent impression. Further, when French scientists transfer the French convention of reporting science in the present tense to their English writing, they seem to be stating general truths, rather than describing their own procedures and findings.

In most cases it is not possible for an author to communicate with the readers of a journal if the author cannot talk to them in the "scientific dialect" of that particular publication. (This is of course also true when you choose a publication channel within one linguistic area.) This requires that the author is fairly well acquainted with the specific journal and knows what types of articles are published and in what format.

Some English-language journals are more sympathetic than others to articles from other countries and cultures. This is possible to find out by doing the following:

- looking at the journal's mission statement to see if it has any policy regarding articles submitted from different countries or cultures;
- checking whether the journal has previously published articles by non–English-language authors;
- checking to what extent the editorial board is international, which may imply a greater understanding of cultural diversity and a more multicultural peer-reviewer pool; and
- contacting the editor to find out if the journal may be interested in your work—pointing out its particular importance and the possible mitigating circumstances of being from an LMIC or non–English-speaking country.

Crossing the Language Border

Montgomery (2004) points out that the linguistic future of the world will be one of diversity, bilingualism, or even multilingualism. An important goal in this world will therefore be "to increase tolerance towards variation in scientific English—to avoid the imperial attitude that one standard must be obeyed" (p. 1335). Until this tolerance is developed, however, authors of scientific articles have to take the language issue seriously.

As noted above, the way in which authors present their results is often crucial to how the editor and reviewers will view the research report. The importance of good English-language usage cannot be over-emphasized. The presentation of the study and the results is particularly important when the topic or setting may seem new and exotic to the editor and reviewers. It is not just a matter of using the right terminology. Many English-speaking editors and reviewers (similar to many French-, German-, or Swedish-speaking editors) will have a rather strict idea of what constitutes good language.

Should one do a professional language check before sending in an article? Although it is expensive and time consuming, the answer is YES. If researchers are certain that they have a good case, a more experienced person has read the article and found it good, and the authors want to publish it in a journal with no resources to help with language editing, it will definitely increase the chances of acceptance. There is also the risk that if the article is considered to be a "borderline case," it will be rejected if there are language problems. However, in rare cases, if the authors know that the journal and the editor have a policy of accepting articles by non–English-language authors and the journal has the resources to do a language-check, it may not be necessary to have perfect English at the time of the first submission. But this is a case where contacting the editor beforehand is definitely worthwhile.

A few words about editing services: in most countries, there are English language manuscript editing services available for academic research papers written by non-native English speakers. These manuscript editors are generally native speakers of English with substantial experience in editing scholarly articles, and many of them are accomplished authors in the field. English editing services usually assure that the most important points, ideas, and opinions are communicated in the appropriate style of scientific writing and using the appropriate vocabulary for the context. The text is also checked for typographical and spelling errors, including punctuation.

Services range from a simple language check through to highly detailed copyediting. Additional options may include formatting according to the particular journal's standards, adjusting the word count to meet journal requirements, and writing a cover letter. Many services use an English language expert to complete a substantive edit first, then pass the text on to a professional English proofreader who makes sure the text flows well and the meaning is clear. Of course, all of the options also raise the price of the service, but even a basic language check can be very useful for teams of non-native English authors, when it can be difficult to maintain a consistent style throughout a document.

Killing Two Birds with One Stone: Dual-Language Publication

Where the topic of the article is such that it would be important to publish both at the national level and in an international journal, the author could consider trying to publish the same text in more than one language. In fact, if authors feel that their results should be considered in the development of local policy, publication of the results in an international journal may very well give the findings more prestige among the politicians of their country. Some addiction journals will agree to publish an article that has already been published in another language or to simultaneously publish the article in several languages.

These practices do not violate ethical codes regarding duplicate publication (see Chapter 14) as long as the editors agree and the simultaneous publication is mentioned along with the source of the original. If there is an interest in presenting the article to several audiences, the general rule for the author is to find out the policy of the journal(s). If the journal is published with open access or provides the option to publish additional material online only, there is a possibility that the same journal can publish an English–language and another language version of the same article. Check this with the editor.

Importance of National and Local Publications

As a researcher, one should not be blinded by the prestige of internationalism but instead try to protect the diversity and applicability of research. The diffusion of relevant research to a national audience fulfils important democratic, social, and health policy aims. Brazil has been prioritizing this as well, and there is good research available in Portuguese but not in English with relevance to policies. (Bastos & Bertoni, 2014; INPAD, 2012). The development of culturally specific research is also important for the global development of addiction research.

Nevertheless, some research may lack universal relevance. Research on specific treatment systems, on special treatment modalities, or on effects of nationally implemented policy measures in LMICs may sometimes be irrelevant outside their national or regional audience. In parallel, some of the research published in the big international journals, based on findings in North America or Europe, may not be relevant in other cultural circumstances or in developing countries.

As long as most of the important databases and indexing systems favor English-language journals and journals from the affluent countries, journals published in LMICs and non-English journals may be regarded as less-prestigious publication channels. However, in some countries, such as Nigeria, there has been a growing acceptance of locally published articles as important parts of a person's academic curriculum vitae The *African Journal of Drug and Alcohol Studies* was set up in response to the number of addiction researchers in Africa having grown and some of the issues of national importance not being of interest to international journals, the only channels for African researchers in earlier times.

The wider acceptance of local publications also recognizes the reality that it is difficult for many researchers to get published in international journals. The number of scientists has increased but not the resources and support—such as libraries and translation services—that are needed to conduct the kind of research and produce the kind of articles that would be interesting for an international journal. This does not mean that the research is not valuable.

For researchers from LMICs, pragmatism in the choice of a publication channel seems essential. As noted above, it can sometimes be problematic to rely on only national or local journals, especially those with few resources, but the situation may be improving.

Conclusions

Addiction problems and their solutions have strong local, national, and cultural characteristics. Addiction research needs to communicate within these milieus. It is important to preserve linguistic and cultural diversity in the communication of scientific findings. Addiction problems are an unfortunate fact of life in many countries and are growing in Latin America, Africa, and Asia. International communication is clearly necessary for the spread of information and can be personally rewarding, as indicated in Box 4.1. The research communities in LMICs need support and encouragement. In a world of increasing globalization, the English-speaking developed world can easily become isolated, not recognizing that it has much to learn from experience in other parts of the world.

In this chapter we have noted some signs that the global balance in science is improving. We know that many international and English-language journals are sympathetic toward publishing research from other countries and linguistic areas (see Edwards & Savva, 2002a, 2002b). The activities within international organizations such as ISAJE will hopefully further increase the awareness of resource, language, and cultural issues among journal editors and the research community in general through fostering networks and striving to change the discriminatory practices of the databases and indexing systems. This is the good

The following quotation from an interview with Professor Mustapha Soueif, an Egyptian psychologist, cannabis researcher, and internationally recognized addiction expert, shows how exciting it can be to confront the challenges of publishing in multiple languages and different cultures (Edwards, 1991):

> I have to be "bilingual" if I care for international readership and acknowledgement. And bilingualism is not an easy job. You cannot reduce it to a pendular movement from Arabic to English and vice versa. Rather, you switch off a whole way of thinking, feeling and mode of expression; and tune yourself to a totally different wave length. At the start of your career you find that this exercise is really tough, and overloaded with frustrating moments. But you accept it the way it is, because you chose to have it this way. Gradually, you attain higher levels of relevant skills; your troubles decrease, yet they never disappear.
>
> Another implication is that you have to accept a double load of responsibilities most of the time; I mean your local duties (the university, the private clinic, sharing in national meetings and writing in periodicals) and international requests (usually meetings and writings). Sometimes you have to turn down a request from one side or the other. But you have to be very careful if you intend to play the two roles with optimum smoothness. It takes creative effort to find points of convergence between both, and it is, therefore, highly rewarding.
>
> A third implication is that gradually your role is redefined for you. You are no more just a local scientist with international resonance. You are transformed into a culture-transmitter or a bridging factor. You are expected to behave as a medium for communication between two cultures. Whenever you cross the fence you should do something useful and interesting to the people on the other side. Of course what you carry with you should always be relevant to scientific endeavour. But it is sometimes peripheral. Yet it proves to be quite instrumental in promoting mutual understanding between investigators trying to transcend national and/or cultural barriers. This is all the more important when it comes to an area like research in drug abuse. (pp. 438–439)

Box 4.1: Professor Mustapha Soueif on "Bilingualism" in addiction publishing.

news for researchers from less-resourced countries and non–English-language cultures.

The bad news is that the competition within research is hardening, strengthening existing hierarchies in the world of science and putting increasing demands on researchers from LMICs. Researchers from these countries face special challenges. General advice and rules of conduct are of limited value. Hard work and a good dose of pragmatism are needed if you want to communicate your research to the appropriate audience and get scientific credit for it.

In this chapter we have pictured the unique challenges faced by addiction scientists who work outside the cultural and linguistic mainstream. It will take a great deal of skill, persistence, and courage to get to the top of your field. But the rewards awaiting you at the summit may be that much greater, because you will have acquired the skill to read the map and orient yourself both in your country of origin and in the world that lies beyond.

Please visit the website of the International Society of Addiction Journal Editors (ISAJE) at www.isaje.net to access supplementary materials related to this chapter. Materials include additional reading, exercises, examples, PowerPoint presentations, videos, and e-learning lessons.

References

Amin-Esmaili, M., Nedjat, S., Motevalian, A., Rahimi-Movaghar, A., & Majdzadeh, R. (2009). Comparison of databases for Iranian articles; Access to evidence on substance abuse and addiction. *Archives of Iranian Medicine, 12,* 559–565.

Babor, T. F. (1993). Megatrends and dead ends: Alcohol research in global perspective. *Alcohol Health and Research World, 17,* 177–186.

Barata, R. B. (2010). International cooperation: Initiatives in graduate studies in public health [Editorial]. *Cadernos de Saúde Pública, 26,* 2008–2009. DOI: https://doi.org/10.1590/S0102-311X2010001100002

Bastos, F. I., & Bertoni, N. (2014). *Pesquisa Nacional sobre o uso de crack: Quem são os usários de crack e/ou similares do Brasil? Quantos são nas capitais brasileiras?* Rio de Janeiro, Brazil: ICICT. Retrieved from http://www.obid.senad.gov.br/portais/OBID/biblioteca/documentos/Publicacoes/329797.pdf.

Burrough-Boenisch, J. (2013). Editing texts by non-native speakers of English. In P. Smart, H. Maisonneuve, & A. Polderman (Eds.), *Science Editors' Handbook* (2nd ed.). Surrey, England: European Association of Science Editors. Retrieved from http://www.ease.org.uk/sites/default/files/1-2.pdf.

Edwards, G. (Ed.). (1991). *Addictions: Personal influences and scientific movements.* New Brunswick, NJ: Transaction.

Edwards, G., & Savva, S. (2002a). Hur icke-engelskspråkiga tidskrifter på ber-oendeområdet arbetar med språkliga frågor: En enkät av ISAJE. [How non-English language journals in the addiction work with language questions: An ISAJE questionnaire.]. *Nordisk Alkohol- & Narkotikatidskrift, 19,* 207–210.

Edwards, G., & Savva, S. (2002b). Hur engelskspråkiga tidskrifter på beroende-området arbetar med språkliga frågor: En enkät av ISAJE. [How English language journals work with language questions: An ISAJE questionnaire.]. *Nordisk Alkohol- & Narkotikatidskrift, 19,* 211–215.

Ezzati, M., Lopez, A. D., Rodgers, A., Vander Horn, S., & Murray, C. J. L. (2002). The comparative risk assessment collaborating group (2002). Selected major risk factors and global and regional burden of disease. *The Lancet, 360,* 1347–1360.

Gosden, H. (1992). Research writing and NNSs: From the editors. *Journal of Second Language Writing, 1,* 123–139.

Holmgren, M., & Schnitzer, S. A. (2004). Science on the rise in developing countries. *PLOS Biology, 2*(1), e1. DOI: https://doi.org/10.1371/journal.pbio.0020001

Ingwersen, P. (2002). Visibility and impact of research in psychiatry for north European countries in EU, US and world context. *Scientometrics, 54,* 131–144.

INPAD (Instituto Nacional de Ciência e Tecnologia para Politicas Publicas do Álcool e Outras Drogas). (2013). *II Lenad: Levantamento nacional de álcool e drogas.* Retrieved from http://inpad.org.br/lenad/resultados/alcool/resultados-preliminares.

Linardi, P. M., Coelho, P. M. Z., & Costa, H. M. A. (1996). The 'impact factor' as a criteria for the quality of scientific production is a relative, not absolute, measure. *Brazilian Journal of Medical and Biological Research, 29,* 555–561.

Maj, M. (2010). World Psychiatry and the WPA task force to promote dissemination of psychiatric research conducted in low and middle income countries. *Epidemiologia E Psichiatria Sociale, 19,* 204–206.

Maisonneuve, H., Berard, A., & Bertrand, D. (2003). International submissions to journals. *The Lancet, 361,* 1388–1389.

Meneghini, R., & Packer, A. L. (2007). Is there science beyond English? *EMBO Reports, 8,* 112–116.

Midanik, L. T. (2004). Biomedicalization and alcohol studies: Implications for policy. *Journal of Public Health Policy, 25,* 211–228.

Monteiro, C. A., & Barata, R. B. (2007). Fórum de Editores Científicos em Saúde Pública [Editorial]. *Revista de Saúde Pública, 41,* 1–2. DOI: https://doi.org/10.1590/S0034-89102007000100001

Montgomery, S. (2004). Of towers, walls, and fields: Perspectives on language in science. *Science, 303,* 1333–1335.

Murthy, P., Manjunatha, N., Subodh, B. N., Chand, P. K., & Benegal, V. (2010). Substance use and addiction research in India. *Indian Journal of Psychiatry,* 52(Supplement S3), 189–199.

NSD. (2014). ERIH PLUS. Retrieved from https://dbh.nsd.uib.no/publiserings kanaler/erihplus/about/index (Accessed October 31, 2015).

Obot, I. (2014). Personal communication, September 15, 2014

Room, R., Jernigan, D., Carlini-Marlatt, B., Gureje, O., Mäkelä, K., Marshall, M., ..., Saxena, S. (2002). *Alcohol in developing societies: A public health approach.* Geneva, Switzerland: World Health Organization.

Saxena, S., Levav, I., Maulik, P., & Saraceno, B. (2003). How international are the editorial boards of leading psychiatric journals? [Correspondence]. *The Lancet, 361,* 6090. DOI: https://doi.org/10.1016/S0140-6736(03)12528-7

Science for Brazil (2014). Web of Knowledge gets Brazilian Link. www. scienceforbrazil.com/web-of-knowledge-gets-brazilian-link/. Accessed October 31, 2015.

Targino, M. G., & Garcia, J. C. R. (2000). Ciencia brasileira na base de dados do Institute for Scientific Information (ISI) [Brazilian periodicals in Institute for Scientific Information (ISI)]. *Ciencia da Informacao Brasilia, 29,* 103–117. Retrieved from http://www.scielo.br/pdf/ci/v29n1/v29n1a11.

Turci, S. R. B., Guilam, M. C. R., & Câmara, M. C. C. (2010). Epidemiologia e Saúde Coletiva: Tendências da produção epidemiológica brasileira quanto ao volume, indexação e áreas de investigação - 2001 a 2006. *Ciência & Saúde Coletiva, 15,* 1967–1976. DOI: https://doi.org/10.1590/S1413-81232010000400012

van Weijen, D. (2012). The language of (future) scientific communication. *Research Trends, 31.* Retrieved from http://www.researchtrends.com/issue-31-november-2012/the-language-of-future-scientific-communication.

Volpato, G. L., & Freitas, E. G. (2003). Desafios na publicação científica [Challenges in scientific publication]. *Pesquisa Odontologica Brasileira, 17*(Supplement 1), 49–56.

Warner, K. E., Tam, J., & Koltun, S. M. (2014). Growth in *Tobacco Control* publications by author from low- and middle-income countries. *Tobacco Control, 23,* 231–237. DOI: https://doi.org/10.1136/tobaccocontrol-2012-050762

West, R., & McIlwaine, A. (2002). What do citation counts count for in the field of addiction? An empirical evaluation of citation counts and their link with peer ratings of quality. *Addiction, 97,* 501–504.

World Bank. (2014). World Development Indicators 2014. Washington DC: World Bank Author.

World Health Organization. (2004). Galvanizing mental health research in low- and middle-income countries: Role of scientific journals. A joint statement issued by editors of scientific journals publishing mental health research. Department of Mental Health and Substance Abuse, WHO, Geneva, January 2004. *Bulletin of the World Health Organization, 82,* 226–228.

World Health Organization. (2007). Research capacity for mental health in low- and middle-income countries. Geneva: Global Forum for Health Research and WHO.

World Health Organization. (2011). Global status report on alcohol and health. Retrieved from http://www.who.int/substance_abuse/publications/global_alcohol_report/msbgsruprofiles.pdf?ua=1.

Zyoud, S. H., Al-Jabi, S. W., Sweileh, W. M., & Awang, R. A. (2014). A Scopus-based examination of tobacco use publications in Middle Eastern Arab countries during the period 2003–2012. *Harm Reduction Journal, 11,* 14. DOI: https://doi.org/10.1186/1477-7517-11-14

CHAPTER 5

Getting Started: Publication Issues for Graduate Students, Postdoctoral Fellows, and Other Aspiring Addiction Scientists

Dominique Morisano, Erin L. Winstanley,
Neo Morojele and Thomas F. Babor

Introduction

In recent years, there has been increasing pressure on graduate and medical students, postdoctoral fellows, and even research assistants and lab technicians to write or co-author scientific publications. Some of this pressure has extended to undergraduates (e.g., Trammell, 2014), often before they have had the opportunity to take a statistics course.

The number of publication credits is frequently a key criterion for students' acceptance into advanced study, postdoctoral opportunities, and internship placements as well as for the receipt of scholarships, fellowships, grants, and employment. For novice academics, publication numbers and authorship order are often at the top of considerations for tenure-track advancement. More competitive universities that value high publication numbers might urge students and junior faculty to compose theoretical papers and review articles or to write reports based on publically sourced unpublished data (e.g., www.apa.org/research/responsible/data-links.aspx) instead of running original studies, which take time and do not always yield publishable results. In some countries, students are advised to publish articles in addition to producing a monograph-style dissertation; in others, they are expected to focus solely on the production of a "compilation thesis" or article-based dissertation that might lead to

How to cite this book chapter:
Morisano, D, Winstanley, E L, Morojele, N, and Babor, T F. 2017. Getting Started: Publication Issues for Graduate Students, Postdoctoral Fellows, and other Aspiring Addiction Scientists. In: Babor, T F, Stenius, K, Pates, R, Miovský, M, O'Reilly, J and Candon, P. (eds.) *Publishing Addiction Science: A Guide for the Perplexed*, Pp. 89–118. London: Ubiquity Press. DOI: https://doi.org/10.5334/bbd.e. License: CC-BY 4.0.

multiple publications. Some students must produce dissertations that are based on published articles (possibly with multiple authors). In any case, for postgraduate trainees and junior academics, authorship is increasingly at the forefront of issues faced in education and early employment.

This chapter presents issues that are particularly relevant to publishing as a graduate student or postdoctoral fellow, but anyone early in her or his publishing career might benefit from reading through the topics covered. The chapter begins with a discussion of general issues related to authorship and then addresses the more specific topic of publishing graduate-level theses. The latter section focuses on the entire process of thesis publication, ranging from issues that might arise before writing one's thesis all the way to eventual postpublication submission to an appropriate journal. Our main sources of information on this topic come from North American and European universities in high-income countries, but the issues and solutions discussed are increasingly relevant to university students in other regions. Accordingly, special attention is provided to the challenges encountered by students or novice investigators in less resourced countries.

General Issues

The challenges of publishing early on the academic trajectory include making decisions about authorship and timetables, navigating ethical dilemmas, and balancing publication pressures with training goals. Yet publications can open doors for both career advancement and financial remuneration.

Authorship

As noted in Chapter 11, authorship of peer-reviewed journal articles is the "coin of the realm" in academic settings, although the ability to write even unpublished reports is a valuable skill in any work situation. For the great majority of graduate students and postdoctoral fellows, early-career authorship will come only from collaboration with faculty members,[1] senior researchers, and supervisors. As such, both mentors and mentees should consider a number of ethical and practical issues that could arise on joint projects (see Chapters 14 and 15 for a discussion of authorship ethics). At the heart of such trainee–faculty (or even employee–supervisor) collaborations lies an inherent power imbalance (Fine & Kurdek, 1993; Gross et al., 2012). Often, the faculty members with whom students and trainees have the most interactions (and thus the greatest chance to do research) are responsible for providing them with recommendation letters and evaluating their work. These faculty members may even be responsible for trainee salaries, as in the case of graduate assistantships or postdoctoral fellowships. Many students and trainees begin with minimal experience and competence in publishing and must rely on faculty support and guidance. Even if students and postdoctoral trainees are consulted during the process of assigning authorship, faculty members generally make the ultimate decisions on where (or whether) students or trainees are placed on the author list. Students who disagree with or misunderstand such decisions might fail to voice their opinions for fear of negatively impacting the ways in which those faculty members will evaluate them.

The academic level of the collaborating faculty member or supervisor could also influence the authorship decision-making process. Senior faculty with established research grants might be more likely to give students or trainees opportunities for first authorship on co-authored publications. With potentially bigger labs or projects and greater numbers of volunteers and research assistants, senior faculty might even provide more chances to publish in general, handing over projects, ideas, and datasets to their mentees. In contrast, junior faculty members are frequently under significant pressure to get their own names on publications in order to earn research grants, advance to higher faculty positions, and gain tenure. As a result, they might have more concerns about sustaining and advancing their own careers than about taking time to help their students or trainees to publish.

Figure 5.1 provides a satirical view of authorship situations sometimes encountered by students who work on publications with more experienced or higher ranked investigators. Although the cartoon is a spoof, many academics would agree that it is uncomfortably close to the procedures witnessed in some research labs, centers, and departments. The procedures for determining student–faculty co-authorship are likely to vary by discipline, institution, and even culture, but they should ideally reflect a dynamic process that evolves as the authors revise and resubmit their article.

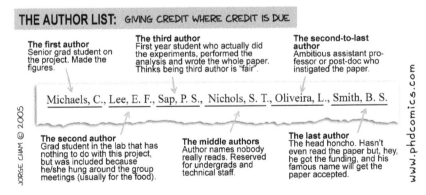

Figure 5.1: Authorship credit comic from "Piled Higher and Deeper" by Jorge Cham (www.phdcomics.com, reprinted with permission of author. All rights reserved.).

Graduate students, postdoctoral fellows, and young professionals working in basic and applied research settings are often uninformed about acceptable procedures for deciding authorship within a given field or discipline. In addition, procedures seem to vary so greatly even within departments that it can be difficult to stay abreast of what constitutes acceptable practice. The availability of specific guidelines is indispensable to establish equal opportunities for student authorship and consistent procedures for student–faculty collaborations. As in the case of the more general issue of authorship (discussed above), there are specific guidelines available that can facilitate this process at some institutions and help prevent problems from arising in the first place. Some examples of these guidelines are discussed below. If they are not readily available at your research center or university, however, it is possible to adopt guidelines from another institution or professional society (see Chapter 11 for an example).

As a rule, graduate students should be the first authors of journal articles based on their thesis or dissertation manuscripts. Many disciplines and institutions enforce this broad principle. For example, the American Psychological Association's *Ethical Principles of Psychologists and Code of Conduct* (American Psychological Association, 2010) explicitly states, "Except under exceptional circumstances, a student is listed as principal author on any multiple-authored article that is substantially based on the student's doctoral dissertation" (Section 8.12). Further, the American Psychological Association indicates that faculty advisors should discuss publication credit with students as early as feasible and throughout the research and publication process. However, the "exceptional circumstances" mentioned highlight a universal gray area, and it is often the case that other factors might complicate seemingly straightforward authorship assignment, for instance when the graduate student's dissertation is based on part of an advisor's grant.

In line with changing times, several institutions of higher learning have posted general authorship guidelines on their websites. The University of Pennsylvania, for example, has developed a broad policy on fairness regarding authorship credit for publications co-authored by graduate students and faculty. A university-wide process for determining authorship sets forth simple principles and an appeal process and requires graduate programs to provide more specific guidelines to reflect interdisciplinary and interdepartmental differences in assigning authorship credit (University of Pennsylvania's Office of the Provost, 2013). Mandating such procedures within each graduate group clarifies expectations about authorship for both students and faculty members. Specific departmental guidelines cover topics such as authorship criteria (specific and general principles regarding the kind of work that warranted a publication credit), whom to consult to resolve disputes, and the issues that faculty should discuss with students when beginning joint projects. Examples of such issues include (a) whether the graduate student will share authorship credit, (b) the expected order of authorship, (c) the division of labor on the project, and (d) when to revisit or review work that is being completed by each collaborating member of the pair or group. The University of Alberta's website hosts a similar set of guidelines around intellectual property and authorship (University of Alberta, 1996).

In general, with the expansion of the Internet as the primary tool of communication in most circles of higher education, online policies appear to be an efficient and user-friendly way of spreading authorship and intellectual-property guidelines to junior investigators with adequate access. Harvard Medical School Office for Research Issues (1999), the University of Toronto (2007), Washington University in St. Louis (2009), and the University of Cambridge (2014), among others, have also provided statements on authorship or intellectual property for members of their institutions—although some are rather brief in nature, they seem to be evolving. University of Pennsylvania and University of Alberta guidelines provide the best models for the development of similar policies in higher learning institutes across the world. Such university-wide policies are an excellent way to keep students, postdoctoral fellows, and faculty members informed about the most fair and equitable procedures to follow in joint-authorship situations.

In what has become a US benchmark article for writings on student–faculty co-authorship, Fine and Kurdek (1993) produced a set of authorship guidelines based on the idea that both faculty and students should meaningfully participate in the authorship decision-making process. Fine and Kurdek recommended that, at the very initiation of joint projects, supervisors and faculty collaborators provide new students and postdoctoral fellows with information about how authorship decisions are made. They also put forth a series of specific and potentially controversial recommendations about student authorship, arguing, for example, that supervisors cannot and should not expect as much from students as from experienced professional colleagues. Instead, the authors suggested that there should be a different standard for the level of professional contribution required by students to attain a given level of authorship credit within a student–faculty collaboration. At the same time, however, they maintained that student contributions must be professional in nature: that is, creative, intellectual, and integral to completion of the paper. Examples of such contributions might include developing the research design, writing sections of the manuscript, integrating diverse theoretical perspectives, developing new conceptual models, designing assessments, contributing to data-analysis decisions, and interpreting results. Other tasks—such as entering data, carrying out statistical analyses specified by the supervisor, and typing a manuscript—might warrant a footnoted acknowledgement, but they would not, according to the authors, deserve authorship credit. Fine and Kurdek suggested that supervisors and students decide early in the publication process what combinations of professional activities would merit a given level of authorship credit for both parties. These decisions might now need to be checked against journal or discipline-specific guidelines and standards, many of which have become more detailed over the years in response to authorship confusion and transgressions (see Chapter 11).

Fine and Kurdek (1993) raised a variety of issues and case scenarios surrounding authorship in student–faculty collaborations that are still relevant more than two decades later. Chapter 11 is a direct response to articles such as this as well as to the diverse but brief and scattered array of individual university guidelines mentioned above. Students, postdoctoral fellows, and other early investigators in the process of article publication should refer often to the general set of very practical authorship guidelines provided in Chapter 11. These guidelines span the planning, drafting, and finalization stages of authorship. Indeed, the chapter is an ideal source for beginning researchers to consult as they try to determine where (or if) they should appear within author lists. It touches on potentially controversial issues, such as what constitutes a "substantive" authorship contribution. For example, if a graduate student has developed, coordinated, and carried out a research project for a mentor or supervisor but did not come up with the original idea, analyze or interpret the resulting data, or participate in the writing of the ensuing manuscript, does he or she deserve to be listed as an author on publications arising from the project? According to the recommendations in Chapter 11, the answer is no, because there is no involvement in the writing process (and to be an author, one must write!). However, one might argue that this student should at least be given the *option* of contributing in a more substantive way to the publication process in order to earn authorship. Students might therefore want to explicitly express their interest in being involved in future publications.

In summary, there is a great amount of room for improvement in the realm of early-career publishing. The process has not yet been clearly documented in terms of student and junior investigator rights, responsibilities, and roles. Although progress has been made in clarifying the issues and formalizing some long overdue policies, much remains to be done at both the level of the academic institution and the level of the individual faculty and trainee. Fortunately, there are plenty of opportunities to learn more about this area to improve the process. The mentorship of a seasoned investigator can provide her or his students, postdoctoral fellows, or other trainees with a golden opportunity to ascertain how publication works. At the very least, the sharing of articles such as this chapter might help to raise awareness of the issues and how to deal with them.

Publishing One's Thesis or Dissertation

Converting the thesis or dissertation into one or more journal articles is a key publishing opportunity for aspiring researchers. Incentives to early publication include building confidence, establishing a pattern of scholarly activity, enhancing student satisfaction, increasing knowledge of the publication process, and advancing or updating the science.[2] Sometimes, early publication affords a novice researcher the opportunity to demonstrate the need for a particular area of research (Robinson & Dracup, 2008). As noted above, there are many incentives to begin publishing early or publishing before the research data "shelf life" has expired, particularly for those who are interested in academic careers (Resta et al., 2010). Given the amount of work that is invested in the preparation of a thesis or dissertation, this is often the ideal place to begin one's publication career, and it is important to be strategic about the development of a publication plan.

When considering a timeline for publication, there are several questions researchers might ask themselves. For instance, "What is my academic trajectory?" Or, "How fast is this area of research developing?" "How much information is available in my content area?" "Is the literature up to date or does it need updating?" "What is the potential real-world impact of my research?" "Does current literature support the need for my research, or do I need to build a published case?" "What audience is most interested in my area of research?" Answering these simple questions could help a novice researcher to develop a successful publication plan both during and after thesis or dissertation completion. The following section describes additional considerations.

Before Writing One's Dissertation: Format Considerations

There are several different doctoral dissertation formats, which vary in acceptability depending on the country and the university in which they are written. Two of the more popular formats are the monograph style (single authored) and the separate manuscript style (multiauthored; Hagen, 2010). Many graduate programs increasingly favor dissertations that depart from the traditional monograph style and that instead facilitate the incremental translation of the dissertation into publishable manuscripts.

The manuscript style of dissertation—although it might have different names—generally requires that chapters be written in article format. For example, at The Johns Hopkins Bloomberg School of Public Health, a student can choose to write a traditional monograph-style (chapter-based) dissertation or a "papers option." The latter format requires that a minimum of three of the dissertation chapters take the shape of publishable manuscripts, with one chapter usually serving as a critical review of the literature and two chapters comprising empirical analyses. To the extent that the papers are "publishable," whether they must be submitted or accepted for publication to earn a degree varies across universities. In the Nordic countries (Denmark, Iceland, Finland, Norway, Sweden), most of the dissertation articles must have been published or accepted for publication before the dissertation can be passed.

Manuscripts may represent the entire chapter or a portion of a dissertation chapter that is supplemented with a synthesis or independent introduction. An example of the purposive changes that may be made to a manuscript to fulfill the chapter requirements include the addition of regional data and epidemiological information, the definition of terms for lay readers, a longer and more in-depth explanation of the phenomenon, the theoretical tenets guiding the proposed study, and a conclusion that illustrates student mastery of the subject.

The extent to which manuscripts need to be interrelated and reflect a single focus of research, as occurs in a monograph-style dissertation, varies across institutions, departments, and advisors. It is, in part, contingent on the clarity of the institutional guidelines provided. Anecdotal evidence suggests that the rules are not concrete. Furthermore, if one is writing a literature-review chapter, it is helpful to keep in mind that many addiction journals do not accept unsolicited review articles and that getting this type of manuscript published could be a special challenge. Literature reviews using a systematic or structured approach are more likely to be published.

If one has the opportunity to choose which dissertation format to take, it is important to consider the benefits and particular challenges of a style that is meant to facilitate the publication process. For example, even if one chooses to write one's thesis in the manuscript style, resulting chapters might still require significant revision if they need to be shortened and formatted later for a particular journal and written with a broader audience in mind than one's dissertation committee (Azar, 2006).

In the Trenches: Writing One's Dissertation with Publication in Mind

While writing the thesis or dissertation, it is helpful to think about whether chapters or sections will eventually be suitable for journal publication. If the answer is yes, then several issues arise that should be addressed sooner rather than later. For instance, if one hopes to publish one's data in a particular journal, it is important to consider the author guidelines during the drafting stage in order to tailor the writing and formatting style of the dissertation toward specific journal requirements. It is also useful to consider the intended audience of that journal early on (see Chapter 3 for issues related to choosing a journal). Even if a particular target journal has not yet been identified, the chapter can be written with the potential audience in mind (e.g., clinicians or policy makers), in a way that can help refine the scope of the manuscript. It is also important to remember that if one publishes data or other study-related material before submitting the dissertation or thesis, one must consider which parts of the published manuscript(s) are eligible for inclusion in the final dissertation. Journals and publishers will often grant permission to students to submit published manuscripts as dissertation chapters, but it is wise to request written confirmation.

Furthermore, for many, a considerable amount of time can elapse between creating initial drafts of the thesis or dissertation and preparing to publish the content in a journal. It is therefore important to maintain adequate documentation of all analyses and datasets. The lengthy dissertation-writing process plus the journal-submission process could result in a situation in which, months or years after data collection, a journal reviewer requests that data analyses be revised or substantially expanded. Although this issue is generally relevant for the authors of any research study, the significant time that it takes to complete the dissertation amplifies the importance of keeping an adequate record of completed work.

In sum, the forward-thinking student will strategically balance dissertation requirements with potential journal submission requirements. This is not always easy. Dissertations typically require a much greater level of detail than most journal manuscripts. This means that significant portions of the dissertation will need to be cut, edited, and fine-tuned for publication. Writing style might also need adjustment, depending on the intended audience (e.g., dissertation committee vs. journal editors, and reviewers vs. the scientific community at large). There are benefits to this conversion exercise, however. The process of transforming dissertations into publishable articles teaches graduate students not only how to summarize research findings in a succinct manner, but also how to communicate to a broader audience than faculty and committee members.

In the long and sometimes dark days of creating one's dissertation with publication in mind, it is key to remember that publication presents multiple rewards. In addition to fulfilling degree requirements and contributing to scientific advancement, all of one's hard work can be directly applied to making progress on the career front. Publication is, after all, the coin of the realm.

Preparing for Publication

Once the dissertation has been approved, and the appropriate celebrations have concluded, the time for publication is nigh.[3] Frequently, suggestions made during the final dissertation defense will be relevant to the initial stages of preparing for publication. During this phase, several issues inevitably will come to the surface.

The first is authorship. As previously discussed, the student should be the first author the majority of the time. In the case of multiple authors, institutional and disciplinary guidelines or even our own recommendations (see Chapter 11) can help to determine authorship order. If committee members are to be invited as potential co-authors, it should be made clear that all authors are required to have made substantial contributions to the journal manuscript itself, as opposed to simply "being a part" of the dissertation-development conversation. Given that many journals now require written statements that specify authorship contributions, this is no longer just a traditional courtesy.

Assignment of authorship is a dynamic process that will depend on the amount of time that has lapsed since graduation, the extent of revisions required for publication, and the context in which those revisions are made. For example, revisions are sometimes required at the final stage of the dissertation-approval process, and it might not be feasible to anticipate the target journal until after graduation. If substantial revisions are requested, the opportunity might arise to seek expertise outside of the dissertation committee. The recruitment of external co-authors can offer several advantages. First, fresh insight might facilitate the process of tailoring a manuscript for a particular target audience. Alternatively, external experts might be able to address weaknesses in a manuscript that fall outside the student's field of knowledge. Sometimes new graduates might recruit the co-authorship assistance of a former labmate or graduate-student peer to make broad cuts in superfluous content that might be difficult for the primary author to do. This offers the added opportunity or benefit of publication experience for a peer.

One should also consider publication of the dissertation itself, with or without an accompanying short-form article. This is a requirement at many European institutions, where dissertations often result in published books. Some graduate programs might provide structured guidance regarding the process of indexing the dissertation, copyrighting dissertation materials, and publishing the dissertation as a complete document. Some university libraries now do this automatically (e.g., McGill University: www.mcgill.ca/library/find/theses). Alternatively, there are an increasing number of low-cost opportunities to publish one's full work online. A sampling of websites offering this possibility is presented in Box 5.1. For example, Dissertation Abstracts Online indexes dissertation abstracts and disseminates them across a wide range of literature search engines. ProQuest Dissertations and Theses allows graduates the option to purchase a permanent link for dissertation abstracts; this can be useful for citation purposes. Other sites offer interested readers the choice to either download or receive a .pdf or paper copy of a dissertation for a nominal fee.

If one is looking to reach the widest audience, writing the dissertation in manuscript style can facilitate the process of achieving one or more first-author publications. Finding the time for even one article can be difficult after graduation, when important life changes (e.g., finding or starting a new job, starting a family, catching up on things that might have been on hold during graduate school) are often inevitably competing for one's time. This is why a postdoctoral position, when available, offers an ideal solution: the very nature of the job often includes the development of publications as a primary goal. Furthermore, depending on the area of research, postdoctoral positions of 1–3 years might not allow sufficient time to be a part of a new project from inception to publication. Entering the position with one's own dissertation provides an immediate publishing goal.

Publication Timelines

Some supervisors and faculty members feel it is important to set formal limits, policies, and procedures regarding the time that students have to publish their

1. UMI (University Microfilms International) Dissertation Publishing: www.proquest.com/products-services/dissertations
   a. ProQuest Dissertations & Theses Global database
   b. American Doctoral Dissertations
   c. Masters Abstracts International
   d. ProQuest Dissertations and Theses—United Kingdom (UK) & Ireland
   e. Dissertations & Theses @
   f. Dissertation Abstracts International/Dissertation Abstracts Online/Comprehensive Dissertation Index
2. OCLC WorldCat Dissertations and Theses (includes manuscripts from OCLC member libraries): http://www.oclc.org
3. Networked Digital Library of Theses and Dissertations: www.ndltd.org
4. DART-Europe (28 countries): www.dart-europe.eu/basic-search.php
5. BNF: Thèses et écrits académiques (France): http://signets.bnf.fr/html/categories/c_011theses.html
6. EThOS (UK): http://ethos.bl.uk
7. theses.fr (France): www.theses.fr
8. Theses Canada Portal (Canada): www.bac-lac.gc.ca/eng/services/theses/Pages/theses-canada.aspx
9. DissOnline (Germany): www.dnb.de/DE/Wir/Kooperation/dissonline/dissonline_node.html
10. Tesionline (Italy): www.tesionline.com/intl/index.jsp
11. Tesis doctorales: TESEO (Spain): www.educacion.es/teseo
12. dissertations.se (Sweden): www.dissertations.se
13. Database of African Theses and Dissertations (Africa): www.aau.org/page/database-african-theses-and-dissertations-datad
14. Networked European Deposit Library (France, Norway, Finland, Germany, Portugal, Switzerland, Italy, and the Netherlands): www.ifs.tuwien.ac.at/~aola/publications/thesis-ando/NEDLIB.html
15. Google Scholar: http://scholar.google.com
16. Amicus (Canada): http://amicus.collectionscanada.ca/aaweb/aalogine.htm

Box 5.1: Online dissertation indexing and publishing resources.

thesis or project data in a scholarly journal. When this timeline is expired, there might be a debate over whether the right to publish the data should be forfeited to the supervisor or members of the dissertation committee. It is a common belief that if work is not published in a timely manner, it is unlikely to be published at all (Rudestam & Newton, 1992).

In most cases, students should have the right to publish their results as first author, even with considerable delays. If the timely dissemination of important scientific findings is at the root of such policies, however, then these procedures might be warranted. Graduate students sometimes lose interest in publishing project data after their theses have been defended (or even before!), and important or interesting scientific results are often buried under more salient tasks at hand (e.g., seeking full-time employment). Regarding specific policies, this is something that supervisors and dissertation committee members should discuss with their students early in the collaboration process. A reasonable solution for the various parties in these cases might be to designate a mutually agreeable time period together and then sign a written agreement that would bind them to it.

One example of an individual professor's policy that was put together and published online is that of Professor Karl Wuensch (2008,[4] East Carolina University). On his website, Wuensch clearly states his policy regarding timeliness of publication for student theses. For example, if the thesis is the student's idea, the student does most of the work (e.g., collects and analyzes the data, writes the manuscript), and the manuscript is prepared within 18 months from the date of the research initiation or one year from the date of the thesis defense, then the student is first author. If warranted by their contributions to the journal manuscript, the thesis director and other committee members might also be listed as authors. However, if the student does not complete the research, including defending and depositing the thesis and preparing the manuscript for submission for publication within the time limits mentioned above, then all rights to use that thesis data revert to the thesis-committee director. Wuensch also indicates procedures for other situations that might arise, for example, if the student-submitted manuscript is not accepted upon initial submission to a journal. Guidelines such as these might also be adapted for postdoctoral fellowship projects.

In the discussion of publication timelines, it is important to remember that exceptions (e.g., illness) can always be considered if one fails to publish within the agreed-upon period—one need not despair. As long as steady progress is shown and good communication among co-authors is in place, the pressure that might come from thesis advisor(s), co-authors, and committee members can be reduced. Sometimes the issue lies not with one's own progress but with getting co-authors to respond in a reasonable amount of time. Although all authors might struggle with the multiple-author publication process, novice writers in particular must learn to develop effective communication strategies, ideally from their advisors. It can be useful to set specific time frames for co-authors with concrete deadlines and frequent email reminders. If response time becomes unreasonable, a direct conversation with these co-authors about their place on the manuscript might become necessary. If motivation or writer's block is an issue, it might be useful to take advantage of some of the strategies presented in Box 5.2.

Publication Contracts and Guidelines

Several attempts have been made to develop formal procedures to address the ethical, practical and logistical issues discussed above. Professor Bruce M. Shore, an educational psychologist and professor emeritus at McGill University (Montreal, Canada), developed a formal supervision contract (Shore, 2014) for use with students. This contract covers matters such as authorship order, publication credit, and general responsibilities of both the advisor and the student within the supervisee–supervisor relationship. As a supervisor, he required that all of his students read, discuss, and sign the contract before agreeing to work with him, and he often raised the issues involved with authorship before projects even germinated. He agreed to share his contract as an example of an advisee–advisor agreement for the purposes of this chapter (see Appendix A). The process recommended in the agreement is refreshing. Regardless of whether a student agrees with the various conditions of the contract, the issues are transparent and open to discussion at the onset of the mentee–mentor relationship.

A similar guideline was developed by graduate students at the University of Connecticut School of Medicine (Cornell et al., 2014; Authorship Rights of Graduate Students, see Appendix B) to protect graduate students working in various areas of health science by clearly defining student–faculty authorship criteria and the ethical responsibilities of each party. The procedures described in the guideline (as well as Professor Shore's contract) can be adopted by department chairs, center directors, student organizations, and individuals to protect graduate students from negligence or mistreatment related to scientific authorship.

Financial Remuneration

Conversations about financial remuneration can arise in the creation of a manuscript. Some faculty and supervisors feel that students or other individuals who are paid as research or graduate assistants should not be given authorship because credit for performed work is being given in the form of a salary. These same faculty members might express that publication credit replaces the need for financial remuneration, because the individuals will ultimately benefit from having their names listed on a paper. Fine and Kurdek (1993) are firm in their position that paying a research assistant or graduate student should not

Even if you love writing, sometimes it takes great effort to put a line down on paper. With an infinite array of potential distractions on the Internet (e.g., social media), especially when one must make use of online resources (e.g., Google Scholar, PubMed) to write, writing time can suffer. Add to that the existence of smartphone apps and offline "distractions" (e.g., work tasks with deadlines, that new novel you can't put down, television, family or household obligations, social invitations), and finding time to write can be nearly impossible. Some potential solutions:

1. Make a writing schedule and stick to it. Mark the time in your calendar, and treat it like you are getting paid by the hour. If an extra incentive is needed, take a cue from behavior-modification experts and give yourself a small reward when you successfully follow through with your writing goals for the day (or even the hour!).

2. Find a great place to write. Many new scholars find that writing at a local cafe or public library is easier than writing at home. Alternatively, designating an area of your home for "writing" might help to keep you on task.

3. Do something about your smartphone/tablet while you write. Put it on "airplane" mode; take it offline; or, at the very least, turn off notifications.

4. Take advantage of free, online writing tools and apps. Do a quick web search for "free writing tools," and you will encounter a bevy of computer- and smartphone-based applications that will allow you to do such things as (a) keep you offline (e.g., "Freedom" app), (b) block you from specific sites (e.g., "Self-control" app), (c) organize your thoughts (e.g., "Evernote" app), (d) monitor writing breaks (e.g., "Time Out" app), or (e) be rewarded or "punished" for progress (e.g., "Write or Die" app). The popular website *The Huffington Post* has even designated an entire section of their site for keeping up to date with the latest writing apps: www.huffingtonpost.com/news/writing-apps. For those without computer access, setting frequent, proximal, and challenging yet achievable short-term goals has been closely linked to achievement success (see Morisano, 2013).

5. Give yourself a few minutes each day to de-stress. Often, our most creative ideas arise when we pull ourselves out of "go mode" and take a moment to sit and think, relax, take a walk, close our eyes, exercise, or meditate (e.g., with Jon Kabat-Zinn at www.youtube.com/watch?v=iZIjDtHUsR0).

6. Keep up-to-date on the latest research by subscribing to relevant listservs such as the one maintained by the Kettil Bruun Society for Social and Epidemiological Research on Alcohol (instructions at www.kettilbruun.org/Listserve.htm). They are often the source of good ideas and occasionally an inspiration for future articles.

Box 5.2: Writing strategies.

substitute for authorship credit, when credit for professional and intellectual contributions is due.[5] This extends to the hiring of consultants to contribute to the research and writing of an article; payment is not a substitute for authorship. The extent of controversy surrounding financial remuneration indicates that this topic should be covered when creating institutional and departmental guidelines surrounding authorship procedures. In light of the authorship criteria discussed elsewhere in this chapter and in Chapter 11, it is clear that neither financial reimbursement nor its absence should be considered in the determination of authorship credits.

The Nitty Gritty: Submitting a Manuscript and Responding to the First Rejection

After carefully choosing a target journal (see Chapter 3 for advice), one should normally write a cover letter to the journal's editor and a brief description of the manuscript. Some journal editors might have sympathy for novice writers when sending written feedback (e.g., by providing more detail), so one's inexperience could be worth noting here. One should be mindful that some journals require specific cover-letter content (e.g., word count, conflict-of-interest statement); therefore, author guidelines must be consulted in advance. These are most often found under Author Guidelines or Instructions for Authors on journal websites, or in the paper copy of the journal itself. Some journal editors (particularly of smaller journals) are also open to receiving presubmission emails to gauge interest in potential submissions; this is worth considering.

Even for the most fastidious researchers and stellar writers, the day will likely arrive when a rejection letter is received. If the rejected work is based on one's dissertation data, the decision can be particularly devastating, given the time and energy invested (and other issues previously discussed). It is important to understand that rejection is simply a part of the writing and publication process—even senior and experienced researchers have manuscripts rejected.[6] It is surprisingly easy to forget that if one is reaching for the stars and submitting to a competitive journal, acceptance rates are low. Even lower ranked journals are increasingly incorporating rigorous standards that might require a decent paper to go through a "revise and resubmit" round or two before acceptance. The most productive step to take post-rejection is to read and incorporate reviewer feedback as much as possible into a new draft, and try, try again (at another journal, unless resubmission is specifically invited). Chapter 12 provides guidance on how to respond to editors' requests for revised manuscripts.

A Word on Predatory Publishers

With the dramatic expansion of open-access and online journals (see Chapter 3), a number of for-profit enterprises have created new "journals" that will publish almost any article submitted for a processing fee ranging from $500 to several thousand dollars (Beall, 2012). The name "predatory publisher" has been applied to this type of business because it involves charging publication fees to authors without providing the editorial and publishing services associated with legitimate journals. Several new addiction-science journals have been launched by these publishers, raising serious questions about their impact on a field that is already plagued by conflict-of-interest threats from the alcohol, tobacco, gambling, and pharmaceutical industries (see Chapter 16).

The characteristics of these journals include rapid acceptance of articles with little or no peer review; aggressive marketing, often using poor grammar and syntax; journal editors with no academic standing in the addiction field; misleading or nonexistent publication metrics (e.g., impact factors, indexing services); and publication fees that are not revealed until after the article is accepted.

It is easy to understand both the frustration of a new investigator who might receive multiple rejection letters and the appeal of an online journal that levies page charges after a cursory review. If early-career scientists or trainees choose to publish in such journals, however, the most likely consequence is to appear to peers, grant reviewers, potential employers, and promotion committees to be naive, unethical, or desperate for authorship credits. Many researchers are not familiar with the complicated and often confusing developments in journal publishing and may be easily scammed and embarrassed. Fortunately, resources on how to protect the integrity of science and avoid these unscrupulous phantom publishing operations masquerading as addiction journals are available, including Jeffrey Beall's (2015) list of predatory publishers (see References for a link). Prospective authors should also consult Chapter 3 of this book and the updated website of the International Society of Addiction Journal Editors (ISAJE; www.isaje.net), which provide a list of journals that subscribe to the Farmington Consensus, a code of ethics for journals and journal editors (Farmington Consensus, 1997).

Special Issues of Relevance to Students and Junior Investigators from Low- and Middle-income Countries

Thus far, this chapter has focused on publication issues that are likely to be most relevant to those from well-resourced countries with an established scientific community in the addiction field. Students and junior investigators in less resourced countries face a number of different issues related to conducting and disseminating research (see Chapter 4 for a discussion of broader issues related to addiction research in developing countries). The following section addresses some of the special challenges encountered by students and novice addiction researchers from low- and middle-income countries (LMICs) as well as those from LMICs who earn their degrees from universities in developed countries and then return home. There is an imperative at both the national and international levels to publish research on addiction issues that is relevant to populations outside of Europe, Australia, and North America. High-quality dissertation research in general has the potential to significantly impact addiction science. Further, individuals from LMICs have especially strong obligations (and pressures) to conduct research and publish the results. In many LMICs, research is used to shape both the policy agenda and prevention/intervention programs. But most of the evidence on what policies and interventions "work" to reduce substance-related harms is based on studies conducted in developed countries. Indeed, the notion that research is limited in LMICs is highlighted by the shocking 10/90 gap statistic, according to which, only 10% of global research spending is directed to health problems that comprise 90% of the world's disease burden (Global Forum for Health Research, 2004).

General Capacity Challenges

The capacity of individuals to conduct and publish research varies considerably within and across LMIC academic institutions. In many university environments, salaries may be low, with both high teaching loads and competing demands. Personal financial constraints might compel academics to undertake other activities, such as seeing private patients or conducting various types of consulting work to supplement their incomes. Academics in LMICs often have minimal staff support and must conduct the bulk of their research work unassisted. In better resourced environments, investigators are more likely to have staff assistance for many of the activities that are required to write and submit papers for publication (e.g., literature reviews, data collection, entry, and data analysis) and for other aspects of research (including grant writing).

Publishing in countries with minimal research infrastructure outside of an academic institution is a special challenge, because writing is often lower on the priority list than tasks that are directly related to conducting the research, running prevention and intervention programs, or moving on to the next project. The final product of research is often a report for a local or national agency rather than a formal journal article. Although reports are an important mechanism for disseminating research findings, redrafting them into journal articles is necessary for the data to reach a broader scientific audience, to influence work in other LMICs, and to contribute to global knowledge. Publishing in a peer-reviewed outlet might also provide the author(s) with helpful feedback and ways to improve the work and thus the contribution.

Converting reports into journal articles under intensive work constraints can be a difficult, albeit surmountable, challenge. ISAJE has developed a writing mentorship program for this purpose (see http://www.parint.org/mentor_1.htm for more information; Miller, 2011). It provides novice researchers with the opportunity to be mentored by senior researchers, which can be useful if the immediate work environment does not provide sufficient opportunities to learn how to write for peer-reviewed academic journals.

Some LMIC researchers might sometimes fear that their work does not meet the standards of certain journals. With the development and use of increasingly sophisticated equipment and statistical techniques in high-income countries, the perception might arise that any research that is not state-of-the-art is not publishable. This is absolutely not the case. As suggested in Chapter 4, LMIC research may provide drug and alcohol policymakers with regionally specific data and evidence-based interventions. When implementing new laws, treatment policies, or programs anywhere, it is imperative that they are culturally appropriate and relevant. Furthermore, it is useful for researchers in Europe, Australia, and North America to have a more global perspective on research and prevention or intervention outcomes when developing their own protocols and policies. Exposing addiction scientists from non-LMICs to researchers from LMICs might lead to important investigative collaborations and cross-cultural research. Some of the most valuable studies of alcohol and drug screening, brief intervention, treatment, and epidemiology were conducted as cross-national collaborations between researchers from LMICs and high-income countries (Humeniuk et al., 2012; Rehm et al., 2010; Saunders et al., 1993). By regularly reading journal articles, attending conferences, and joining international research societies, LMIC researchers can gain exposure to diverse international research and build the confidence, skills, and connections that could lead to opportunities for international collaborative research.

Research Topics

Although there is still a significant underrepresentation of LMIC publications in scientific journals, improvements have been observed in recent years (Large et al., 2010; Warner et al., 2014). Large and colleagues demonstrated that the proportion of psychiatric publications from LMICs, as identified via PubMed, increased from 8.0% in 1998–2002 to 12.5% in 2003–2007. Similarly, Warner and others reported an increase in LMIC research publications in a leading addiction journal (*Tobacco Control*), from 10.1% in 1992–2006 to 30.9% in 2007–2011.

The relative lack of studies emerging from many developing countries in a multitude of research areas, however, provides ample topics for publication. Recent graduates have an easy publication target: their dissertations or theses. Academics will likely conduct new research. Further, people in government agencies, clinical settings, and nongovernmental organizations, who may not have access to original data, might consider alternative publication routes such as narrative or systematic reviews that involve synthesizing the results of multiple research studies on a specific subject. The Cochrane Collaboration site (www.cochrane.org) can be consulted for information on potential review topics and systematic-review writing procedures. Similarly, it is possible to write case studies, letters, or policy and opinion pieces, all of which can stimulate public debate and influence policies.

Selecting an Appropriate Journal

As indicated in previous chapters (e.g., Chapter 3), there are many addiction journals, but the selection of the "right" journal can present special challenges for those from LMICs. A number of competing considerations might influence the choice.

First, in both developed and developing countries, many academics are under pressure to publish in "high-impact" peer-reviewed journals (see Chapter 3 for a discussion of impact factors). In South Africa, for example, academic institutions receive government subsidies based on the number of peer-reviewed publications produced in journals that have been accredited by the Department of Higher Education and Training. Moreover, in many LMIC institutions, the academic evaluation of faculty and their potential for career advancement is dependent on their publication record.

Second, authors might be faced with having to choose between publishing in a high-impact international journal that furthers their research careers or publishing in a low-impact local journal that reaches the public health audience of interest. In some cases, a middle-ground solution can be reached (e.g., publishing one's papers in both types of journals under agreed-upon conditions; see Chapters 3 and 4 for a discussion).

Third, the topical foci or missions of different journals must also influence one's choice. Some journal reviewers and editors might not have an interest in studies of non-American or non-European populations. Advance familiarization with the contents of the journal under consideration can help to gauge the likelihood that an LMIC health or addiction issue would engage the journal's editors and readership.

Finally, one could consider publishing in an open-access journal, which is usually accompanied by payment (although it is important to take heed of the predatory publishers previously mentioned!). Advantages to authors might include increased accessibility and citations, which contribute to researchers' rankings and assessments. However, submission or publication costs can be high and difficult to justify in the case of limited research funds.

Language

Many times a manuscript is rejected by a journal not because of the quality of the research but because of the authors' failure to express their ideas clearly. For authors whose first language is not English, translating one's work for English-language journals can be difficult. Writing in English can even be a challenge for individuals who have attended English-language academic institutions and who have written their theses or dissertations in English. Converting the dissertation or thesis to the shortened format required for most journals can add to the difficulties of working in a second or third language and can lengthen the time to publication. To manage such language constraints, it is advisable to invite a native English speaker to serve as a co-author and help with editing, as long as she or he meets all the key authorship criteria. International conferences and meetings can be a good forum for networking with potential co-authors. Alternatively, authors may consider using English-language editing services, which usually entail a fee (e.g., http://webshop.elsevier.com/languageservices/languageediting, http://wileyeditingservices.com/en/english-language-editing).

Access to Literature and the Internet

In numerous academic and other institutions in LMICs, access to journal articles, books, and other relevant literature is a major challenge that hinders research, writing, and publishing. Paper copies of articles and other literature often have to be ordered via slow, costly, and unreliable interlibrary loan systems. Furthermore, many academics do not have easy access to online journals because (a) they have unreliable Internet connections, (b) their institutions do not own subscriptions to the required journals, or (c) they might be unaware of free or reduced-cost options for accessing journal articles. In 2002, the World Health Organization and a number of major publishers established the Health Inter-Network Access to Research Initiative (HINARI) to directly address such difficulties. HINARI provides free or reduced-cost online-journal access to health workers and researchers from local, not-for-profit institutions in many LMICs. More information about the initiative, including eligible countries, instructions for access, and related initiatives is available on the HINARI website (www.who.int/hinari/en).

The lack of consistent and reliable Internet access also causes problems at the online article-submission stage for authors from LMICs. This process can be lengthy even for those with good Internet connections. Establishing collaborations with researchers who have better access to these resources might, in some cases, help to address this challenge.

Challenges of Rejection

As noted above, it is quite common for manuscripts to be rejected for publication after initial submission. Papers from non-European and non–North American settings are sometimes rejected because the reviewers or editors are not aware of the significance of the research in its cultural or local context. In such cases, authors may exercise their right to appeal the rejection if they believe it is based on the editors' or reviewers' lack of appreciation of the importance of the topic. It might also be useful to precede submission with an email to the journal editor about the topic and its importance before sending it in.

Comment: Be Optimistic

Despite the significant challenges for novice scientists from LMICs, there are advantages to the relative lack of existing research for those just setting out on their research and publication careers. One might be able to claim truthfully that the research has never been conducted or replicated outside of the developed world. Furthermore, the presence of numerous academics from LMICs who continue to be prolific despite the under-resourced settings in which they work provides evidence that many of the aforementioned difficulties are surmountable.

Conclusions: Take the Long View

A career in addiction science is not for everyone, but it can be very rewarding for those who have the motivation and the aptitude (Edwards, 2002). The best way to begin is to attempt publication of one's thesis or dissertation, work closely with one or more well-published investigators, employ the writing strategies discussed, and find a place for postdoctoral research or clinical training. The writing process from student to postdoctoral fellow to junior researcher is generally the same, although the level of autonomy increases with each transition. Greater autonomy is usually accompanied by more security regarding one's place in the publication process and an increased ease in negotiating authorship order. Further, full-time research scientists are not the only ones who enjoy the rewards of publishing. Those who work in clinical settings, government agencies, and other organizations often find that while journal publications are not rewarded by their employers, neither are they likely to be discouraged. The pros of publishing one's work usually outweigh the cons.

In this chapter, some basic guidelines have been outlined for inexperienced authors. Although there is no magic formula for guaranteed publication, finding a mentor, learning to persevere in the face of rejection, and never ceasing to believe in addiction science are key elements to the process.

> Please visit the website of the International Society of Addiction Journal Editors (ISAJE) at www.isaje.net to access supplementary materials related to this chapter. Materials include additional reading, exercises, examples, PowerPoint presentations, videos, and e-learning lessons.

Notes

[1] For the general purposes of streamlining and efficiency, the term *faculty member* or *faculty* (as an adjective) is used throughout this chapter to represent any kind of higher education advisor, supervisor, teacher, or researcher who might otherwise be called a researcher, a lecturer (junior or senior), a professor, etc. It should be noted that, depending on the country and/or institution, different terminology may be used.

[2] Let us not forget that this is the true purpose of scientific publishing!

[3] For students at some institutions, the dissertation articles must be published before the dissertation can be approved, and publication must therefore be prepared at an earlier stage.

[4] See http://core.ecu.edu/psyc/wuenschk/Help/ThesisDiss/thauth.htm

[5] And of course, those faculty members are usually getting paid as well—publication is an expected part of the job.

[6] Including all of the authors of this chapter!

References

American Psychological Association. (2010, June 1). *Ethical principles of psychologists and code of conduct: Including 2010 amendments.* Retrieved from http://www.apa.org/ethics/code/index.aspx.

Azar, B. (2006, March). Publishing your dissertation. *gradPSYCH Magazine.* Retrieved from http://www.apa.org/gradpsych/2006/03/dissertation.aspx.

Beall, J. (2012). Predatory publishers are corrupting open access. *Nature, 489*(7415), 179. DOI: https://doi.org/10.1038/489179a

Beall, J. (2015, January 2). *Beall's list of predatory publishers 2015.* Retrieved from http://scholarlyoa.com/2015/01/02/bealls-list-of-predatory-publishers-2015/.

Cham, J. (2005, March 13). The author list: Giving credit where credit is due. *Piled Higher & Deeper: A Grad Student Comic Strip.* Retrieved from http://phdcomics.com/comics.php?f=562.

Cornell, E., Doshi, R., Noel, J., & Rusch, L. (2014). *Authorship rights of graduate students.* Unpublished guideline prepared for the Department of Community Medicine and Health Care, University of Connecticut School of Medicine, Farmington, CT.

Edwards, G. (Ed.). (2002). *Addiction: Evolution of a specialist field.* Oxford, England: Blackwell Publishing.

Farmington Consensus. (1997). *Addiction, 92,* 1617–1618.

Fine, M. A., & Kurdek, L. A. (1993). Reflections on determining authorship credit and authorship order on faculty-student collaborations. *American Psychologist, 48,* 1141–1147.

Global Forum for Health Research. (2004). *The 10/90 report on health research 2003–2004.* Geneva, Switzerland: Author. Retrieved from http://announcementsfiles.cohred.org/gfhr_pub/assoc/s14789e/s14789e.pdf.

Hagen, N. T. (2010). Deconstructing doctoral dissertations: How many papers does it take to make a PhD? *Scientometrics, 85,* 567–579.

Gross, D., Alhusen, J., & Jennings, B. M. (2012). Authorship ethics with the dissertation manuscript option. *Research in Nursing & Health, 35,* 431–434.

Harvard Medical School Office for Research Issues. (1999, December 17). *Authorship guidelines.* Retrieved from http://hms.harvard.edu/about-hms/integrity-academic-medicine/hms-policy/faculty-policies-integrity-science/authorship-guidelines.

Humeniuk, R., Ali, R., Babor, T., Souza-Formigoni, M. L. O., de Lacerda, R. B., Ling, W, McRee, B., Newcombe, D., Hemraj, P., Poznyak, V., Simon, S., & Vendetti, J. (2012). A randomized controlled trial of a brief intervention for illicit drugs linked to the Alcohol, Smoking and Substance Involvement Screening Test (ASSIST) in clients recruited from primary health-care settings in four countries. *Addiction, 107,* 957–966.

Large, M., Nielssen, O., Farooq, S., & Glozier, N. (2010). Increasing rates of psychiatric publication from low- and middle-income countries. *International Journal of Social Psychiatry, 56,* 497–506.

Miller, P. M. (2011). Introducing the ISAJE-PARINT Online Mentoring Scheme. *Journal of Groups in Addiction & Recovery, 6,* 272.

Morisano, D. (2013). Goal setting in the academic arena. In E. A. Locke & G. Latham (Eds.), *New developments in goal setting and task performance* (pp. 495–506). New York, NY: Routledge.

Rehm, J., Baliunas, D., Borges, G. L. G., Graham, K., Irving, H., Kehoe, T., Parry, C. D., Patra, J., Popova, S., Poznyak, V., Roerecke, M., Room, R., Samokhvalov, A. V., & Taylor, B. (2010). The relation between different dimensions of alcohol consumption and burden of disease—an overview. *Addiction, 105,* 817–843.

Resta, R. G., Veach, P. M., Charles, S., Vogel, K., Blase, T., & Palmer, C. G. (2010). Publishing a Master's thesis: A guide for novice authors. *Journal of Genetic Counseling, 19,* 217–227.

Robinson, S., & Dracup, K. (2008). Innovative options for the doctoral dissertation in nursing. *Nursing Outlook, 56,* 174–178.

Rudestam, K. E., & Newton, R. R. (1992). *Surviving your dissertation: A comprehensive guide to content and process.* Newbury Park, CA: Sage.

Saunders, J. B., Aasland, O. G., Babor, T. F., de la Fuente, J. R., & Grant, M. (1993). Development of the Alcohol Use Disorders Identification Test (AUDIT): WHO collaborative project on early detection of persons with harmful alcohol consumption--II. *Addiction, 88,* 791–804.

Shore, B. M. (2014). *The graduate advisor handbook: A student-centered approach.* Chicago, IL: The University of Chicago Press. DOI: https://doi. org/10.7208/chicago/9780226011783.001.0001

Trammell, A. (2014, October 14). *The benefits of publishing as an undergraduate.* Retrieved from https://publish.illinois.edu/ugresearch/2014/10/14/the-benefits-of-publishing-as-an-undergraduate/.

University of Alberta Faculty of Graduate Studies and Research Council. (1996, November 15). *Intellectual property policies: Guidelines for authorship.* Retrieved from https://uofa.ualberta.ca/graduate-studies/about/graduate-program-manual/section-10-intellectual-property/10-2-guidelines-for-authorship.

University of Cambridge. (2014, November). *Good research practice guidelines (Section 9): Dissemination and publication of results.* Retrieved from http:// www.research-integrity.admin.cam.ac.uk/sites/www.research-integrity. admin.cam.ac.uk/files/good_research_practice_guidelines_11.14.pdf.

University of Pennsylvania's Office of the Provost. (2013, February 15). *Fairness of authorship credit in collaborative faculty-student publications for PhD, AM, and MS students.* Retrieved from https://provost.upenn.edu/policies/ pennbook/2013/02/15/fairness-of-authorship-credit-in-collaborative-faculty-student-publications-for-phd-am-and-ms-students.

University of Toronto. (2007). *Intellectual property guidelines for graduate students & supervisors.* Retrieved from http://www.sgs.utoronto.ca/currentstudents/ Pages/Intellectual-Property-Guidelines.aspx.

Warner, K. E., Tam, J., & Koltun, S. M. (2014). Growth of Tobacco Control publications by authors from low- and middle-income countries. *Tobacco Control, 23,* 231–237.

Washington University in St. Louis. (2014, November 21). *Policy for authorship on scientific and scholarly publications.* Retrieved from https://wustl. edu/about/compliance-policies/intellectual-property-research-policies/ scientific-scholarly-authorship/.

Wuensch, K. (2008, November 30). *Thesis authorship.* Retrieved from http:// core.ecu.edu/psyc/wuenschk/Help/ThesisDiss/thauth.htm.

Appendix A. Example of a Research Advisor-advisee Contract
(Excerpted from Shore, 2014)[a]

Mutual Expectations Regarding Research Advising
High Ability and Inquiry Research Group
Department of Educational and Counselling Psychology,
McGill University

These notes are designed as guidelines to facilitate positive and mutually beneficial student-advisor relationships and to avoid problems on matters such as authorship and credits on publications, the extent of participation in activities other than the Thesis, Research Project, or Special Activity, and future access to data collected in the course of our work together. Some of the activities described below may be conducted in groups. Where these notes hinder rather than help, they should be amended to meet mutually acceptable needs, in general or as occasions arise.

A. Advisor's Responsibilities
   1. Meet regularly with students and be contactable at other times.
   2. Arrange substitute advising during extended absences.
   3. Advise on course selection.
   4. Assist in the preparation for comprehensive or oral examinations.
   5. Help prepare conference and journal presentations based on work done in the program and assist with applications for support to attend suitable conferences at a reasonable distance and on whose programs students earn a place.
   6. Help apply for funds to cover direct research costs and to provide stipends to full-time students.
   7. Provide feedback within a mutually agreed time-frame on written work submitted for review.

B. Students' Responsibilities
   1. Regularly pursue work and keep the advisor informed of progress or problems.
   2. To a mutually agreed degree that respects other responsibilities and priorities, contribute to advancing team activities that further the common good of all of us working together—e.g., workshops for teachers, parent contacts, library orders, data bases, maintaining bibliographies and mailing lists, convening meetings, maintaining computers and supplies. These tasks will be equitably distributed.
   3. Join in the preparation of conference presentations and publications on research and other activities done with faculty members.

4. With appropriate guidance, prepare a draft version of the thesis or major report, normally within 3 months of its final presentation for master's degrees, or 6 months for doctoral degrees; after that point the advisor may take over such preparation and the order of authorship may be changed (within CPA, APA and McGill authorship guidelines).
5. Apply for scholarships and bursaries, especially FQRSC, McGill, and SSHRC (where eligible) [this list of funding sources should be amended to match local availability].
6. Participate to a mutually agreed extent in teaching-related activities such as the TA course.
7. Take a professional role in one's discipline by undertaking at least one student or regular membership in an appropriate professional or academic organization.
8. Keep at McGill a copy of raw data, coding sheets, instruments, and subject-identification data.
9. Upon graduation, leave with the advisor a printed copy of the main research report, and an electronic copy in modifiable form (e.g., not PDF) of any data and the text of the thesis or project.
10. Use Microsoft Word and APA [or other, as appropriate] style for written submissions.
11. Report annually in writing on progress and contributions (department and university forms).
12. Regularly attend and participate in research-team meetings.

C. Joint Responsibilities
1. Give full credit for the contributions of others and to research funding in all products.
2. Assign authorship according to the latest APA publication guidelines. (For example, if a thesis topic or report is entirely the student's original contribution, then the advisor's contribution is due a footnote. Shared scientific responsibility calls for co-authorship, with the student as first author on the main points of the student's research of those for which the student took primary creative responsibility, and the advisor as first author on any specific subpoints which the advisor contributed or a broader study of which the student is part.)
3. Both have unlimited access to the data collected on or about the topic of a thesis or project during the time worked together, plus any other that may be agreed to, giving due credit to its origin either by footnote or reference to previous publications.

D. Degree Covered by this Agreement
Check-mark all that apply [and revise this list as needed for your institution]:
☐ PhD Thesis or Dissertation
☐ MA Thesis
☐ MA Research Project

☐ MEd "Special Activity" Project
☐ Undergraduate Honors Thesis
☐ Independent Graduate Student Project
☐ Independent Undergraduate Student Project
☐ Other (specify): _____
☐ Not for formal credit

E. Comments, Additions, or Special Notes [expand this space as required]

F. Signatures
We agree to work together in an advisory relationship in accord with the above guidelines.

Advisor	Date	Student	Date

Printed Name	Printed Name

One copy for each.

Note:
[a] This sample contract was also reproduced in: Shore, B. M. (2014). *The graduate advisor handbook: A student-centered approach.* Chicago, IL: The University of Chicago Press (in the series *Chicago Guides to Academic Life*). DOI: https://doi.org/10.7208/chicago/9780226011783.001.0001

Appendix B. Authorship Rights of Graduate Students[b]

It is agreed that...

1. Graduate students are a vulnerable population with regard to authorship issues in scientific publications because of their junior status in the academic hierarchy.
2. Graduate students rely on principal investigators, faculty members, and other individuals in positions of power for funding and for access to research opportunities and data.
3. Graduate students rely on principal investigators, faculty members, and other individuals in positions of power for successful completion of any graduate program.
4. Graduate students who participate in research studies often fulfill necessary roles and provide vital support toward the completion of research projects conducted by teams of faculty, students, and staff.

5. Principal investigators, faculty members, and other individuals in positions of power can influence, directly or indirectly, positively or negatively, the credit given for work done by students following the successful completion of a research study.
6. Authorship credits are often important for graduate students' careers.
7. Students may be given inappropriate and unethical authorship credits to enhance the student's chances of success. Conversely, students may be denied appropriate and ethical authorship credit.
8. There is little recourse for a graduate student should a principal investigator, faculty member, or other individual in a position of power negatively influence deserved authorship credit.
9. A set of rights and guidelines to protect graduate students and to define faculty–student authorship criteria are needed.
10. The rights and guidelines listed in the sections "General Research Studies" and "Dissertation or Thesis Research" listed below shall be adopted to protect graduate students from negligence or mistreatment and to define graduate student authorship.

General Research Studies

1. A graduate student who has participated in a research study conducted by a faculty member who is affiliated with graduate student's program or who supervises the graduate student has the right to be invited to become an author on any report, abstract, journal manuscript, or other document developed based on the results of the study, provided the student has completed sufficient training.
   a. Study participation may include, but is not limited to, the following: recruitment of study subjects, providing an intervention, data collection, data entry, questionnaire coding, supervision and training of study personnel, writing of the research protocol, or the provision of other technical services.
   b. Authorship is defined as providing a major contribution to a report, abstract, journal manuscript, or other document including, but not limited to, the following: writing the final version of the submission, designing the study, interpreting the results, study coordination, statistical analysis, laboratory analysis, data management, or providing informative advice on study design and analysis.
   c. Sufficient training may include, but is not limited to, the following: completion of specific coursework, knowledge of the subject matter, or knowledge of the study design. The extent of training is to be agreed upon prior to the student's involvement in the research study and occurs between the student, the study's principal investigator, and/or the student's major advisor.

2. A graduate student's role in the drafting of a report, abstract, journal manuscript, or other document, as well as possible authorship position, is to be discussed prior to the first draft of a report, abstract, or journal manuscript.

3. Financial compensation, whether through graduate assistantships or by other means, is not a replacement for authorship credit.

4. Acknowledgement is not a replacement for authorship credit.

5. A graduate student's role on a report, abstract, journal manuscript, or other document shall not change without notifying the student, allowing the student to respond to the notification, and agreement of all co-authors.

6. A graduate student has the right to refuse authorship on a report, abstract, journal manuscript, or other document for any reason.

7. If a disagreement over authorship occurs between a graduate student and a principal investigator, the graduate student may appeal to the Director of their graduate program or the Chair of the department with which the principal investigator is affiliated to appoint an unbiased arbitration committee to resolve the conflict. This committee will be comprised of three individuals and will consist of at least one student.

8. The principal investigator or any other faculty member shall not penalize a graduate student by eliminating future authorship opportunities, removing study responsibilities, assigning an excessive workload, withholding monetary compensation, or imposing any other punishment, directly or indirectly, should the student disagree with the principal investigator over authorship or invoke independent arbitration.

9. These guidelines shall apply for an agreed upon amount of time after the student graduates, changes institutions, or otherwise is no longer affiliated with the graduate program. The time limit shall be agreed upon by the student, the study's principal investigator, and/or the student's major advisor.

Dissertation or Thesis Research

1. Research and analyses conducted by a graduate student for the purposes of fulfilling doctoral dissertation or master's-thesis requirements is considered the property of the graduate student, regardless of who is listed as principal investigator on funding, regulatory documentation, or other documentation.

2. A graduate student has the right to first authorship on any report, abstract, journal manuscript, or other document that is created based on the results of dissertation or thesis research conducted by said graduate student.

3. The principal investigator listed on funding, regulatory documentation, or other documentation that supports a graduate student's dissertation

or thesis research shall in no way impede, and will support, said graduate student in creating a report, abstract, journal manuscript, or other document.

4. Data generated from dissertation or thesis research will revert to the principal investigator if, and only if, a graduate student has not produced a first draft of a report, abstract, journal manuscript, or other document within a previously agreed upon time window.

   a. If no window is agreed upon, then the data generated from dissertation or thesis research shall not revert to the principal investigator under any circumstances.

   b. If the first draft of a report, abstract, journal manuscript, or other document is not produced by the student within the previously agreed upon time window, the principal investigator must include the graduate student in the drafting of a report, abstract, journal manuscript, or other document using the guidelines specified in the "General Research Studies" section, unless the graduate student agrees to be excluded from the process.

5. A graduate student has a right not to publish, and not to have published, dissertation or thesis research.

   a. A graduate student may invoke this right at any time prior to, during, or after the previously agreed upon publication window, unless the previously agreed upon window has already been exceeded, the graduate student has been included in the authorship process, and the results have already been published in a peer-review journal; or the graduate student has previously agreed to be excluded from the process.

6. If a disagreement over authorship occurs between a graduate student and a principal investigator, the graduate student may appeal to the Director of their graduate program or the Chair of the department with which the principal investigator is affiliated to appoint an unbiased arbitration committee to resolve the conflict. This committee will be comprised of three individuals and will consist of at least one student.

7. The principal investigator or any other faculty member shall not penalize a graduate student by eliminating future authorship opportunities, removing study responsibilities, assigning excessive workload, withholding monetary compensation, or by imposing any other punishment, directly or indirectly, should the student disagree with the principal investigator over authorship or invoke independent arbitration.

Note:
[b]This guideline was developed by Erin Cornell, Riddhi Doshi, Jonathan Noel, and Lisa Rusch in April 2014, when they were graduate students at the University of Connecticut in the Graduate Program in Public Health.

CHAPTER 6

Addiction Science for Professionals Working in Clinical Settings

Richard Pates and Roman Gabrhelík

Introduction

This chapter is aimed at doctors, psychologists, social workers, therapists, and other staff in the health sector, social care sector, and criminal justice system (e.g., prisons, probation) working in addiction. It is also written for workers in the nongovernmental (non-statutory or "third") sector with some professional training or expertise. These clinical workers often are the first to identify new trends in substance use, effects, problematic consequences, and problems that may support or hinder rehabilitation. Therefore, clinicians can play an important role in research. In many developing countries or in countries without a history of alcohol and other drug research, clinicians may be the only people who are able to document problems. At the same time, they also have a duty to identify and collect this information and distribute it. This chapter will discuss what sort of research might be suitable for clinicians, how to approach it, where to publish, and pitfalls in addiction research and publishing. The purpose is to encourage professionals who work in the field of addiction, not primarily as researchers, but as clinicians who have conducted work or research projects that could be worthy of publication. This chapter also provides instruction on how clinicians can collaborate with researchers.

Historically, clinicians have played an important role in research. It is worth remembering that the early pioneers in alcohol and other drug research were often doctors such as Trotter, Rush, and Huss (in alcohol research) as well as Dole and Nyswander (in research on the use of methadone in the treatment of

---

heroin addiction). It is also of note that, today, many of the people working at the top of large research institutes and public health bodies such as the World Health Organization have clinical backgrounds in psychology and medicine.

Although a research component is included in many (or most) undergraduate and postgraduate clinical courses, it is sometimes seen as a process that must be passed before qualification rather than as an exciting opportunity to expand a professional role. In many professional fields, the number of well-trained staff who never do research or publish anything after they have qualified is surprising (e.g., Jowett et al., 2000; Salmon et al., 2007), especially given that the work clinicians perform, whether in the statutory or non-statutory sector, is usually based (or should be based) on proven results and methods founded on research-related best practices.

This lack of willingness to undertake research or to publish research results may result from lack of confidence, opportunity, or willing collaborators. But as will be seen in this chapter, there are plenty of opportunities and subjects appropriate to study systematically in the clinical setting. Although this chapter is not meant to teach research methods, it is aimed at those who have previously had some research training and who have had the opportunity to undertake research projects. It is also aimed at those interested in evaluating their work or investigating some aspect of their work that may be worthy of publication.

We cannot take for granted that the majority of professionals have the skills for conducting scientifically sound studies. To conduct a research study using appropriate design, adequate measures, and correct statistics can sometimes be difficult. Further, there are additional problems of trying to publish the findings in peer-reviewed journals.

Early addiction practice was based on a clinical approach (problems were observed, described, and explained), and this slowly started to shift toward empiricism (allowing for testing hypotheses through observation and experiment). More recently, an evidence-based approach to addiction services has been promoted and widely accepted. Evidence based practice (or applied addiction science) means that the nature and method of addiction services is based on findings from research studies. The level and quality of clinical work is quantified. Quantified results serve as an evidence of effectiveness or ineffectiveness of any interventions provided.[1] In practice, this means a range of things, including treatment of addiction problems, prevention of relapse, and provision of aftercare and other post treatment interventions aimed at helping those in recovery get back to a regular lifestyle. Over time, addiction professionals began to ask questions about the effectiveness of the methods being used in the treatment of addiction problems, the prevention of relapse, and the provision of aftercare and other post treatment interventions aimed at reintegrating the person into daily life. As a consequence, interest in appropriate interventions grew. Professionals from the field started to search for new ways to achieve better results in less time but with a longer duration of action.

Studying the effective factors in addiction services and monitoring the benefits of different interventions became the domain of research.

Why bother Doing Research?

For many clinicians, the idea of undertaking research may seem to be yet another demand on their time and not part of their job. But clinicians should always be asking whether what they do is effective and the best practice. As will be discussed later, research can take many forms in terms of evaluating interventions, trying to find the cause of a problem, studying individual cases or reviews of a subject area. Many of these areas may be too complex and involved for the professionals in clinical practice to undertake, but there are some types of research that are well within the capability of clinical staff. Examples of this are research into brief interventions with alcohol and tobacco, which has had an impact on clinical work.

Many benefits can be derived from taking part in research. There is an intrinsic satisfaction in undertaking a good piece of research, especially if it produces results that may affect your work and make it more effective. There is also the respect that you will earn from your colleagues. But most importantly, performing research can help to further your career. Even if your work has been mainly clinical, having publications on your résumé or curriculum vitae will do no harm and will probably enhance your career. Future employers will respect your endeavors into research.

What Research is Appropriate for You?

Choosing a research project that is suitable is very different if you are working in a clinical field rather than in an academic institution. In a clinical field, you will need to choose a research subject that permits access to participants (if it is a person-based project) and something that is manageable in the context in which you are working. Many clinical services perform regular audits of their work, and these are already simple forms of research. Of course, if this type of research is undertaken, it needs to be more rigorous than a standard audit and should conform to a research protocol.

Sometimes research questions may come from your search for a solution to a problem—you find that little work has been published in that subject area. In the 1990s, when the first author (R.P.) wanted to find treatments for compulsive injecting (needle fixation), a literature search revealed just one article published 20 years before that described three cases. This nevertheless led to a number of research collaborations in a clinical setting in which the problem was studied and psychological theories and treatment options were developed (Pates & Gray, 2009; Pates et al., 2001).

If the work you do routinely is common practice and already described in the literature, then it is unlikely to be of interest to journals. However, if you are doing something innovative or have noticed unusual results, this may well be worth formalizing and investigating. If you are planning innovative work, this should be investigated carefully following proper designs and ethical considerations.

The late Griffith Edwards, a great champion of addiction science and someone who was influential in encouraging junior researchers to publish their work, made an interesting observation that many clinicians will recognize. In a book of his to be published posthumously (Edwards, in preparation), he asked the question of where addiction research ideas came from. He observed that clinical research often comes from something said by a patient but also noted that the clinician "must have ears with which to listen. It is often too easy to ignore what patients may be saying by believing that expertise lies with the expert! He went on to describe a situation in which a patient of his commented that he (Edwards) had previously given the patient very bad advice: Edwards had told the patient that, to become sober, the patient would need a lengthy in-patient stay—the current practice at that time. The patient said he did not need that sort of help, would not accept it, and that it would mean the end of his business if he chose that path.

Inspired by this man, Edwards went on to conduct a comparison trial of in-patient versus out-patient treatment and found, at the 12-month follow-up, there was no significant difference between the two groups. This evidence helped to overturn the conventional consensus at the time—that in-patient treatment for a significant drinking problem is essential for recovery. This is a good example not only of the need to listen to patients but also of the need to challenge conventional ideas in places where they may be rigidly held.

In additional to quantitative reports, some journals will accept case reports or series of case studies (see also Chapter 8 on qualitative research), in which unusual findings may be reported (e.g., uncommon manifestation of diseases, "off-label" uses of medication, previously unreported effects of medications, unexpected effects of treatment). These studies can be of great interest because you may be the first to report a phenomenon—only make sure you are seeing and understanding cause and effect. These can add to the literature in an incremental but important way. It is often in clinical settings that these unusual practices come to notice, which could be the beginnings of a phenomenon or just unusual outliers in the field.

In addition to working within a centre there is the opportunity to work with other professionals doing similar work. This might entail being part of a multi-centre trial, in which a number of treatment centres work on the same project to increase the numbers of people being treated and provide greater statistical power to the analyses. A multi-centre trial also allows for comparison across sites and thus increases the generalizability of findings. This sort of trial is usually expensive because it needs coordination, usually from a research center.

This can be exciting work but requires a lot of extra effort to ensure that the interventions are the same in each center and that all the protocols are being followed in the same way.

Another type of investigation clinicians can do is historical research conducted by extracting data from case notes. For example, the first author of this chapter (R.P.) wanted to examine whether outcomes had changed in the clinic in which he was working from the establishment of the service 20 years previously to the present. This was performed by asking a number of questions that he formulated based on case notes and by taking a cohort of the first 200 patients registered with the clinic to establish things such as morbidity, mortality, recovery, and loss to the service. These data were then compared with data obtained from another cohort some 15 years later. This study evaluated a span of time when changes in practice were occurring in service delivery, and it was important to see if outcomes had improved. The study results actually had important consequences in terms of delivering services and learning lessons from practices that were found to be too rigid.

Good quantitative research is worth pursuing if the topic is original and not just repeating previous research. But, of course, many topics that have been researched are the product of an original idea that was investigated and then later research added to the findings and expanded it. In this way, individual studies become a body of research. Sometimes it is worth investigating a previously published research topic by adding a new dimension or helping to generalize a finding through the study of a different group. It must be borne in mind that, if the study is using a control group for comparison, it would be unethical to withhold a recognized treatment from the control group, even in the interests of the research.

Qualitative research is becoming more common in the addiction field. Twenty-five years ago, it was difficult to get qualitative research published because it was often not seen as "proper" research. That view has changed, and qualitative research is becoming more common. The advantage of conducting qualitative research is that you can investigate questions more deeply and follow up information that comes out of the research. It is often undertaken with fewer participants than quantitative research but still requires a rigid methodology and the same safeguards. (See Chapter 8 in this book for a full discussion on carrying out qualitative research.)

One major difference between working in a clinical setting and undertaking academic research is that, often in randomised controlled trials, there is a set of exclusion criteria that is used to remove what may be confounding factors for research. The problem for clinicians is that the people they treat are not subject to exclusion criteria. Storbjörk (2014) has written about this in a large piece of research on alcohol problems with 1,125 participants. She asked the following question: If 10 of the most common exclusion criteria were operationalized and applied to this group, what would be the percentage of real-world problem alcohol users excluded from her study and how would this exclusion, bias treatment outcomes. She found that 96% would have been excluded by at least one exclusion criterion. She found that on average, participants fulfilled 2.56 of the less exclusive criteria (eg unemployed or homeless) and 3.99 of the more exclusive criteria (Currently medicated for psychiatric problems or overdose recently). The percentage of treatment seekers excluded because of not meeting the less exclusive individual criteria ranged from 5% being excluded for lack of education to 80% excluded for past or current addiction treatment. The importance of these results is that if our clinical work with real-world populations is informed by biased results, we will not see the same clear results that are published in some academic journals.

One example of this is in research undertaken in the United Kingdom on the treatment of amphetamine problems by substitute prescribing. This is now a common practice in the United Kingdom. However, one of the exclusion criteria has always been the presence of comorbid mental health problems, specifically because heavy use of stimulants such as amphetamine can produce paranoia and psychosis. Carnwath and colleagues (2002) challenged this by a piece of retrospective research examining the case notes of eight patients with schizophrenia who had been prescribed dexamphetamine for co-existing amphetamine dependence. The authors commented that the patients with co-existing problems had poorer treatment outcomes, often did not comply with treatment plans, and had frequent periods of hospitalization. However, they found that, in four of the eight cases examined, the prescription of dexamphetamine led to good progress in terms of both substance use and mental health. In two cases, progress was more equivocal although there had been some benefit, and two cases were deemed to be treatment failures but the condition of the patients was no worse at the end of treatment than at beginning. There was greater adherence to neuroleptic regimes, and none of the patients suffered an exacerbation of their psychotic symptoms as a result of treatment. This is an example of where exclusion criteria for being part of the trial were ignored and good results followed.

It is also true that, although randomised controlled trials are seen as the "gold standard" for research, use of a randomized controlled trial sometimes may be unethical if it means depriving one group of potentially advantageous treatment. An example of this can be seen in a research design in which needle exchanges are established in one city and not in another to measure the incidence of new viral infections among injection drug users. This, of course, would be entirely unethical and would have other methodological problems, unless there were only enough resources to establish programs in one of the cities.

How to get Started

Before starting on a project, you should discuss it with other colleagues to get their approval and cooperation. If this is seen to be feasible, then a thorough research protocol should be written with a description of the scientific need for the study (a literature search and an explanation of your hypothesis), methods of recruiting your sample of participants, methods of measurement, intervention, and statistical analysis.

If you have any doubts or questions, discuss them with colleagues or other people who are active in the field that you wish to conduct research in. More-experienced colleagues are often interested in what you might be doing and will be happy to answer questions and make suggestions about your line of research. Establish a coordinating committee that can provide advice and discuss the project as it progresses. This committee can include members of your department, but it is often useful to have someone from outside to ask the awkward questions and raise points you might not have thought about before. Another option might be to seek collaboration with doctoral students and postdoctoral students. Doctoral students and postdocs may offer their time, knowledge, and skills while supervised by their mentors.

Always make sure all the staff involved in the unit are aware of the research, understand the process, and have any queries answered satisfactorily. These may be the people who refer suitable subjects for your research or whose cooperation you may need to get to the project running smoothly.

Any research that involves human or animal subjects will require ethical approval. Where to obtain this will vary from country to country, but usually universities or major health centers will have a standing ethics committee. An application to the ethics committee will have to follow its standards and will possibly involve a personal appearance in front of the committee during which you will answer questions, provide assurances, and discuss potential changes to the research protocol.

Research usually requires extra funding. Such funds may be obtained as research grants, obtained as small grants from the employing authority, or absorbed in the normal running costs of the unit. Some research may be conducted in house with no extra costs by putting in place research protocols that allow other staff and colleagues to know what is being done. You will still need to be thorough and objective in your research, but it can be undertaken as part of clinical work. Investigators working in academic institutions will routinely be applying for research grants and will know the main funding bodies available in their field. These are likely to be less familiar to clinicians, but research funds are available from small charitable bodies as well as national funding bodies (e.g., the U.S. National Institute on Drug Abuse) and major organizations (e.g., the Gates foundation) who have a huge commitment to solving major world social and health problems. To be approved, it is important that you are working in an area covered by the funding body's activities and that, when you complete the application form, you answer all the questions and explain exactly what you are doing and why.

Make sure you have identified someone experienced in statistics who may be able to guide you on statistical techniques. Collaboration with a statistician from the beginning, when writing the project proposal, is encouraged (e.g., when focusing on patients, power sample analyses should be calculated before conducting research or when choosing appropriate data-collection tools). This is also true for someone experienced with quantitative methods when conducting qualitative research

As an example, an on-going project in the Czech Republic was conducted in therapeutic communities for users of illicit drugs. Research activities are relatively infrequent in these facilities because of many contextual reasons (e.g., low capacity of individual facility, low interest by staff, no uniform treatment models). Within last few years, a new, interesting research problem has emerged (not only) in the context of therapeutic communities: attention-deficit/hyperactivity disorder as a comorbid factor and risk factor for significantly higher treatment drop-out and reduction of treatment effect (Miovský et al., 2014). This interesting and important issue was formulated and clarified through a systematic discussion and series of meetings with staff within a two-year preparatory phase. The Czech team decided to invite the National Association of NGOs and its working group of therapeutic communities to participate. After a selection procedure, they contracted particular therapeutic communities, trained the staff in data-collection methods, and supervised the data-collection procedure. Particular communities were direct partners of the study and had participated since the beginning. To stay within the study budget and make the study manageable, however, the original concept had to remain limited because of potential travel costs and technical complications linked to the difficulty of testing all new clients for attention-deficit/hyperactivity disorder (which is an unpredictable and irregular procedure). Nonetheless, it is also a good example of how to create, through networking, a very attractive opportunity for extensive and sophisticated clinical research with a large number of clients.

Who should be in My Article-Writing Team?

Conducting a good-quality research project requires knowledge, skills, and enthusiasm combined with high levels of persistence. Writing a scientific article is, however, a discipline on its own. Many colleagues who are involved in the data-collection phase of your research will not participate in the actual writing of the article (s) for various reasons (e.g., because of a low interest in writing, lack of confidence or time). It is often the case that data are available but that there are only a limited number of people who are willing to write an article based on it. You may end up writing the actual manuscript on your own. To avoid this situation, you may want to start an early search for collaborators who will help you to write and submit articles to save time.

In the previous section, we suggested that a statistician be part of your research project. With the advent of modern statistical packages for your computer, it is often simple to run the statistics, but frequently people are using inappropriate statistics for the problem. It is important to get this right before you start. When your article is reviewed, your statistical techniques will be examined. If you have used the wrong technique, the article will be rejected. This will either mean you have wasted much time and effort or that it will take a lot more time to rework the statistics—which may of course then produce different outcomes. Similarly, preparing a high-quality qualitative article is difficult without the appropriate experience of a good qualitative researcher.

Where Should I Publish?

Choosing the right journal to which you may submit your finished article should be done with care. Chapter 3 of this book discusses this and should be consulted. There are many journals that focus specifically on addiction, but, in addition to these "addiction specialty journals," scientists and practitioners who work in the field also have a "mother" profession or discipline in which they have been trained (e.g., medicine, psychology). These disciplines also publish many journals in their fields, and these journals may publish articles on addiction. There are also journals published in countries in which information may be more local and more relevant to national or local populations. Therefore, there is a wide variety of potential journals to which you may submit your manuscript.

You must consider, therefore, whether the subject is of national or international importance. If the subject is mainly of interest to people in your country, it may be more appropriate to submit to a national journal. International journals may judge whether an article is of international interest and may not accept an article that is more local. However, it may be that a subject that appears to be local in scope becomes of interest to experts in many another parts of the world. Addiction is a worldwide problem, and practices spread. One example of this is that the use of water pipes to smoke tobacco is very common in the Middle East. Therefore, this form of substance use may be seen to be local. However, the practice does have great potential health risks, and the effects of the diaspora of refugees from this region to many other parts of the world will also export this practice and the concomitant health risks. Both authors and editors need to bear this sort of situation mind.

Before submitting, check the impact factor and acceptance rate of the journal. This can be found in Chapter 3, Table 3.1, in this book. Typically, journals with higher impact factors have lower acceptance rates. If you are submitting to a high-impact-factor journal, it will be more difficult for you to get your article accepted unless it is of high originality and good-quality science.

Furthermore, check the instructions for authors either in the relevant journal or on the website to ensure that the journal accepts the type of work your article describes. Make sure when you submit your article it conforms to the standards of the journal. Follow the instructions for authors regarding word length, style of referencing, and formatting of tables and figures.

What are the Pitfalls to Doing Research in this Setting?

There are many potential pitfalls in doing research in a clinical setting. Yet, if you are well prepared, you may avoid most of these. As has been mentioned above, you need set a clear research question that you want to answer, plan your research design, plan your methodology, decide on your statistical techniques, and make sure you get ethical approval. One important way to ensure the research will be finished and finished correctly is to involve your colleagues. If you share your work with them, they are more likely to cooperate, identify study participants for you, and highlight problems you may not have considered. One exercise that you can do is to write an abstract of the research without the results. In doing this brief exercise, you can set out the methodology, subjects, research question, and statistical techniques.

One of the difficulties in conducting research in clinical areas occurs when there is an ethical conflict between using your clients for research and whether the research or the clinical needs take priority. One must always place clinical need above research interest. Sometimes it is better to undertake research and clinical work in different locations to keep your clinical interests and scientific interests apart

Choosing the right sampling technique is a crucial step that affects the whole study. If sampling is not done appropriately, the results may be flawed, irrespective of how well the study was conducted overall. When you are ready to start your research and wish to recruit study participants, you may find that there were many people who had the problem you are researching at the time you decided to do the research, but, once you start recruiting, they often seem scarce! This is a phenomenon noted by clinical researchers. Therefore, be prepared to go wider to recruit your participants by perhaps involving another agency or advising colleagues in similar facilities to yours that you are trying to recruit.

If you are running a trial with a control group, make sure that your control group is a genuine control and match the experimental group in every way possible, including matching by demographic features and definitions of the problem being researched. Too often, a reviewer on a journal will see that the control group does not match the experimental group and will reject an article on that basis

Another important aspect of doing your own research is the choice of appropriate data-collection tools. It is always better to choose standard, standardized, and well-recognized scales, questionnaires, and other types of measures as opposed to those developed on one's own.

It is not always easy to get research published. But there are some things you can do to increase your chances of getting an article accepted in a well-respected journal. It is well to note the following points:

- *Scientific writing skills* take a long time to acquire, and, with every article produced, these skills improve. Endurance and enthusiasm is the key. Also, collaboration with someone who already possesses these skills is encouraged.
- *Scientific journal language* is specific and differs among fields. For the beginner, it may be difficult and timely to write densely, specifically, and clearly. What may help is to read published articles to become familiar with the language style that is used and to ask someone experienced to "polish" the article.
- Scientific *literature availability* may be a problem for those working in smaller clinical facilities with smaller budgets. Their libraries simply may not be able to purchase access to journal full texts. You may want to invite for collaboration someone from an academic setting or a research facility with access to journal subscriptions. Also, you may ask the study authors for an author's copy. For more options how to search for scientific literature, see Chapter 7.
- *Time* between having completed the research and actually having an article accepted for publication may take months. The approximate time for receiving feedback from a journal is three months. Always try to plan ahead. You can save time by doing literature searches during the data-collection phase. Try to publish outcomes of the pilot phase of your research.
- *Rejection* of an article is common and every author has an experience of receiving negative feedback from the journal on his or her article. Always remember that most rejected articles may be improved based on the feedback that is usually sent together with the letter from the editorial office. Try to learn from the unsuccessful attempts, and do not allow pride or bitterness overcome you.
- *Fighting frustration* should be one of the skills you develop. Research and scientific publishing are very demanding, but getting your article published is very rewarding. All the pain pays off once you see your name connected with an important contribution to the field.

Chapters 7 and 8 in this book will help when you write up your work for publication. Read them carefully and follow the advice, because this will increase your chances of publication.

Serving as a Reviewer

Once you publish your first article, the chances increase that you will be asked by a journal to review someone else's manuscript. You may want to accept for the following reasons:

- reviewing an article will help you see and understand what the journal expects from authors (based, for example, on the reviewer guidelines and other requirements) and what the processes are inside the journal's "black box";

- reading a manuscript from a reviewer's position may help you adopt critical scientific thinking that you may later use when you write your own article; and
- reviewers are an "endangered species," and journal editors need competent, expert volunteers when arranging independent evaluation

Conclusion

Undertaking research as a clinician can be rewarding both for its intrinsic value and its ability to provide answers to many of those troubling clinical questions. It is also valuable for career development and can be an enhancement for a department or unit. It will take more time and effort but adds variety to the working week. You need to follow proper protocols, gain ethical approval, and obtain advice from colleagues or, if necessary, a local academic department. Do not try to take short cuts or believe that, because you are working in the reality of the treatment setting, you know better than academics. Nothing is worse than spending a lot of time and effort on a project for it then failing because of a lack of thorough preparation. Good luck with your research!

Acknowledgements

We offer many thanks to Richard Saitz for his helpful suggestions in our writing of this chapter.

Please visit the website of the International Society of Addiction Journal Editors (ISAJE) at www.isaje.net to access supplementary materials related to this chapter. Materials include additional reading, exercises, examples, PowerPoint presentations, videos, and e-learning lessons.

Note

[1] The widely advertised evidence-based approach is viewed by some as too narrow and formalistic. Hjørland (2011) promotes what is called research-based practice that, besides taking into account quantitative approaches, also considers as legitimate theoretical work and qualitative research.

References

Carnwath, T., Garvey, T., & Holland, M. (2002). The prescription of dexamphetamine to patients with schizophrenia and amphetamine dependence. *Journal of Psychopharmacology, 16*, 373–377. DOI: https://doi.org/10.1177/026988110201600414

Edwards, G. (in preparation). *Seeing Addiction* (Book).

Edwards, G., & Babor, T. F. (Eds.). (2012). *Addiction and the making of professional careers*. New Brunswick, NJ: Transaction Publishers.

Hjørland, B. (2011). Evidence based practice: An analysis based on the philosophy of science. *Journal of the American Society for Information Science and Technology, 62*, 1301–1310. DOI: https://doi.org/10.1002/asi.21523

Jowett, S. M., Macleod, J., Wilson, S., & Hobbs, F. D. R. (2000). Research in primary care: Extent of involvement and perceived determinants among practitioners from one English region. *British Journal of General Practice, 50*, 387–389.

Miovský, M., Čablová, L., Kalina, K., & Šťastná, L. (2014). The effects of ADHD on the course and outcome of addiction treatment in clients of therapeutic communities: Research design. *Adiktologie, 14*, 392–400.

Pates, R. M., & Gray, N. (2009). The development of a psychological theory of needle fixation. *Journal of Substance Use, 14*, 312–324. DOI: https://doi.org/10.3109/14659890903235876

Pates, R. M., McBride, A. J., Ball, N., & Arnold, K. (2001). Towards an holistic understanding of injecting drug use: An overview of needle fixation. *Addiction Research & Theory, 9*, 3–17. DOI: https://doi.org/10.3109/16066350109141769

Salmon, P., Peters, S., Rogers, A., Gask, L., Clifford, R., Iredale, W., Dowrick C., & Morriss, R. (2007). Peering through the barriers in GPs' explanations for declining to participate in research: The role of professional autonomy and the economy of time. *Family Practice, 24*, 269–275. DOI: https://doi.org/10.1093/fampra/cmm015

Storbjörk, J. (2014). Implications of enrolment eligibility criteria in alcohol treatment outcome research: Generalisability and potential bias in 1- and 6-year outcomes. *Drug and Alcohol Review, 33*, 604–611. DOI: https://doi.org/10.1111/dar.12211

SECTION 3

The Practical Side of Addiction Publishing

CHAPTER 7

How to Write a Scientific Article for a Peer-Reviewed Journal

Phil Lange, Richard Pates, Jean O'Reilly and Judit H. Ward

All the chapters in this book speak to our aspirations to contribute to addiction science and to have a role in the scientific life of this field. In large part, this role comes through being published in peer-reviewed journals.

Susan Savva (personal communication)

Introduction

A career in addiction science is largely built on reputation and the (perceived) quality of publications that a researcher (or a team of researchers) produces. If these publications are numerous and of high quality, they may lead to research funding and advancement. To gauge the contribution of a researcher to addiction science, fellow researchers may consciously or unconsciously compute the number of worthwhile publications that a colleague has produced in relation to the number of years he or she has published. The greater speed of release for journal articles when compared with books—typically months versus years—means that those who wish to influence their field of study need to publish in peer-reviewed journals to quickly communicate their research results.

This chapter offers the novice author a step-by-step guide to prepare an article for publication. Annotated bibliographies and references listed at the end

How to cite this book chapter:
Lange, P, Pates, R, O'Reilly, J and Ward, J H. 2017. How to Write a Scientific Article for
a Peer-Reviewed Journal. In: Babor, T F, Stenius, K, Pates, R, Miovský, M, O'Reilly, J
and Candon, P. (eds.) *Publishing Addiction Science: A Guide for the Perplexed*,
Pp. 135–153. London: Ubiquity Press. DOI: https://doi.org/10.5334/bbd.g. License:
CC-BY 4.0.

of this chapter suggest further readings worth consulting about specific problems. This chapter begins with the proviso that a good manuscript written by a graduate student or a junior investigator may be highly praised by faculty and colleagues and yet fall short of being publishable. Indeed, editors regularly receive poor manuscripts that are accompanied by a letter from a graduate student saying that his or her professor recommended submission. Yet the praise from a professor or colleagues does not obviate the need for novice authors to scrutinize every aspect of their text to see that it conforms to the demands of a scientific article.

Here, we offer suggestions on how to use the style guide for the journal of your choice (for which there is additional information in Chapter 3), explain how to use a publication manual, and offer step-by-step guidance on the writing process itself. We also offer advice about working with colleagues, writing strategies, and maximizing the worth of your article for your selected journal. Some of the steps mentioned here are described in more detail, and sometimes with a valuable differing viewpoint, in Chapter 12.

This chapter is written for readers who have completed graduate or postgraduate education and have completed a research project that they want to publish in a peer-reviewed journal but who are unsure of some of the basic steps in preparing the manuscript for submission. This chapter is also appropriate for readers who already are proficient in another field of science but want to add articles in addiction science to their list of publications. For this scientist, we advise caution: Terms may have different meanings for the layperson than for addiction scientists. For example, the word *recovery* connotes in the popular press and in everyday life that someone has undergone a course of clinical treatment or perhaps an affiliation with Alcoholics Anonymous. But in addiction science, *recovery* means achieving precise behavioral goals or a given score on a measure and by a given point in time. There are enough such special concepts built around everyday language that scientists new to the addiction field are advised to gather a group of colleagues to advise their research from the beginning.

We assume here that the reader is already competent in writing a scientific article. This chapter aims to fine tune competence in writing rather than to teach the basics of science writing. At the other end of the continuum, researchers whose articles are already often accepted in the journals of their choice will likely find little of interest here. Authors from developing or non-English speaking countries may wish first to read Chapter 4, which explores some of the special challenges encountered by researchers from developing and/or non-English speaking countries.

A successful publishing career means writing for a scientific audience, and authors may have to submit a number of manuscripts to various journals to discover how to do this in a way that results in a high percentage of accepted articles. An early decision researchers must make is whether to work alone or with colleagues. You can work in isolation from colleagues and hope to learn from rejection letters and from harsh peer reviews (see Chapter 12). Or, you can build an informal team of fellow scientists who are both critical and supportive and who will read and comment on your manuscripts. This is often a quicker, more efficient, and more stimulating path. If you are new to a center or department and you want to sort out quickly who will be supportive of your aims versus who may be less than helpful (e.g., those who have reputations for being always harshly critical or for promising and then failing to read and critique manuscripts), ask people you trust this question: "If you were writing on my topic of _____, whom would you trust to help critique your work in a helpful way?" A novice author can learn much from established authors by passing them drafts for their assessments and their recommendations for getting published.

For a younger or inexperienced writer it may be sensible to check on the acceptance rate of the journals (see Chapter 3) and go for one with a higher acceptance rate. In this way the chances of your paper being accepted are greater.

Writing a scientific article for a peer-reviewed journal can be a creative and enjoyable act. Some people write beautifully and effortlessly, whereas others feel as though they are sweating out each word. But, over time, authors with both writing styles can make successful contributions to addiction science.

This chapter presents one way to write such an article—it is not the only way, of course, but it does offer the advantage of a clear step-by-step method that helps you to plan ahead. If you follow these steps, you will finish with a manuscript worth submitting to the journal of your choice (providing of course that the original science is sound). At the end of this chapter, we also present an annotated bibliography describing other approaches to preparing scientific manuscripts for peer review.

Being methodical, let us start with a checklist.

When you have decided on where to submit your paper make sure you read thoroughly the instructions to authors and follow them precisely. Virtually all journals will now only accept submissions electronically. This may be daunting for the first time a paper is submitted but it makes the process much easier for the journal and the author.

Check the Style Guide for Your Journal of Choice

Each journal has its own specific style configuration, and, to be accepted by a journal, you must write to *its* requirements, not those of another style format and not to your own personal preferences. To do this, have all information on all of the parameters required for the journal that you have (initially) chosen (see Chapter 3 for more information). Many journals offer a one-page style guide. But even the minimal style guides for undergraduate articles issued by university departments typically run to many pages, so clearly a lot will have been left out of a journal's one-page summary. The *Publication Manual of the American Psychological Association* has 66 pages on style alone (American Psychological Association, 2010, pp. 21–86). Much can be said for simply sitting down and reading at one go these 66 pages for a quick and complete overview of essential topics that are left out of most brief style guides. Read these American Psychological Association chapters and you will emerge an enlightened initiate knowing what topics to be sensitive to even if you must use a different style guide than this manual. The journal you are submitting to may have other style parameters that will affect your article, such as the preferred length of the manuscript and its abstract; gender-neutral or other styles of preferred language; the maximum number, length, and style of footnotes or endnotes; and the maximum size of tables.

Your journal of choice may require or recommend the use of reporting guidelines, depending on the type of paper you intend to write. Even if a journal does not require the use of reporting guidelines, it is worth following or at least consulting a systematic guideline to establish a framework for your paper. There are hundreds of such guidelines in existence, helping researchers to produce accurate, complete, and reliable reports. Table 7.1 outlines some common guidelines.

An additional guideline that was developed in 2016 is SAGER (Sex and Gender Equality in Research), a comprehensive procedure for reporting sex and gender information in study design, data analyses, results and interpretation of findings: http://www.equator-network.org/reporting-guidelines/sager-guidelines/

A brief warning about tables and figures: Journals may not specify the size limits on tables and figures, yet these parameters have a huge effect on what information you can include in them and how you organize your writing. Beginning researchers have a tendency to send wider, longer, and less-interesting tables than seasoned researchers. To create tables that will fit the page in those cases where the journal gives no guidance, (a) estimate the typeface in the table when compared to the textual typeface in the journal and (b) build model "trial" tables (one row, the number of columns needed, longest possible data lines per table cell) that would fit within a typeset page. Then build your tables. This alone may save you from immediate rejection or the work of rewriting the text and reorganizing the table. If you have tables that require more than one page, check the journal to see if it publishes tables of that size or check with the editor. Editors have horror stories of good articles that arrive with huge tables that could never fit on a page. (The tricks authors use to create such large tables include using tiny typefaces, margins of less than a centimeter, and rows that run off the edge of the page and the monitor as well as carrying on for several pages with landscape orientation while submitting to a journal that does not accept that format. Do not consider any of these, because you will only infuriate the editor.)

The other problem is tables and figures that are excessive in number or size. These all take up large amounts of space, and this may be a consideration for acceptance of the article. Include only those tables with direct relevance to the article and those that help the comprehension of the work.

Acronym	Full name	Application	Description	Link
CARE	Case Reports	Case reports	Provides a flow diagram for systematic collection of data when seeing a patient or doing a chart review.	http://www.care-statement.org/
CONSORT	Consolidated Standards of Reporting Trials	Randomised controlled trials	Recommendations for reporting randomized trials. There are several extensions, including abstracts, cluster trials, pragmatic trials, and N-of-1 trials.	http://www.consort-statement.org/
PRISMA	Preferred Reporting Items for Systematic Reviews and Meta-Analyses	Systematic reviews and meta-analyses	Minimum criteria for reporting in systematic reviews and meta-analyses. There are extensions for abstracts, equity, protocols, individual patient data, and network meta-analyses.	http://www.prisma-statement.org/
STARD	Standards for Reporting Diagnostic Accuracy	Descriptions of studies involving diagnostic accuracy or validity	Essential items that should be included in every report of a diagnostic accuracy study.	http://www.equator-network.org/reporting-guidelines/stard/
STROBE	Strengthening the Reporting of Observational studies in Epidemiology guidance	Observational studies in epidemiology, including cohort, case-control studies, and cross-sectional studies	A checklist for articles reporting observational research. There is a draft STROBE checklist for abstracts.	http://www.strobe-statement.org/
TIDieR	Template for Intervention Description and Replication	Descriptions or evaluations involving interventions	A guide for writers to describe interventions in sufficient detail to allow their replication.	http://www.consort-statement.org/resources/tidier-2
TREND	Transparent Reporting of Evaluations with Nonrandomized Designs	Surveys and longitudinal studies	A checklist to guide standardized reporting of nonrandomized controlled trials.	http://www.cdc.gov/trendstatement/

Table 7.1: Guidelines commonly used in reporting health research.*
*For information on more than 280 reporting guidelines, visit The Equator Network: http://www.equator-network.org/.

Editors agree that far too many authors ignore the crucial step of reading and following the journal's submission guidelines. Ask yourself, "Am I 100% confident that I have followed *every* one of even the smallest details in the journal's guidelines?" If your silent answer to yourself is, "Hmmm, certainly yes, probably 90% or 95%," then your next step is to conclude that this is not good enough: Go back and fix those few items so that they are correct.

The bottom line: Read and follow the journal's instructions.

Box 7.1: The importance of journal guidelines.

Check the journal's style guide for requirements governing the presentation of figures and make sure that they fit within the journal's page parameters and technical requirements. There is a danger in looking to old copies of a journal to assess table and figure design. If you cannot get a current copy online or at a university library, write to the editor explaining the situation, and the editor—surely pleased at your concern—will likely send a sample copy. Figures are often easily sized by click-and-drag formatting to fit a given space within the correct margins.

Do a Thorough Literature Review

The literature review is a crucial portion of your article. Many beginning researchers have problems with the scope and structure of the literature review. By studying examples of good literature reviews, you can improve your understanding of current standards. See also Chapter 9 on how to write systematic reviews. Wikipedia offers an introduction to the basic points of literature reviews (http://en.wikipedia.org/wiki/Literature_review). Kathy Teghtsoonian offers a useful didactic example explaining alternatives in a review of the literature on smoking (http://web.uvic.ca/spp/documents/litreview.pdf). An example of a thorough literature review article that serves as a model for shorter reviews within an article—with exemplary background, definitions of terms and variables, treatment conditions, and results—is this article on quasi-compulsory drug treatment in Germany by Stevens et al. (2005). (But avoid the one-sentence paragraphs frequent in this otherwise fine review. Most editors and reviewers hate one-sentence paragraphs and complain about even one or two.) Cochrane Group reviews also deserve your attention. Not only may a review from the Cochrane Group spark improvements in your research, but reading a collection of reviews can also help you to develop a model for your work. See http://www.health.qld.gov.au/phs/documents/cphun/32103.pdf.

Reviewers will be much more familiar with the literature than you are, and, therefore, your literature review needs to be informed and critical, not naive and accepting of all that is cited. One way to improve your literature review is with a step-by-step approach. Have these materials handy:

- all the relevant literature needed to establish the theory or hypothesis that you will examine (it will help you to outline your article and to see what background or literature reviews you need for each section);
- all relevant literature for each of the measures that you have used (the initial article describing each measure and crucial articles describing challenges, alterations, refinements, including statistics on validity, reliability, and all other relevant attributes); and
- all the data needed for your methods, procedures, and results sections (a good way to assess if you need more literature for a given section is to ask yourself, "If I were challenged to support why I chose this [measure, method, statistic], what literature supports my choice?").

If you are writing about qualitative research for a journal that publishes little of your specialty, be sure to have the latest work on rigor in qualitative research and link it solidly to your work, because the probability is high for a rough ride from reviewers who know little about qualitative research and who may be more biased than they realize. ("I have seen a few good qualitative papers, but very few," they tell me.) Also, please read Chapter 8, which explains how to write about qualitative research.

Writing Step #1

Contact your chosen journal with a working title and abstract, ask if your article is of interest and relevant to the journal's mandate, and ask any awkward questions (. . . flexibility on article length? average time for the peer-review process?). Now is the time to learn if your article is acceptable to this journal, not after you have spent days writing an article to a specific format when that journal is unlikely to accept it. If the answer is favorable, you are ready to start writing. If the response is unfavorable, look for another journal. Alternatively, you might consider asking knowledgeable colleagues what journal(s) they feel are the best choice(s) for your article.

Writing Step #2

Now settle down to write for colleagues and your posterity your unique contribution to addiction science. Here are a few specific guidelines for each section of your article:

Title: You should know the overall writing style of your chosen journal well enough to know intuitively what is a suitable title for your article. If in doubt, (a) read the table of contents of several issues to get a feel for their style of titles and (b) make up a couple of possible titles and ask for reactions from colleagues who know this journal well.

Mistakes to avoid: Trendy and cute titles soon look trivial and dated. An editor may allow such a title (especially if rushed), but years from now it will look embarrassing in your curriculum vitae when reviewers read it to determine if you deserve research funding.

Abstract: The abstract summarizes how you carried out your research and what you learned. It is increasingly common and often requested that you use a structured abstract (objective, methods (or) design, sample, results, and conclusion). For example, BMJ (n.d.) requires structured abstracts within a sound framework: objectives, design, setting, participants, interventions, main outcome measures, results, and conclusions.

Mistakes to avoid: Do not go beyond what is established in your article: Offer no nonsignificant results, no speculation. Do not use telegraphic style (e.g., omitting articles and other parts of speech to achieve brevity) unless allowed by the journal. Do not go over the abstract size limit set by the journal.

Introduction statement: A good introduction tells the reader why the article is important in terms of the problems to be investigated, the context for the research question, what place this research question has in understanding addictions, and what is original about the endeavor.

Mistakes to avoid: Do not simply describe the substance or behavior under study. Authors who see this as sufficient too often feel that the problem substance or behavior itself implies what research is needed. This is almost never true. At no point should the volume of loosely related information make the reader feel lost and wonder, "Why is all of this information here?" Avoid archaic arguments that have been resolved or that are not pertinent to your

---

A frequent mistake made by beginning researchers is to not make clear to the editor and reader what is the *original* contribution of an article. It is easy to forget that scientific journals exist only to publish original knowledge. Describe the originality of your research analyses in your initial letter to the editor to see if there is interest in your article so that if the article later appears on the editor's desk, he or she will remember it for the innovative understanding that it offers. For the reader's benefit, your original contribution(s) should be clear from the title (if possible), mentioned in the abstract, and described in the introduction and in the discussion (and/or conclusion).

Box 7.2: The importance of originality.

article, even though you may have spent months researching these and you have a fascinating solution to the debate. Avoid formulaic first lines: A sentence such as "Access to legalized gambling has increased greatly in the last two decades" begins at least one third of articles on gambling. An occupational hazard of editing is to receive by the dozen manuscripts with opening lines such as "Alcoholism (or drug dependence or tobacco use) is a significant public health problem." The editor's eyes glaze.

Literature review: The literature section of a dissertation is an entire chapter. For an article, it should briefly summarize only the most important references that lead directly to understanding the importance of your article and the methods used. Keep the topics of your literature review grouped so that the flow is logical and the reader does not have to move back and forth. Move from the general subject to the more specific studies relevant to your research question. For detailed guidance on which articles to cite, refer to Chapter 10 (Use and Abuse of Citations). For detailed information about how to use state-of-the-art search technologies to locate articles relevant to your literature review, see Appendix A. When your draft is completed, compare it with the literature reviews in your journal of choice.

Mistakes to avoid: If several authors have been involved in writing the literature review, then it is likely to be too long and detailed, because each author tends to add what he or she knows are essential works. Keep the review concise.

Method: After readers have gone through this section, they should know the research methods in such detail that they could replicate the study in full with another sample. One way to check the completeness of this section is to have colleagues read it and ask them to verify if they could carry out this research project wholly from the methods section. If there are previously released articles using the same methods—whether your article or those of others, and especially if the method is described in more detail elsewhere—then you should cite these. This may allow you to shorten the method section.

Mistakes to avoid: If some aspect of your methods is suboptimal, it is better to mention it here with the comment "see the limitations section" and then be straightforward in the limitations section. Do not try to hide or disguise poor methods; reviewers will pounce on them. If your research involves randomized control trials, editors may refer you to the CONSORT Statement promoting high standards and uniform methods: http://www.consort-statement.org.

Results: Here you describe the outcome(s) from your research. Double check that each novel finding mentioned in the discussion is reported here.

Mistakes to avoid: This section especially lends itself either to over-writing (excessive detail beyond what is needed for analysis, excessive weight given to nonsignificant results) or to under-writing (cursory attention to important aspects and variables). Avoid reporting results as "approaching significance"; if they are not statistically significant, do not quote them as a near result. A mistake to avoid here is opening the results section with a description of the sample or an analysis that is more relevant to the methods, such as the validity of your measures. Start your results section with the main findings. Beginning researchers often take up too much of their manuscript with nonsignificant results; be ready to drop a result that colleagues or reviewers suggest is unimportant, even if it seems like a wondrous and magical thing to you.

Discussion and/or conclusion(s): Describe how your specific results fit into the world of addiction science. You may address issues raised in the literature review, address policy issues, or raise new questions that are either unaddressed or rarely addressed by others.

Mistakes to avoid: A little speculation is allowed, but limit it and ask your supportive colleagues what they think. Restrict your discussion of your future research plans to a line or two. Some authors like to end with the trite conclusion "More research is needed." It always is. If you wish to write in this vein, be as specific and creative as possible in tracing what original work needs to be done and what interesting hypotheses it will test.

Limitations: Describe in brief detail the suboptimal aspects of your research. This newish trend has come as a result of demand for more transparency in research publishing. Junior authors are often afraid that being open about the limitations of their research will create prejudice against an article. In fact, the opposite is true. Senior researchers (i.e., editors and reviewers) will see flaws in your work that you will likely not see. Reviewers and the editor ask only that you acknowledge limitations. To do so is not a sign of weakness in you or your approach, but much to the contrary: It shows that you are an author who is on top of what are best practices and that you are a person who sees the need for better methods (as opposed to one who stumbles along pleased with his or her inadequate work). In concise, simple, and unapologetic language, describe the shortcomings that kept your work from being optimal. Some journals allow an author to note limitations throughout the text (i.e., not as a subheading toward the end of the article). You may wish to check to see if your journal of choice allows or prefers this alternative.

Mistakes to avoid: Do not be ingratiating (e.g., do not apologize, promise to avoid these mistakes in the future, or offer excuses), for this creates the impression of servility. You are not groveling You are only signaling to your peers that you know what is better practice in research.

References: It is easy to forget that the function of references is to allow any reader to retrace the evidence you cite. Electronic sources that become unavailable threaten this openness. You must check that all the references in the text are cited in the reference section and that all the references in the reference section are cited in the text. Too often, authors neglect to check this, and these mistakes may be found by reviewers. You should be completely fluent in the minute details of proper reference style for your chosen journal. Too many errors tell the editor that an author has been careless, and this suggests carelessness perhaps elsewhere

Mistakes to avoid: Verify if translation of foreign language titles is required. If it is, translate foreign-language titles even in the first version you send to the editor.

Appendices: If your journal of choice seems not to have published appendices, then check with the editor to see if they are allowed. Appendices represent an excellent solution to the problem of presenting background information (e.g., legislation, policy statements, questionnaires and measures, speeches, protocols) that is too long for the body of the article. They are also easy for a reader to skip: a blessing. Online, some journals allow for the posting of appendix materials such as video and sound files, and URL access, as well as more traditional yet space-consuming items that are difficult or impossible to include in print journals. Note: Such data may not have peer-review status if not evaluated by the reviewers.

Mistakes to avoid: Omit appendices that you feel are relevant to the article but that colleagues feel are not pertinent.

Writing Step #3

You have written this first version early enough to allow you to circulate it to several colleagues whom you can trust to read it and to offer prompt and fair critiques. Once you have their feedback, consider if their assessments warrant rewriting before submitting it to your chosen journal.

Writing Step #4

Submit your article to the editor. It might be useful to read Chapter 12 on manuscript preparation at this point. Bon voyage on this first step in becoming a contributor to the world of addiction science.

Writing Step #5

Your article has been accepted for review (whether minimal or extensive) and has come back with the reviewers' and the editor's comments. This would be a good time to consult Chapter 12, which describes referees' reports and how to respond to them. If you decide the referees' criticisms are too severe for you to answer, then write the editor to tell him or her so and provide your precise reasons for not revising your article. This accomplishes several good things to your benefit: (a) It labels you as someone who takes editing a journal seriously, who knows his or her goals, and who does not let work slide; (b) it signals to editors how serious the criticisms were and may lead them to discuss options with you; and (c) they will remember you as someone who did not leave them hanging and wondering if that article was ever coming back.

If you decide to revise your article, you have several choices. Authors should not see themselves as helpless in the face of reviewers' comments. To reassure authors of their rights, we at our journal send the following paste-in text to even experienced researchers.

> As we tell all authors, a reviewer's comments are *not* orders that have to be carried out. To the contrary, for each point that a reviewer has made, an author has these three options:
>
> (a) discuss/debate/refute a reviewer's comment(s),
> (b) rewrite the text in response to a comment(s), or
> (c) a combination of these so that an author both discusses/debates/refutes a reviewer's comment(s) and rewrites to accommodate some comments by a reviewer.
>
> In many of the articles that you see in print, there are several points that appear just as authors intended, because they debated and defended their approach as written. As editor, I sometimes very much give the author the benefit of the doubt.

The last point in answering the reviewers' comments is practical but often over-looked. Be crystal clear in accounting for how you responded to each point made by each reviewer. It is a good idea to provide in a letter to the editor the responses you have made point by point to the reviewers comments and to use track changes in the text of the article.

If your article is rejected, then carefully read the critiques and see if you feel that submitting it to another journal seems a wise step. If so, be sure to format it thoroughly to that journal's style and revise it in response to the reviewers' criticisms. It is worth remembering that if your article is rejected and you submit it to another journal, it may be sent to a reviewer who has already rejected it.

Writing Step #6

Once your article is accepted, you may have little more involvement until the editor or publisher sends you the proofs to check. When the proofs arrive and you see how the nuances of your careful writing style have been altered, it is easy to feel lonely and unappreciated. But please respect that copy editors know well what is more readable and credible to the target audience. If you have a hard time deciding on whether to accept a change or not, a criterion is to ask yourself is, "Has my meaning been respected or has it been changed?" If it has been respected, then let it be as edited and trust the copy editor. If you read your article a year-later, you will usually see the wisdom of the copy editor's changes.

Publishing Dissertations

Most postgraduates who have successfully completed a master's thesis or doctoral dissertation will want to have their findings published. (Chapter 5 treats this topic in more detail.) It is important to remember that these dissertations are usually much longer and more detailed than will be required for publication as an academic article. Think carefully about how many articles your dissertation can be split into: Often a doctoral dissertation has enough material for three or four articles. Do not replicate exactly the methodology or literature review (this will be seen as self-plagiarism; see Chapter 14), and keep the methodology as simple as is necessary to explain what you did. Often the methodology in dissertations is much more comprehensive than is required for an academic article, keep it to what is needed to explain your procedure. Editors will get frustrated when presented with an unedited dissertation and may reject it before sending it for review.

When writing up a dissertation for publication it is important to bear in mind who should be included as authors (see Chapter 11 for discussion of how to assign authorship credits) and appropriate acknowledgement of supervision etc.

Conclusion

When your first addiction article is published, you will have made a contribution to the addiction sciences and to the public arena where the dialectics between what is, what could be, and what will be are in struggle. A proverb: some Inuit say that a man can be only as good a hunter as his wife's sewing will let him be. In the addiction sciences, the effectiveness of our research, treatment methods, policies, and advocacy can be only as good as the literature that we publish.

For Further Reading

Boxes 7.3 and 7.4 describe resources for improving your scientific writing in general (writing style and motivation issues) and in particular areas, respectively. If they do not contain a work specific to your needs or the books are unavailable, try searching your local university or professional library using terms such as *scientific writing* or *publication manual* in a title or subject search.

Yet another technique is to find the library classification codes (call numbers) at your nearest university for books on writing psychology and biomedical science (e.g., in academic libraries using Library of Congress call numbers, they are mostly among the books labeled with H61 (social

Alley, M. (1996). *The craft of scientific writing* (3rd ed.). New York, NY: Springer.
- Lengthy chapters on building competence and curing shortcomings.

Greene, A. E. (2013). *Writing science in plain English*. Chicago, IL: University of Chicago Press.
- A short, focused guide presenting twelve writing principles based on what readers need in order to understand complex information, including concrete subjects, strong verbs, consistent terms, and organized paragraphs.

Matthews, J. R., & Matthews, R. W. (2014). *Successful scientific writing: A step-by-step guide for the biological and medical sciences* (4th ed.). Cambridge, England: Cambridge University Press.
- Step-by-step advice helps researchers communicate their work more effectively. The fourth edition has been updated to provide more guidance on writing and organizing each part of the manuscript's draft.

Rogers, S. M. (2014). *Mastering scientific and medical writing: A self-help guide* (2nd ed.). New York, NY: Springer.
- A compact guide with exercises as solved problems; good for overcoming specific writing handicaps. It also addresses issues troublesome to authors of a non-English language origin. This second edition answers questions resulting from new developments in scientific communication.

Silvia, P. J. (2007). *How to write a lot: A practical guide to productive academic writing*. Washington, DC: American Psychological Association.
- This breezy guide is especially good for authors who realize that their writing style needs improvement or who have been told that a component of their article (e.g., abstract, introduction, method, or discussion,) misses the point of what it should communicate. Journal articles have 23 pages of coverage in this book.

Strunk, W., & White, E. B. (1999). *The elements of style* (4th ed.). New York, NY: Longman.
- Still one of the best and shortest writing guides, easily read and absorbed. Those learning English find its clarity and brevity helpful. The 1918 edition by Strunk is available for free as an e-book from Project Gutenberg at http://www.gutenberg.org/ebooks/37134.

West, R. (2002) A checklist for writing up research reports. *Addiction, 95*, 1759-61.
- This is an advanced, comprehensive guide to scientific writing prepared by the Editor of one of the leading addiction journals.

Box 7.3: Annotated bibliography of scientific writing: basic problems of writing style and motivation.

Goldbort, R. (2006). *Writing for science.* New Haven, CT: Yale University Press.
- This book offers detailed chapters cover every type of science writing by using numerous examples. The author discusses how to approach various writing tasks as well as how to deal with the everyday complexities that may get in the way of ideal practice.

Gustavii, B. (2003). *How to write and illustrate a scientific paper.* Cambridge, England: The Cambridge Press.
- This work is oriented to the biological and medical sciences. It is the clearest and most succinct work that we found among all such works at our local university. A marvel of clarity and utility. It is also full of relevant URLs for up-to-date information.

Huth, E. J. (1990). *How to write and publish papers in the medical sciences* (2nd ed.). London, England: Williams and Wilkins.
- This compact work offers practical advice on how to make decisions about what to write and what to leave out for both novice and experienced researchers. A highly readable source.

Miller, J. E. (2005). *The Chicago guide to writing about multi-variate analysis.* Chicago, IL: University of Chicago Press.
- This work shows how specific the aids available to scientific authors are. The book is a mini-course in writing about numbers (i.e., statistical analysis).

Schimel, J. (2012). *Writing science: How to write papers that get cited and proposals that get funded.* Oxford, England: Oxford University Press.
- This book is built upon the idea that successful science writing tells a story. The author discusses every aspect of successful science writing, from the overall structure of a paper or proposal to individual sections, paragraphs, sentences, and words

Box 7.4: Annotated bibliography of scientific writing: focusing on standards for scientific articles and specific scientific areas.

sciences), Q158 (biomedical sciences), R119 (biomedical sciences, and T11communication)), and then scan the shelves in those sections for books that did not come up in your title or subject search. Some would call this a strategy of desperation, but half of the books in the annotated bibliographies below were found this way.

Finally, most academic libraries offer so called LibGuides, i.e., special research guides on scientific writing that are not just for students. Advanced guides include a collection of links to invaluable print resources in house and links to authoritative and reputable online options on the Internet. Here are a few examples, all from the USA:

- Michigan State University – http://libguides.lib.msu.edu/medwriting
- Duke University – http://guides.mclibrary.duke.edu/scientificwriting
- Wilkes University – http://wilkes.libguides.com/scientific_writing
- University of California San Diego – http://ucsd.libguides.com/psyc
- Bowling Green State University – http://libguides.bgsu.edu/techwriting

Acknowledgements

The authors are grateful for contributions from Susan Savva, Ian Stolerman, Kerstin Stenius, Sheila Lacroix, Thomas Babor, and Gerhard Bühringer.

> Please visit the website of the International Society of Addiction Journal Editors (ISAJE) at www.isaje.net to access supplementary materials related to this chapter. Materials include additional reading, exercises, examples, PowerPoint presentations, videos, and e-learning lessons.

References

American Psychological Association. (2010). *Publication manual of the American Psychological Association* (6th ed.). Washington, DC: Author.

BMJ. (n.d.). *Resources for authors: Research.* Retrieved from http://resources. bmj.com/bmj/authors/types-of-article/research

Savva, S. (2007, July 25). Personal communication.

Stevens, A., Berto, D., Heckmann, W., Kerschl, V., Oeuvray, K., van Ooven, M., Steffan, E., & Uchtenhagen, A. (2005) Quasi-Compulsory Treatment of Drug Dependent Offenders: An International Literature Review, *Substance Use & Misuse, 40,* 269–283.

Appendix A. How to Locate Articles Relevant to Your Literature Review

No matter how easy it seems to Google your topic, the scholarly article you are writing deserves a more in-depth literature search than Google or even Google Scholar provides. On the other hand, it would be very time consuming to check individual journals for relevant articles, even though in certain cases the majority of the pertinent articles seem to have been published in a handful of journals. The purpose of scientific databases is to aggregate all publications from a variety of journals in a single database on a particular topic, such as PubMed and Medline on biomedical and health science and PsycInfo on psychology. These databases abstract and index every article published in the journals in their coverage, making the scientific content easily discoverable through literature searches. Since currently there is no single and comprehensive database in the field of addiction science, expect to spend a significant amount of time searching scientific databases with various scopes. Please see the more general discussion of relevant databases and abstracting & indexing services in Chapter 3.

A literature search can also serve as a great start to conceptualize the topic of your article, since in order to run your search in a database, you will first have to produce a list of search terms. Searching is a skill that can best be learned with the help of a professional searcher. Before you start your literature search, please consult your librarian on the latest trends and, if possible, schedule a one-on-one session to find out which databases are available at your institution and what search strategies would work the best in those resources.

Choose the Right Database

The first step of the search process is choosing the appropriate databases. The best way to start is by reading the description of a database to define the type, scope, and coverage of the resource. For example, the Rutgers Alcohol Studies database is a collection of bibliographic records for books, book chapters, journal articles, government documents, conference papers, and dissertations. Although you will not have access to the full text of any document, you can use the reference to find it elsewhere. The Rutgers Alcohol Studies database is a very comprehensive database; discontinued in 2007, it can be considered an excellent source of articles written before 2007. Other useful resources include the Alcohol and Alcohol Problems Science Database, or ETOH, discontinued in 2003, and the CORK database, updated until early 2015.

Because there are currently no comprehensive databases for the addiction field, resources such as Medline or PsycInfo will usually provide the best results at the beginning of your literature search on any addiction science-related topic. Searching a major database also comes with an additional bonus: if your institution subscribes to the journal and has an article linker software application in place (most academic libraries do), you will have instant access to the full text of those articles.

Build Your Search

Search interfaces vary depending on the platform your institution provides (e.g. Ovid or EBSCO). Spending 45–60 minutes with your local librarian can save you precious time to locate the most important features of the databases and can allow you to focus on the search strategies. In a nutshell, it is highly recommended to use the "Advanced" search option, if possible, in any database on a platform where you perform your search in multiple search boxes (e. g. EBSCO platforms, Academic Search Premier). It's important to be comfortable using the Boolean operators, truncation, and wildcards, and familiar with the concept of controlled vocabulary, mapping, and the thesaurus. Each database defines its own preferred terms; for example, Medline, PubMed, and PubMed-Central (reiterations of the same collection in slightly different formats) use Medical Subject Headings called MeSH terms, the controlled vocabulary the US National Library of Medicine uses for indexing articles. Another notable collection of terms is the Library of Congress Subject Headings used by academic libraries, book publishers and Academic Search Premier, a software application originally designed to allow similar titles to be placed physically close to each other on the shelves of brick-and-mortar libraries. For example, "marijuana" is the preferred term in PsycInfo, while Medline uses "cannabis" as an index term. A keyword search usually searches the full text, for example, searching for the word "ganja" as keyword anywhere in an article. A useful feature of the Ovid platform is "mapping" your term to these preferred terms to achieve a high precision of search results. The Ovid platform also prompts you to build a search line-by-line (or term-by-term), resulting initially in an alarmingly high number of hits. Then, using the Boolean operators AND and OR, you can modify and combine your search with additional terms in as many ways you want to filter articles. Each database offers a variety of filters, such as date range, populations and document types, which are essential in the search process.

This comprehensive search strategy will retrieve the relevant articles that were indexed by a subject heading or a descriptor matching your concept. Other searches may target certain parts of the articles the database defines as searchable, such as the author, title, abstract, and keywords, usually in a single search box. This type of search is perfect to locate known items, i.e., to find an article written by an author knowledgeable about your topic, or to retrieve the full text of an article discovered earlier. It should be noted that sometimes old-fashioned methods, such as "footnote chasing" or finding a good review article on your topic, may result in unexpected breakthroughs in your literature search. Many novice searchers take screenshots of their most successful search strategies for future reference or documentation, since most databases do not allow you to save your search and return to it.

Benefit from Citation Management Software Applications

It's a good idea to save the results of each search, i.e., the bibliographic records and/or the full texts of the articles, in a citation management software application. Many authors rely on these applications, such as the proprietary EndNote and RefWorks, or the open source Zotero.

These applications serve multiple functions in the process of conducting research and sharing results in a publication. They are integrated with most of the platforms content providers use for databases and individual journals so that researchers can immediately download the metadata, including links to the full text, of several articles retrieved during the search. They can then share them with collaborators, and can finalize which ones to cite in the article to be published. The in-text citation function, such as Write-N-Cite in RefWorks, allows the author to insert placeholders in paragraphs that serve as the basis of the reference list at the end of the article in the format required by a particular journal, such as APA first author/year or numerical style. Most major citation styles are built into most citation management software applications as output styles. Authors who create their own lists and folders of articles to be cited will benefit from the convenience of creating a list of references, endnotes, or footnotes with one click of a mouse. Should the article be rejected, there is no need to reformat the in-text citations and the entire bibliography to match the required style of another journal. All that needs to be done is to change the output style in the citation management software.

CHAPTER 8

How to Write Publishable Qualitative Research

Kerstin Stenius, Klaus Mäkelä, Michal Miovský and Roman Gabrhelík

Introduction

Conducting and publishing qualitative research requires the same principal skills as quantitative research. In addition, there may be special challenges for qualitative researchers. They may have to overcome prejudice and communication barriers within the scientific community. This chapter provides advice to authors who wish to publish their research in a scientific journal. The chapter starts with some remarks on the special characteristics of the processes of qualitative study that can affect the reporting of the results. It then identifies the common criteria for good qualitative research and presents some evaluation principles used by editors and referees. Finally, it offers practical advice for writing and publishing a qualitative scientific article.

In *quantitative* research, the observations typically follow a systematic scheme whereby the classification of the observations is already determined to a large extent when the data collection starts. This makes it possible to gather large data sets for numerical analyses, but the understanding of the findings will be restricted by the concepts on which the collection of data was based. One can argue that in *qualitative* research, in which the observations (e.g., texts, sounds, behaviour, images) are usually fewer, the researcher's preconception of a social phenomenon does not determine the research results to the same extent as in quantitative research (Sulkunen, 1987) Qualitative research thus is often used

How to cite this book chapter:
Stenius, K, Mäkelä, K, Miovský, M and Gabrhelík, R. 2017. How to Write Publishable Qualitative Research. In: Babor, T F, Stenius, K, Pates, R, Miovský, M, O'Reilly, J and Candon, P. (eds.) *Publishing Addiction Science: A Guide for the Perplexed*, Pp. 155–172. London: Ubiquity Press. DOI: https://doi.org/10.5334/bbd.h. License: CC-BY 4.0.

to study social processes or the reasons behind human behaviour (Sulkunen, 1987), or as Wikipedia puts it: The **why** and **how** of social matters more than the **what, where,** and **when** that are often central to quantitative research.

Qualitative addiction research focuses on topics that range from historical processes to treatment outcomes. Qualitative research is used increasingly to answer questions about alcohol and other drug policy, including rapid assessment of policy developments (e.g., see Stimson et al., 2006). It is used to study program implementation and to evaluate various policy measures (e.g., see Miovský, 2007; Miovský & Zábranský, 2003). Furthermore, ethnographers have used qualitative methods to increase the understanding of patterns of substance use in various population groups (e.g., see Lalander, 2003).

There is also an important and growing interest in combining qualitative and quantitative research into so-called mixed-methods research, notably within evaluation and intervention research in the clinical and policy fields (Creswell & Plano Clark, 2007). The combination of qualitative and quantitative methods can deepen the understanding of processes, attitudes, and motives. There is frequent discussion in theoretical mixed-method studies of the relations between various kinds of knowledge and the actual procedure of combining qualitative and quantitative methods (Creswell & Tashakkori, 2007). Box 8.1 presents criteria for good mixed methods articles.

Despite what we believe is an increasing interest in qualitative research, many journals do not publish qualitative studies. In addition, many editors of addiction journals have noted that qualitative manuscripts are more likely to present the editors with problems and are more often declined for publication than are quantitative research reports. Some of the problems are related to how the articles are written.

In the addiction field, there is no journal dedicated exclusively to qualitative research, and in many journals articles must follow a strict standard format. Qualitative articles tend to break with that format, putting special demands

---

a) The study has two sizeable data sets (one quantitative, one qualitative), with rigorous data collection and appropriate analyses, and with inferences made from both parts of the study.

b) The article integrates the two parts of the study, in terms of comparing, contrasting, or embedding conclusions from both the qualitative and the quantitative strands.

c) The article has mixed-methods components that can enrich the newly emerging literature on mixed methods research.

Box 8.1: Criteria for good mixed-methods articles.
Source: Creswell and Tashakkori (2007).

on the reader. Another problem for a comparatively small research field such as addiction research is that it is difficult to find referees who are competent to evaluate qualitative methods and analyses. A qualitative article thus runs the risk of being reviewed by someone who not only is unqualified but also may be prejudiced against qualitative research. For all of these reasons, qualitative researchers have to be particularly professional in their writing.

The Challenges of Publishing Qualitative Research

Qualitative methods can be used for pilot studies, to illustrate the results of statistical analysis, in mixed-methods studies, and in independent qualitative research projects (c.f. Denzin & Lincoln, 1998). This chapter will focus on the last category: original research reports that use qualitative methods. We will emphasise the similarities and considerable overlap in the evaluation, and effective presentation, of both qualitative and quantitative research.

The first and foremost aim of all social research, quantitative as well as qualitative, is to present a conceptually adequate description of a *historically specific* topic, subject, or target. In qualitative research, the determination of the subject is as important as the choice of a population in a statistical study. The description of the subject is always, in both types of study, a theoretical task because it requires a conceptually well-organised analysis.

The processes of classification, deduction, and interpretation are in their fundamental aspects similar for both qualitative and quantitative research. Quantitative analyzing operations, however, are more clear-cut than qualitative operations. Furthermore, the various steps of quantitative research can be more clearly distinguished than can those of a qualitative study. The first issue is that, in qualitative work, the collection and processing of data are more closely intertwined than in a quantitative study. Especially when the researchers personally collect the data, they will not be able to avoid problems of interpretation during the collection phase. A specific issue in some qualitative research is that the methods used can change during the study, depending on interim results. It is a challenge to explain in a short article why this has happened, and why one has used a different method in the final phase of the data acquisition than in the previous parts; or why one changed a classification scheme and encoded the data in a different manner. The researchers must also carefully consider their relations with the study objects. Many qualitative reports discuss at length the character and psychology of the process of data collection, but are less careful in describing what happened to the interview tapes afterwards. Were they transcribed in whole or in part, how was the resulting stack of papers handled and sorted out? In qualitative research, these data processing explications may be necessary to render credibility to the analysis.

A second issue is that qualitative analysis is not restricted to an unambiguously demarcated data set in the same way that a quantitative study is. Good researchers may keep a detailed field diary and make notes of all discussions and thus produce a corpus to which they limit their analysis. Nevertheless, during the analysis phase, they may recall an important detail that they had not recorded in their notes but must take into account in the analysis. The qualitative researchers have to describe this analytical process in an honest and convincing way.

There are several basic factors that make the publication of qualitative research harder and different from standard journal article models of quantitative research (Miovský, 2006):

- The research design may be less strictly defined from the beginning of the research project, and it is not unusual to have design changes as new questions arise and new findings are considered. Redesigning necessitates an especially thorough and sometimes lengthy methodology section to explain those changes.
- Qualitative research uses many different theoretical frames (phenomenology, constructivistic approaches, hermeneutics, etc.) that affect data selection, methodology, and presentation. This variance is also to some extent found within quantitative research. But because analysis and reporting are more closely intertwined in qualitative research, the differences in theoretical perspectives become even more important. As an author, you will have to argue even more clearly for the choice and sufficiency of your data and their scientific significance.
- Compared with quantitative research, qualitative research uses different concepts of research validity (e.g., credibility), with different theoretical backgrounds (Whittemore, Chase & Mandle, 2001) and different views on correct sampling methods and the representativeness of data (Patton, 1990). Some sampling strategies combine qualitative and quantitative perspectives (e.g., respondent-driven sampling). Qualitative-oriented research can be performed with a single case study but also with sampling methods such as snowball sampling or respondent-driven sampling, which can combine traditional probability sampling methods with qualitative-oriented methods. It can be a challenge to describe these data sets and the data collecting methods, as well as why and how they were used, within the length limits usually applied to research reports.

All these factors present authors with a set of practical difficulties, not only because of technical page limits but also because there are not many reviewers with insight into qualitative methods and analysis. Scientific publishing has also gradually become more streamlined, with a lot of written and unwritten habits and rules that are usually based on quantitative approaches and methods. A qualitative researcher must be prepared to tackle these obstacles.

Evaluation Criteria for Qualitative Analysis

There are some differences between the evaluation of qualitative and quantitative research. The replicability of a qualitative study cannot be formulated as a problem of reliability, and the accuracy of a qualitative interpretation cannot be compared with the explanatory power of a statistical model. In the following paragraphs, we propose three main criteria for evaluating qualitative studies: 1) significance of the data set and its social or cultural place; 2) sufficiency of the data and coverage of the analysis; and 3) transparency and repeatability of the analysis. Since in qualitative research the analyses and reporting are very closely intertwined, these criteria are as relevant to researchers and authors as they are to reviewers and editors.

1. Significance of the Data Set and its Social or Cultural Place

The researchers should be prepared to argue that their data are worth analyzing. It is not easy to identify criteria for the significance of data. One precondition can, however, be presented: the researcher should carefully define the social and cultural place (contextualising) and the production conditions of the material.

The production conditions can be discussed at several levels. When the data consist of cultural products, their production and marketing mechanisms should be considered. Texts produced by individuals should be related to their social position. Furthermore, the situational aspect of the data production and the researcher's potential influence on the data should be evaluated. The relationship of cultural products to people's everyday life depends on the production and distribution network. Weekly magazines and movies represent the ambient culture at a number of levels. When doing comparisons over time, it is important to bear in mind that the social and cultural place of one and the same genre may vary from decade to decade.

In international comparisons, it is important to be able to exclude demographic variation as a factor causing differences. If we wish to identify the distinct characteristics of Finnish A.A. members' stories, we should make sure that we do not compare Finnish farmers with American college professors. The criterion for selecting the target group is not demographic but cultural representativeness.

Additionally, people speak of the same things in different ways on different occasions, and it is the task of the researchers to decide which discourse they want to study and argue for their decision in the article. Informal interviews are often advocated instead of questionnaires on the grounds that they will produce more genuine information. But, on the other hand, an in-depth interview is a more exceptional situation for a present-day person than completing a questionnaire. Possible effects of the power structures and gender relations present in every social situation should be considered in the discourse analysis, since it could affect the outcomes of the qualitative research.

Study of the variations of discourse, i.e., the incorporation of the production conditions into the study design, can be rather laborious. Members of A.A. emphasize various sides of their story according to the composition of the audience, and depending on whether they talk at a closed or an open A.A. meeting. Furthermore, the life story will change in relation to how long the speaker has been in A.A. Even when variation cannot be incorporated into the actual study design, it is important to consider and discuss the conditions under which the material was produced and their place in the potential situational variation of the discourse.

2. Sufficiency of Data and Coverage of Analysis

For statistical studies, we are able to calculate in advance the extent of data needed to estimate the parameters accurately enough for the purpose of the analysis. We have no similar methods for estimating the extent of qualitative data required. We usually speak about data saturation: data collection can be terminated when new cases no longer disclose new features (Strauss & Corbin, 1998). The difficulty here, of course, is that the limit is not always known in advance, and the collection of data is rarely a continuing process that can be terminated or extended at will.

Only in very special cases can you base your analyses on just a handful of observations. In most cases, you will need to be certain that you cover the variation of the phenomenon you are studying. On the other hand, a loose but useful rule is that one should not collect too much data at a time. It is better to analyze a small data batch carefully first and only then determine what additional data will be needed. To divide the analyses into smaller parts also helps to produce manageable results for a publishable report.

It is often advisable to group the collection of data according to factors which may prove important as production conditions. The goal is not to explain the variation but to make sure that the data are sufficiently varied. For example, it would be helpful to stratify the collection of A.A. members' life stories according to the members' social position, sex, age, and length of sobriety (Arminen, 1998). The only difficulty is that we will have no advance knowledge of which characteristics will decide the type of life stories; the stories may depend more on drinking experiences than on external circumstances, and within A.A. there may be various narrative traditions which have an influence on the life stories.

Proper coverage of the analysis means that the researchers do not base their interpretations on a few arbitrary cases or instances but on a careful reading of the whole material. Qualitative reports are often loosely impressionistic because the excessive amount of material has made it unfeasible to analyze it carefully enough.

3. Transparency and Repeatability of the Analysis

Transparency of the analysis means that the readers are able to follow the researcher's reasoning and that they are given the necessary information for accepting the interpretations—or for challenging them. The repeatability of an analysis means that the rules of classification and interpretation have been presented so clearly that another researcher applying them will reach the same results. We may identify three ways of improving the transparency and repeatability of qualitative analysis and the report: 1) enumerating the data; 2) dividing the process of interpretation into steps; and 3) making explicit the rules of decision and interpretation.

The best method to decrease arbitrariness and increase repeatability is to enumerate all units on which the interpretation is based. To do this an analytical unit must be specified and it should be as small as possible. In other words, do not choose a movie or a group discussion but rather choose a scene, a statement, or an adjacent pair. The identification of the unit of analysis is in itself part of the process of interpretation.

The process of interpretation and analysis can never be fully formalized. It is above all a question of working step by step so that the process of interpretation can be made visible to both the researchers themselves and the reader.

Qualitative analysis is of necessity more personal and less standardized than statistical analysis. Thus, it is even more vital that the reader is given as exact a picture as possible of both the technical operations and the chain of reasoning that have led to the reported results. The reader must not be left at the mercy of the researcher's intuition alone. The demand for transparency in qualitative research is of crucial importance.

Editors' and Referees' Assessment of Qualitative Research Reports

A discussion of the evaluation criteria for peer review of qualitative research can start with evaluation principles for quasi-experimental research or natural experiments. The *American Journal of Public Health* published an evaluation system for these types of study (Des Jarlais, Lyles & Crepaz, 2004) entitled TREND (Transparent Reporting of Evaluations with Nonrandomized Designs). TREND was designed specifically for research results in which the randomisation principle was somehow restricted. The criterion of *transparency*, which is central to this evaluation system, emphasises a detailed description of all steps and procedures, as well as a detailed justification of the choice and manner of application of the individual methods and theoretical background (see also Mayring, 1988, 1990).

Mareš (2002) analysed quality criteria for research using pictorial documents and summarised the findings with the concepts of *completeness* (how well the data capture the phenomenon examined), *transparency* (the accuracy, clarity, and completeness of the description of the individual phases of the study), *reflexivity* (the ability of researchers to reflect on their different steps and measures during the study and how the investigators may have influenced the research situation), and *adequacy* of interpretation and *aggregation* of contradictory interpretations (the identification and weighting of alternative interpretations and other validity-control techniques).

Des Jarlais, Lyles & Crepaz (2004, pp. 363–365) have drawn up a 22-item list to serve as a general assessment guide for authors and evaluators. Box 8.2 shows some of their requirements and recommendations.

Additional recommendations proposed by Gilpatrick (1999) and Robson (2002) are summarized in Box 8.3.

---

a) An article should be provided with a structured abstract (as a minimum: background, aims, sample, methods, results).

b) The sampling should be described and justified, including an explanation of criteria used.

c) The theoretical background of the entire study, or individual methods, should be described, to show that the sample and data collection were consistent with the study's theoretical background.

d) The context (setting) in which the study was carried out should be described. The authors must describe the characteristics of the field in which the study was carried out, and what made it different from other settings.

e) A detailed description of the research intervention should be included, and of how study participants responded during that intervention.

f) A detailed description of the analytical methods applied, how they were used, including the tools used for minimising bias; and a validation of the results should be presented.

g) A description of the manner of data processing (e.g., technical aspects and procedures) is needed.

h) Description of outcomes and their interpretation are obviously necessary. This includes a discussion of limitations (contextual validity of results), and an analysis of how the design of the study reflects these limitations.

Box 8.2: Assessment Criteria for Qualitative Studies.
Source: Des Jarlais, Lyles & Crepaz (2004).

a) The research issue and the research questions and goals derived from it, should be properly presented.

b) The goals should be contextually embedded and put into a theoretical framework, with an analysis of the present state of knowledge.

c) The author should argue for the importance of their study against this background (e.g., what questions or issues the results should contribute to, how they will move the field forward).

d) Control tools (e.g., research logs, control points) should be reported and how ethical problems were handled (e.g., use of informed consents, careful adherence to research protocols, manner of preparing the research team to manage risky or problem situations).

Box 8.3: Evaluation Criteria for Qualitative Studies.
Sources: Gilpatrick (1999) and Robson (2002).

The qualitative paper, both in its entirety and in its constituent parts, will be evaluated by and large according to the same criteria and expectations as those applied to a quantitative report.

Practical Advice for Writing a Publishable Qualitative Article

A good way to start the process of improving both your writing skills and your chances of publication is to become familiar with the common reasons why editors reject qualitative articles (see Box 8.4), and then carefully read some examples of well-written qualitative articles (see Box 8.5).

Based on our experience as journal editors, referees, and researchers, we now present nine recommendations for potential authors of qualitative articles.

1. Consider the Format and Structure of Your Article

When you get acquainted with various addiction journals, you will realize that qualitative articles can look very different depending not only on their topic but also on where they are published. You can choose to target a specific journal and try to follow closely the format used in that publication. But if you want a greater choice of potential journals for your manuscript, and in particular if

- The author has not related the study to earlier (international) literature.
- The research question is not clearly stated.
- The structure of the article is not clear or does not respond to the expected structure of articles in the journal.
- Theories, methods, and data analyses are not consistent.
- The central concepts are not clearly presented or used in a consistent way.
- The methodology is poor.
- The size of the data set is not defended in a convincing way.
- The data set is not sufficiently contextualised, or there is a clear selection bias.
- The data collection is poor and lacks validity control.
- The methods and analyses are not explained clearly enough, which may lead the referees and the editor to regard the article as too descriptive and the analyses based too much on intuition.
- The author makes unsound conclusions or unfounded generalisations.
- Ethical rules are violated or ethical issues are not mentioned or adequately discussed.
- The text is too long.

Box 8.4: Common reasons why editors decline qualitative articles. Source: Drisko (2005).

you are not a very experienced researcher, it may be wise to choose a traditional structure for your research report.

2. Begin with the Abstract

Most addiction journals require the authors to write very short abstracts, covering background, aims, data and methods, results, and discussion. It is a good idea for the author of a qualitative article to write a preliminary abstract at an early stage of the writing process to ensure that the text will be coherent and logical.

3. Choose a Title that Corresponds to the Content

The title of an article is very important. Drisko (2005) gives the following advice: present the research question reshaped into the manuscript title.

Amos, A., Wiltshire, S., Bostock, Y., Haw, S., & McNeill, A. (2004). 'You can't go without a fag . . . you need it for your hash' – a qualitative exploration of smoking, cannabis and young people. *Addiction, 99*(1), 77–81.

Demant, J., & Järvinen. M. (2006): Constructing maturity through alcohol experience – Focus group interviews with teenagers. *Addiction Research and Theory, 14*(6), 589–602.

Herd, D. (2005). Changes in the prevalence of alcohol use in rap song lyrics, 1979–97. *Addiction, 100*(9), 1258–1269.

Maher, L., & Hudson, S. L. (2007). Women in the drug economy: A metasynthesis of the qualitative literature. *Journal of Drug Issues, 37*(4), 805–826.*

Maeyer, J. D., Vanderplasshen, W., Camfield, L., Vanheule, S., Sabbe, B., & Broekaert, E. (2011). A good quality of life under influence of methadone: A qualitative study among opioid-dependent individuals. *International Journal of Nursing Studies, 48*, 1244–1257.

Miovský, M. (2007). Changing patterns of drug use in the Czech Republic during the post-Communist era: A qualitative study. *Journal of Drug Issues, 37*(1), 73–102.

Phillips, D., Thomas, K., Cox, H., Ricciardelli, L. A., Ogle, J., Love, V., & Steele A. (2007). Factors that influence women's disclosures of substance use during pregnancy: A qualitative study of ten midwives and ten pregnant women. *Journal of Drug Issues, 37*(2), 357–376.

Please visit the website of the International Society of Addiction Journal Editors (ISAJE) at www.isaje.net to access supplementary materials related to this chapter. Materials include additional reading, exercises, examples, PowerPoint presentations, videos, and e-learning lessons.

Box 8.5: Examples of well-written qualitative articles.

A title that indicates what you are interested in will generate more readers who really are interested in your research—and probably more citations of your article (see Chapter 10). Sometimes it is possible to formulate the title so that it also describes what kind of data you have used. A title should not promise too much or be too fancy. If the title of the article is "The commercial discourse on alcohol," the reader expects that the theoretical contribution will be substantial. If it is "An analysis of alcohol marketing" and you deal only with beer advertisements in a short period in Greece, the reader may be disappointed.

4. State the Research Question Early and Clearly

It is a common failure in qualitative reports to embed the research question so deeply in the text that the reader cannot find it. The best way to avoid this is to include, at the beginning of your manuscript, a subtitle called "Research question" or "Aim of the study." An alternative is to present the question at the end of the background or introduction section.

It is not unusual for the reader of a qualitative article to find several different, sometimes even contradictory, research questions presented throughout the various sections of the article: one question in the introduction, another in the methods and data section, and a third in the discussion (Drisko,2005). Even if the research process in qualitative research is often more unpredictable than in quantitative research and you gain new insights during the research process that will affect your perspective, the aim of a research report is as a rule to report not on this exploratory process but instead on specific findings answering a specific question. The reader does not want to be taken through the whole story of the researcher's mistakes and new choice of questions. Focus on a single clear question that will orient the reader's interest and prepare him for the text to come. It may be that your research project will in fact be able to answer many questions. Perhaps then you should consider producing several shorter and focused articles, rather than trying to squeeze it all into one text.

If possible, phrase the research question in a way that reflects the scientific ambition of the study: Is it an article that explores a topic, aims at discovering a new social phenomenon, presents a new perspective, seeks to raise consciousness about a problem, evaluates a project, or tests a theory (Drisko, 2005)?

5. Conduct a Thorough Review of Earlier Research

A good review of earlier research on the topic is essential for your claim that you are contributing new knowledge. It also shows that you want to take your place in the research community and engage in serious dialogue with other researchers. If the referees find that you have overlooked important literature, particularly if it is their own work (and since qualitative addiction research is a small field, you will often have a referee that has contributed to your topic), or that you have misinterpreted earlier studies, they will read your study with skepticism. Do not limit yourself to literature from your own country, but be sure to cover what has been written from your own culture.

The literature review should not be solely descriptive. Use it to position yourself in relation to other researchers and to demonstrate that you are doing something new. What conclusions about your questions can already be drawn from earlier research? State why you think earlier studies have missed a particular aspect of the topic or have taken a perspective that can be complemented with a new one. Alternatively, say why and in what way you want to use an approach or develop a line of thought presented by someone else.

When you have presented a good review of earlier research, you will also have defended your theoretical and methodological position and your choice of data. Be certain to choose the right body of literature with theoretical relevance for your question. If you are studying gender differences in advertisements for tobacco, be sure to cover the literature on gender and media: do not focus exclusively on what we know about gender differences in smoking patterns.

A thorough review, in which you position yourself clearly, also offers a practical way to avoid unfavorable referees. If you state that you disagree with X who has not taken Y into account, the editor will probably not send your text to X, to avoid a conflict of interest. Since the number of possible referees available to the editor usually is limited, this is an important consideration.

6. Present Enough Information in the Methods and Data Section

According to Drisko (2005), inadequate methods are among the most common reason for qualitative articles being declined by editors. It is important to justify the choice of methods. If you want to be really convincing, explain your choice in relation to alternative methodologies. If you use several methods, explain how they complement each other. For instance, it is not enough simply to state that you use focus group interviews and a post-structuralist text analysis: You should describe how and why you use them

Remember that many readers of addiction journals will not be familiar with qualitative methods. Therefore, you must describe the content of the method quite explicitly. Show that the research methods are suitable for the purpose of the study. It is important to convince the reader that you have used your method(s) systematically and on the entire data set. This includes the consistent use of crucial concepts.

You must argue that the size of the sample is sufficient for your purpose. As noted above, a small sample is one of the factors that raises skepticism among readers of qualitative research. How extensive is your data set? How many interviews with how many persons? How many meetings or observations? Position the sample clearly but without being too wordy: Try to focus on the essential features that will help an uninitiated reader to understand what you are analyzing and what the sample represents.

It is important to explain why your data set is the most illustrative and useful to answer the question you are posing. Be careful to describe how you picked your sample. What criteria did you use? Can you compare the data set with other alternatives and why did you choose this one? Describe the important variations within the data set (e.g., age and gender distributions) so that the reader gets a good picture of it. If you have used only a part of the data you have collected within a project, describe the rest of the data briefly to illustrate the context or refer to another, already-published, article in which these data are presented.

For the interpretation and transparency of your reasoning, it is crucial to describe how the data were produced and collected and how these conditions may have influenced the data. What special conditions, for example, come into play if you collect data from A.A. members, for whom anonymity is important? Do they affect the research participants' willingness to be interviewed or how they talk during an interview? Tell the reader how (or whether) you presented the study to the participants. If you used focus groups, describe the groups' dynamics.

Describe carefully each step in the analysis so that the reader can accept your conclusions—or argue against them. A good rule is to present the analysis of one observation/item/response in detail. Describe your interpretations during the analysis in a systematic way and in small identifiable steps. Show the fruitfulness of your concepts. Show how you argued for saturation and how you handled diversity and contradictions in the data.

A thorough description of how the data were handled is also important. It should be clearly stated, for instance, how and whether the interviews were transcribed, coded, and grouped.

7. Link the Results to the Research Question

The presentation of the results is easiest for the reader to follow if the structure is directly linked to the research question, moves in logical steps according to the theory and method, and consistently uses the concepts presented earlier in the article.

Present your data in a systematic way in the body of the text, so that quotations, field notes, and other documentations are easily identifiable. The reader must be certain, for instance, whether you are using direct citations or analyzing interpretations of what the observed or interviewed persons said. The citations or other illustrations must be clearly contextualised. For observational material, state whether you collected the data yourself or if you used data collected by someone else.

Give enough raw data (e.g., direct citations) but not too much. Avoid very short quotations. If you run out of space, ask the editor if you can use online appendices for additional material. Do not refer in the results section to data that you have not already presented in the data and methods section; if you state that you are going to use interviews, do not refer to observations in the results section. If the results are contradictory, declare that fact openly and explain how this may have occurred and what it may mean.

If you use grounded theory, you should be able to present a theory as a result. Descriptive statements are not enough. The theory should be a product of the analyses and not just confirm or illustrate earlier theories (Glaser & Strauss, 1967).

8. In the Discussion Section, Restate Your Main Findings and Relate Them to Earlier Research

The structure of the discussion in a qualitative article can follow the same structure as in quantitative research reports. After a very short summary of your research question (check that it is the same as in the introduction) and the motivation for your wish to explore it, you can repeat in one sentence the main result of your study.

Following this, discuss how your findings relate to earlier research: Do they fill out the picture of what we already know or possibly challenge or even contradict earlier findings? In this section, you can also, if possible, refer to earlier quantitative research. In what way has your study been important for the research community or for a larger audience? Can the results change the picture of similar phenomena in other cultures? Discuss the extent to which the findings with this data set are relevant to the understanding of other situations. What are the concepts that can be transferred to other settings?

As noted in Chapter 12, a good discussion will also contain a consideration of the limitations of your study. What problems with the sample and data collection restrict the possibility of getting a full answer to your research question? With what other data could the answer have been more complete? Could you have used an additional or alternative method?

Finally, consider giving recommendations for further research that will improve knowledge about the topic you have studied.

9. And Finally, Some General Advice

First, it is sensible for qualitative as well as for quantitative researchers to save their good data for scientific articles. Many qualitative researchers publish their results as reports, sometimes in series that will have limited distribution, or as longer articles in monographs. If you want to spread your findings to a larger audience, it is often more efficient to publish one or more articles in a scientific journal.

Second, choose the right journal—a crucial success factor if you want to get your article published. The first step is to choose among either an addiction journal; a journal for qualitative research; or a scholarly journal for sociology, anthropology, history, etc. (see Chapter 3).

If you choose an addiction journal or a disciplinary journal, find out if it accepts qualitative reports. Table 8.1 presents a list of English-language addiction journals that publish qualitative research. Non–English-language journals as a rule accept submissions of qualitative articles. Check if the journal has particular demands on article length that will make it difficult for your submission to be accepted. Look at the editorial board anddetermine whether it includes members who are familiar with qualitative methods. Finally, look at the content

Addiction	International Journal of Drug Policy
Addiction Research and Theory	Journal of Addictions Nursing
Addictive Behaviors	Journal of Alcohol and Drug Education
African Journal of Drug and Alcohol Studies	Journal of Drug Education
Alcohol and Alcoholism	Journal of Drug Issues
Alcohol Research and Health	Journal of Ethnicity in Substance Abuse
Alcoholism Treatment Quarterly	Journal of Gambling Issues
American Journal of Drug and Alcohol Abuse	Journal of Smoking Cessation
Contemporary Drug Problems	Journal of Social Work Practice in the Addictions
Drug and Alcohol Dependence	Journal of Studies on Alcohol and Drugs
Drug and Alcohol Review	Journal of Substance Abuse Treatment
Drugs: Education, Prevention and Policy	Journal of Substance Use
	Nordic Studies on Alcohol and Drugs
European Addiction Research	Substance Abuse Treatment, Prevention, and Policy
Harm Reduction Journal	Substance Use and Misuse
International Gambling Studies	Tobacco Control

Table 8.1: English-language journals that publish qualitative articles.

of the journal: To what extent do they publish qualitative articles? Bear in mind that many addiction journals are open to various research methods, even if those journals have a predominantly quantitative orientation.

Finally, consider if it would be good to suggest a suitable referee for your article. Some journal editors may find it difficult to identify experienced referees for your manuscript. As an author, you can always suggest someone whom you would like to review your text, without, of course, any guarantee that the editor will follow your advice.

Conclusions

In this chapter, we have emphasised that the similarities between conducting and writing up quantitative and qualitative research are greater than the differences. We have presented some quality criteria, particularly for qualitative research, discussed criteria for evaluation of journal articles, and given some practical advice to authors.

Publishing qualitative research is as least as challenging as getting quantitative reports accepted. However, it is apparent that the addiction field as a whole is increasingly coming to realise the value of qualitative studies. We believe that, in the future, there will be an even greater interest in good qualitative research and a growing demand for mixed-methods studies. Those who have dug themselves down into the qualitative or quantitative trenches will emerge and start communicating with each other, for their own and everyone's mutual benefit.

Acknowledgements

The authors thank Tom Babor, Phil Lange, Tom McGovern, Peter Miller, Jean O'Reilly, and Betsy Thom for valuable comments on earlier versions of the text.

Please visit the website of the International Society of Addiction Journal Editors (ISAJE) at www.isaje.net to access supplementary materials related to this chapter. Materials include additional reading, exercises, examples, PowerPoint presentations, videos, and e-learning lessons.

References

Arminen, I. (1998). *Therapeutic interaction. A study of mutual help in the meetings of Alcoholics Anonymous,* Vol. 45. Helsinki: The Finnish Foundation for Alcohol Studies.

Creswell, J. W., & Plano Clark, V. L. (2007). *Designing and conducting mixed methods research.* Thousand Oaks, CA: Sage.

Creswell, J. W., & Tashakkori, A. (2007). Developing publishable mixed methods manuscripts. *Journal of Mixed Methods Research, 2*(1), 107–111.

Denzin, N. K., & Lincoln, Y. S. (Eds.). (1998). *The landscape of qualitative research: Theories and issues.* Thousand Oaks, London, New Delhi: SAGE Publication.

Des Jarlais, D. C, Lyles, C., & Crepaz, N. (2004). Improving the reporting quality of nonrandomized evaluations of behavioral and public health interventions: The TREND statement. *American Journal of Public Health, 3*(94), 361–366.

Drisko, J. (2005). Writing up qualitative research. *Families in Society: The Journal of Contemporary Social Services, 86*(4), 589–593.

Gilpatrick, E. (1999). *Quality improvement projects in health care.* London, Thousand Oaks, New Delhi: SAGE Publications.

Glaser, B. G., & Strauss, A. L. (1967). *The discovery of grounded theory: Strategies for qualitative research.* New York: Aldine.

Lalander, P. (2003). *Hooked on heroin: Drugs and drifters in a globalized world.* London/New York: Berg Publisher.

Mareš, J. (Ed.). (2002). Sociální opora u dětí a dospívajících II [Social support of children and adolescents]. Hradec Králové: Nukleus.

Mayring, P. (1988). *Qualitative inhaltsanalyse: Grundlagen and techniken.* Weinheim: Deutcher Studien Verlag.

Mayring, P. (1990). *Einführung in die qualitative socialforschung.* München: Psychologie Verlag Union.

Miovský, M. (2006). *Kvalitativní přístup a metody v psychologickém výzkumu* [*Qualitative approach and methods in psychological research*]. Praha: Grada Publishing.

Miovský, M. (2007). Changing patterns of drug use in the Czech Republic during the post-Communist era: A qualitative study. *Journal of Drug Issues, 37*(1), 73–102.

Miovský, M., & Zábranský, T. (2003). Kvalitativní analýza dopadu nové drogové legislativy na drogovou scénu z perspektivy pracovníků zdravotnických zařízení a významných poskytovatelů služeb uživatelům nelegálních drog [Impact of new drug legislation on drug scene from the perspective of health care professionals working with drug users: Qualitative analysis]. *Čs. psychologie, 47*(4), 289–300.

Patton, M. Q. (1990). *Qualitative evaluation and research methods.* London, Thousand Oaks, New Delhi: SAGE Publications.

Robson, C. (2002). *Real world research.* Oxford: Blackwell Publishing.

Stimson, G. V., Fitch, C., Des Jarlais, D., Poznyak, V., Perlis, T., Oppenheimer, E., Rhodes, T., & for The Who Phase II Drug Injection Collaborative Study Group. (2006). Rapid assessment and response studies of injection drug use. Knowledge gain, capacity building, and intervention development in a multisite study. *American Journal of Public Health, 96*(2), 288–295.

Strauss, A., & Corbin, J. (1998). *Basics of qualitative research: Techniques and procedures for developing grounded theory*, Second Edition. Thousand Oaks, CA: SAGE Publications.

Sulkunen, P. (1987). *Sosiologian avaimet [The keys to sociology].* Porvoo, Helsinki, Juva: WSOY.

Whittemore, R., Chase, S. K., & Mandle, C. L. (2001). Validity in qualitative research. *Qualitative Health Research, 4*(11), 522–537.

CHAPTER 9

How to Write a Systematic Review Article and Meta-Analysis

Lenka Čablová, Richard Pates, Michal Miovský and Jonathan Noel

Introduction

In science, a review article refers to work that provides a comprehensive and systematic summary of results available in a given field while making it possible to see the topic under consideration from a new perspective. Drawing on recent studies by other researchers, the authors of a review article make a critical analysis and summarize, appraise, and classify available data to offer a synthesis of the latest research in a specific subject area, ultimately arriving at new cumulative conclusions. According to Baumeister and Leary (1997), the goal of such synthesis may include (a) theory development, (b) theory evaluation, (c) a survey of the state of knowledge on a particular topic, (d) problem identification, and (e) provision of a historical account of the development of theory and research on a particular topic. A review can also be useful in science and practical life for many other reasons, such as in policy making (Bero & Jadad, 1997). Review articles have become necessary to advance addiction science, but providing a systematic summary of existing evidence while coming up with new ideas and pointing out the unique contribution of the work may pose the greatest challenge for inexperienced authors.

How to cite this book chapter:
Čablová, L, Pates, R, Miovský, M and Noel, J. 2017. How to Write a Systematic Review Article and Meta-Analysis. In: Babor, T F, Stenius, K, Pates, R, Miovský, M, O'Reilly, J and Candon, P. (eds.) *Publishing Addiction Science: A Guide for the Perplexed*, Pp. 173–189. London: Ubiquity Press. DOI: https://doi.org/10.5334/bbd.i. License: CC-BY 4.0.

What is the Relevance of a Review?

General definitions are one thing; the practical benefit of writing reviews is another. Why would a novice author/researcher engage in this activity? Why is it important? What benefits can it bring? First, it provides the authors with a general understanding of the subject matter they study as part of their area of expertise. Each field of study has its own terminology, and the more specific a topic is, the greater the terminological differences that may be found among authors. It is therefore important to produce a good description and critical appraisal of existing evidence concerning the topic being explored. Another objective is to integrate the findings generated by different studies into a meaningful body of evidence. The process of writing a review article will help the authors obtain a unique perspective on the issue and assist them in processing the results from many investigators into a consistent form. It will then be possible to summarize the results and interpret the existing evidence in a new light. To increase one's chances of having a review article accepted for publication, it is useful to address topical issues in a given field or areas of research featuring a number of heterogeneous and controversial studies where a consistent approach is needed.

What is a Review?

It is difficult to provide a single definition of a review. Indeed, each journal uses its own—slightly different—definition of a review study. For example, the journal *Adiktologie* defines a review article as a "cogent summary of topical issues; the author's own experience is not the underlying theme of the paper. The maximum extent is 16 pages, with not more than 50 bibliographical citations. References to recent literature (not more than five years old) should prevail" (Gabrhelík, 2013). *Addiction*, meanwhile, simply states that "reviews draw together a body of literature to reach one or more major conclusions" and allows review articles to contain up to 4,000 words with no limit on bibliographic citations (Society for the Study of Addiction, 2015).

Despite these limitations, clear distinctions can be made between the types of reviews that can be drafted. The traditional type of review is a narrative literature review, which assesses the quality and results of a selection of literature using implicit criteria (Culyer, 2014). The conclusions of traditional narrative reviews are often based on subjective interpretations of the literature and may be biased in unsystematic ways. Importantly, narrative reviews are essentially nonreplicable.

In contrast, scientific journals often require reviews to be systematic in nature. Systematic reviews use explicit literature search strategies, inclusion and exclusion criteria, and criteria for determining the quality and reliability of study findings. Systematic reviews are replicable and the conclusions drawn by authors more easily verified.

A systematic review that does not include an evaluation of study findings (i.e. performs only a systematic search using explicit inclusion and exclusion criteria) is referred to in this chapter as a hybrid narrative review. Hybrid narrative reviews provide authors greater freedom to interpret and integrate study results and conclusions compared with systematic reviews but still allow the reader to determine the authenticity of the author's findings. These reviews are particularly important for theory development and problem identification, especially when the peer-reviewed literature may be incomplete and when important studies may not use rigorous experimental or longitudinal designs.

Meta-analyses are a step beyond systematic reviews; they require a quantitative analysis of previously published findings.

The following sections discuss the steps involved in creating systematic reviews and meta-analyses. Although not explicitly mentioned, much of the information applies to hybrid narrative reviews as well. Because traditional narrative reviews are no longer viewed favorably, they will not be discussed. It is strongly recommended, however, that before writing any article, authors should first choose a journal to which to submit their research because of the subtle differences in journal manuscript definitions. Authors should study thoroughly the guidelines for authors and keep them on hand to reference while writing the article. This may save a great deal of time spent on final revisions or even make them unnecessary.

Main Steps to Successful Systematic Review

It is useful to observe the following procedure when designing and writing a systematic review. If the intention is to arrive at a systematic classification of evidence, a well-considered and highly structured procedure should be used. Structure is a crucial requirement, and some specific tools (e.g., PICOS: participants, interventions, comparators, outcomes, and study design) can make this more manageable (Smith et al., 2011). Below, we describe the specific steps involved in creating a systematic review and meta-analysis, using the development of a previously published review as an example of good practice. The following recommended strategies are based on the published systematic review (Čablová et al., 2014).

Aim of the Review

The aim of a systematic review is set in the same way as in an original research study; the article must contribute something new to the given research field. The specific aim should correspond with the research questions. It may be, for example, "to provide a systematic review of the results of studies published from 2000 to 2012 that investigate the specific relationship between the level of parental control and alcohol use among children and adolescents." Alternatively, it may be "to classify parenting strategies in relation to alcohol-using children aged 12–15" or "to make a critical appraisal of recent studies of the emotional bond in young adults who use cannabis."

The aims are typically stated in the last paragraph of the introduction. The aims then determine the choice of the specific procedure used to search sources and process and present the results. In the concluding section of the study, it should be stated whether and to what extent the aims have been fulfilled.

Inclusion of Research Questions

In a review article, the research question is included and expressed in the text, formulated as the problem: the topic and the focus of the work. It can be thought of as a spiral that provides logical connections among the parts of the article; that is, different parts build on and follow up on each other in a logical pattern. In terms of a systematic review, the research question must correspond with the objectives of the study and be aligned with the methodology, which is particularly relevant for the identification of data sources (the literature search) and the determination of study inclusion and exclusion criteria. It represents an imaginary starting point for the selection of key words and other parameters that are looked for in the relevant studies. As an example, we can use an article investigating the quality and type of emotional bonds in young adults who use cannabis and its (implicit) research question: "Can an insecure emotional bond be associated with a higher rate of cannabis use among young adults?" or: "Is there a relationship and difference between the lifetime prevalence of cannabis use among young adults and the individual types of insecure emotional bond?"

Identify Data Sources—Quality Literature Search

The primary and most important data sources are electronic databases, typically accessed through university libraries. Because access to specific papers may be limited as a result of financial constrictions, the levels of access granted to students and staff will depend on the resources of the university subscribing to the journals. Thus, you may find that although you can get into a number of databases, you may be able to access only a few full texts (as the others require payment) and have mostly abstracts available, which may not be sufficient for systematic reviews. This is dealt with in more detail in the next point.

In the field of addictology, we recommend to use following databases:

- Web of Science: http://www.webofknowledge.com
- Medline/PubMed: http://www.ncbi.nlm.nih.gov/pubmed
- EBSCO: http://search.ebscohost.com

- SCOPUS: http://www.scopus.com
- ProQuest Central: http://search.proquest.com/index
- PsycARTICLES: http://www.apa.org/pubs/databases/psycarticles/index.
  aspx

Nevertheless, databases and full-text studies are not the only data sources. It is also possible to include conference presentations if the conference abstracts have been published. At the same time, some journals could have a problem with these types of publications because they did not undergo a standard peer-review process. Also, a quality literature search should not disregard print sources, such as monographs; articles in peer-reviewed, non-indexed journals; handbooks and manuals pertaining to the relevant topic; graduate theses; and dissertations. These could be included into a category "Records identified through other sources" in the PRISMA (Preferred Reporting Items for Systematic Reviews and Meta-Analyses) study flow diagram (see below).

We recommend keeping scrupulous notes on the articles read, either using Endnote or a separate database of references. This is relevant to all research but particularly to reviews.

Determine Selection Criteria

The relevant publications, the results of which are to be processed, are selected according to the classification criteria that follow.

- **Year of publication**—designating the period that is under study—may be used as the first criterion.
- **Number of citations of the article**—this information can be found in databases, most often under the heading "Times cited." Articles with a greater number of citations report on more prestigious research.
- **Key words**—they reflect the terminology used in the given field and also help identify the most relevant studies.
- **Relevance of the article**—online databases may turn up a number of articles but, unfortunately, because of the potential overlap of key words and other parameters, some works may be totally inconsistent with the focus of the review. It is therefore necessary to look through each publication—in most cases the abstract will be enough—and exclude any irrelevant studies.
- **Type of publications**—although you may typically work with original and review studies only, specific topics may require the use of information from annual reports, research reports, or guidelines. It is therefore important to state these factors in the description of the procedure.
- **Study design**—as far as research studies are concerned, these may be further divided into subcategories: for example, reviews versus original works or, with clinical issues in particular, cross-sectional versus longitudinal.

- **Language of the publications**—the languages that currently predominate in science are English and Spanish, with Chinese emerging as a significant language of science (in addition to English, Web of Science databases provide the option of searching studies in Chinese).
- **Sociodemographic environment**—it is useful to describe the sociodemographic environment in which the research was conducted because it is a relevant factor that may influence the review's results. Thus, the review needs to take this into account when presenting the research results.
- **Funding source and conflicts of interest**—last but not least, the funding source of a study and other conflicts of interest may influence how the results are interpreted. As explained in other chapters, significant biases in study reporting have been uncovered when the funding source or authors have a financial stake in the results of the study.

Entered into a database or observed when working with hard-copy sources, these criteria make it possible to focus the work on the research question and the aim of the study you have laid down. Finally, all these criteria/indicators will be considered and interpreted in the subsequent discussion section.

Process of "Data Collection"

The complete literature search process needs to be recorded and documented. When evaluating systematic reviews, peer reviewers pay special attention to the means used to collect the "data" (i.e., specific publications) for the analysis. There are specific methods that can be applied for this purpose, with the PRISMA study flow diagram being the most frequently used one in contemporary science (Higgins & Green, 2008; Moher et al., 2009). Figure 9.1 shows the PRISMA study flow diagram used in the systematic review (Čablová et al., 2014).

Explanation of the Specific Items in the Prisma Study Flow Diagram

The first item, *Records identified through database searching,* shows the number of publications found in databases on the basis of the selection criteria. The item *Additional records identified through other sources* refers to the number of publications found in information sources other than those available online (these are typically print documents, such as research reports, handbooks, and manuals). Another step involves the elimination of duplicate articles. If you work with multiple databases, it is very likely that the same publication will be selected several times. Such duplicates should therefore be removed. This process is very easy if you use a citation manager. When using EndNote, for example, this can be achieved by simply activating the "Find duplicates" function.

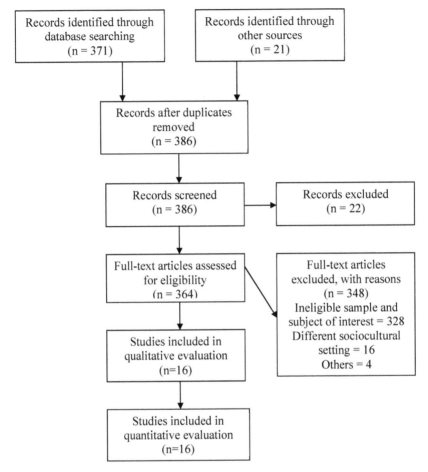

Figure 9.1: PRISMA study flow diagram.
Source: *Čablová et al. (2014, p. 4).*

Then you can focus on the articles. The item *Records screened* indicates the number of publications that remained after the exclusion of duplicates and publications rejected after you have read the abstracts. The number of articles eliminated on the basis of the examination of their abstracts is indicated in the *Records excluded* box. On the other hand, articles for which the full text is available (these should make up as large a proportion of the initial set of records as possible) are assessed in the next step and their final number is given under *Full-text articles assessed for eligibility.* When reading through the studies, you should continue to bear in mind the selection criteria (ideally, with a checklist on your desk) and watch carefully for them being met in the studies under scrutiny. If a more rigorous design is applied, you can also create a table specifically for the selection and assessment of publications. If you come across articles that do not meet the selection criteria, you should state the reasons for such ineligibility and the respective number of studies; see the item *Full-text articles excluded with reasons*. The last figure shows the final number of articles included in the study. This example contains two alternatives—*Studies included in qualitative evaluation* and *Studies included in quantitative evaluation*—but one item only, for example, *Studies included in quantitative evaluation,* is also possible. For more information about the PRISMA study flow diagram method, including further illustrations of the procedure or the PRISMA checklist that helps in keeping a record of the process, visit http://www.prisma-statement. org/statement.htm.

Interpretation of Results

The results of the studies you have obtained will be further summarized in a structured form—ideally a table—according to the classification criteria. It is advisable to compare the qualitative and quantitative perspectives of the studies when processing the results. (Although meta-analysis is not always the goal, it is useful to take quantitative as well as qualitative approaches into account.) When using a quantitative point of view, you can follow the number of studies that used a longitudinal versus cross-sectional design, how many studies applied a standardized methodology versus a methodology developed specifically for the purposes of the study, or how many studies had their samples of participants well balanced in terms of representativeness and how many did not. On the other hand, a qualitative perspective makes it possible to look for broader aspects of the works and fine subtleties in the results that have been ascertained.

There are a number of available tools that can serve as a guide when examining study methodologies and results. The Consolidated Standards of Reporting Trials (CONSORT) statement provides a standardized way to report and interpret the results of randomized clinical trials (Schulz et al., 2010). The primary tool is a 25-item checklist that contains questions on how the trial was designed, the data analyzed, and the results interpreted. The Strengthening the Reporting of Observational studies in Epidemiology (STROBE) and Transparent Reporting of Evaluations with Nonrandomized Designs (TREND) statements are similar checklists for studies using observational study designs (von Elm et al., 2007; des Jarlais et al., 2004). If a more quantitative analysis of study design is desired, the recommendations of the Grades of Recommendation, Assessment, Development, and Evaluation (GRADE) working group may be used (Atkins et al., 2004). These recommendations contain a point system that can be used in combination with the CONSORT, STROBE, or TREND statements to further differentiate among studies. Although useful, the results of using these tools should not be considered as absolute but as guides toward determining the weight that a study's conclusions should be given. In addition, systematic reviews should always be attentive to sex and gender issues, as described in the SAGER Guidelines (Heidari et al., 2016).

Interpretation should always be based on the results and findings specified in a given study; you must refrain from adding any conclusions of your own, because the principal rule is to preserve and express the original author's idea as precisely as possible. When formulating the ideas and working with other review studies, you should always look up the primary source and interpret its results. Other review studies may serve as an inspiration in classifying your results rather than being their source, functioning rather as "background material."

Any copyright rules should be observed when making citations. You should strictly avoid using findings presented by the original authors in their research as your interpretations; if at all, you can resort to a secondary citation, which in itself may appear rather awkward. Therefore, you should seek to be as accurate as possible and restate the author's original argument, looking up other relevant works on the topic that you will cite in the same way. In addition, it is necessary to be attentive and socially sensitive when interpreting the results of studies from different sociocultural settings; you should be careful not to make unreasonable generalizations and ensure that the results are always interpreted in terms of the given social context. This may involve engaging in some additional research but, particularly in the social science field, this extra effort is an element that has a major impact on the final product. In Table 9.1 we present an example that illustrates the processing of the results in a published systematic review (Čablová et al., 2014). The left hand column lists the studies according to authors and year, which corresponds with the standard identification of citations in text. The selection criteria applied to the studies under consideration are indicated in the heading line. The reader thus has a chance to see the results of the work in aggregate and in a clearly structured way without having to wade through a lot of text.

Discussion and Conclusion—was the Aim Really Achieved?

Once the results have been processed and interpreted, what is probably the most challenging part comes next. For one thing, you may be quite tired by now, because the previous systematic procedure was rather demanding in terms of attention and endurance, and now you need to think about the results and compare them with the conclusions drawn by other relevant studies and with each other. In particular, this requires you to bring a new perspective to the subject matter under study, singling out and discussing most salient finding from the results. Importantly, the discussion should compare and evaluate the results against other relevant research projects rather than against the presentation of the author's opinions on the issue. Each idea or result presented in the

Research studies	Country	Study design[1]		Age category[2]			Number of respondents	Parental involvement[3]		Parenting styles[4]					Methods[5]	
		L	C	1	2	3		Y	N	1	2	3	4	5	1	2
Bahr & Hoffmann (2010)	USA		+	+	+	+	4,938		+	+	+	+	+		+	
Barnes et al. (2000)	USA	+			+	+	506	+		+					+	
Burk et al. (2011)	USA	+				+	362	+		+						+ CRPR
Choquet et al. (2008)	France		+	+	+	+	16,532		+				+		+	
Clausen (1996)	Norway	+			+	+	846		+	+	+	+	+		+	
Total		8	8	6	12	14		6	10	13	9	9	7	3	12	6

Table 9.1: Example of processing and presenting the results.
[1] L = longitudinal study, C = cross-sectional study.
[2] 1 = younger children (9–12 years); 2 = older children (13–15 years); 3 = adolescence (16–22 years).
[3] Y = yes; N = no.
[4] 1 = authoritative style, 2 = authoritarian style, 3 = permissive style, 4 = neglectful style, 5 = other.
[5] 1 = questionnaires newly developed for the purposes of the research, 2 = standardised questionnaires.
Source: Čablová et al. (2014, p.5).

article needs to be properly cited, too. The conclusion consists of a practical evaluation of the study; it should not contain any new findings or evidence. Its purpose is briefly to summarize the results and the contribution of the study as a whole. Although this can pose a formidable task to an inexperienced author, it is important to practice the skill of communicating your own views concisely.

The conclusion often includes recommendations (resulting from the study) for further research and tips for practice. It is also advisable to highlight the unique contributions of your review. In technical terms, it is recommended to study carefully the instructions for authors provided by the journal in which you want to submit the article for publication. Although some journals require the discussion and conclusion to come in two separate sections, others prefer to have them combined. The latter requires a slightly different structure, and it is helpful to be familiar with the format requirements before writing the article.

The Most Frequent Pitfalls

When trying to pursue as systematic and transparent a procedure as possible, you can encounter several problems. We have already mentioned the potential problem with differences in terminology used by the authors who publish research on a given subject in the field. To prevent confusion, it is recommended that you read a reasonable number of articles pertaining to your topic and look for the terminology they use. Databases may be helpful in this. The Web of Science platform, for example, features a "related records" function, which may be used to search for similar articles on a certain topic. You may be confronted with a range of often competing theoretical approaches or backgrounds used by the authors to explore the subject matter in question. Because the literature search may be a challenging and time-consuming task, you may need to allow some time to study the relevant concepts thoroughly (for which the studies you have identified may not provide all the answers, requiring you to do further reading), as well as to reflect on such differences in your own conclusions and interpretations. Other differences may be found in the methodology applied by the studies under scrutiny. There are authors who work with standardized methods and their results can be subjected to a simple and valid comparison; on the other hand, there are authors who use their own methodology and whose results are thus difficult to measure. Another aspect that will consume time is the elimination of duplicate records, because researchers sometimes publish the results of the same study in several parts, divided into various subtopics to meet the foci of different journals. A mechanical "remove duplicates" function cannot do all the work. It is necessary to be alert and watch out for any relevant correlates.

Another problem that may be encountered when comparing results between studies is the difference in the number of study participants. Many studies do not use a representative sample of participants, and great differences in their sizes may strongly affect study generalizability. You may also face your own limitations, particularly regarding the inclination toward a selective choice of studies, where certain studies may not be included, either deliberately or inadvertently. Because citation bias may significantly compromise the results, you should try to avoid it at all costs if you want to arrive at a conclusion that is relevant to the field. If you fail to do so, it is most likely that reviewers will discover such a bias, as it is their job to examine related studies in the given area of research.

The last aspect to consider during the interpretation process is the statistical versus clinical significance of studies. In a large number of cases, you will find results that are not reflected in clinical practice, despite being significant. Therefore, it is important to maintain contact with clinical practitioners (or consult other experts) and be able to compare the results with real life. You can then formulate how these significances correlate in the conclusion.

For addiction science, the critical evaluation of systematic reviews is quite important. It is the key to the correct interpretation of selective data from particular studies, it provides background for comparing findings, and it can help to identify potentially disproportionate or inhomogeneous interpretations of findings. It has always been a sensitive issue in the context of publishing addiction science because of potential conflicts of interest, and the history of the field contains examples of published papers in which researchers intentionally distorted data. The tendency to interpret data in a different way and present specific points of view can be a potential source of bias (Bero & Jadad, 1997). For example, there are many examples of contrasting study findings in the area of tobacco policy depending on whether the study was or was not sponsored by the tobacco industry (Glantz, 2005).

Meta-Analysis

Meta-analysis is a form of systematic review that combines findings from a number of studies to create aggregate effect sizes. To do this, the size of the effect is calculated and indexed. This can be used for a number of purposes in addiction science, including the effects of an intervention (e.g., the use of naltrexone and acamprosate for treating alcohol use disorders [Maisel et al., 2013] or the impact of smoking bans on restaurants and bars [Cornelson et al., 2014]) and epidemiology (e.g., substance use among street children [Embleton et al., 2013]) or seroconversion of hepatitis C in relation to shared syringes [Pouget et al., 2012]). By aggregating the effects and applying a statistical analysis, a better understanding may be obtained for some of these research questions.

This is a complicated and time consuming process, probably not best undertaken by inexperienced researchers, but it may add greatly to the better understanding of science and aid treatment providers and policy makers. The process is not dissimilar to that described above in terms of selecting articles for systematic reviews but requires a more complicated analysis. There are also similarities with primary intervention trials, in which one focuses on how well an intervention works. However, in a meta-analysis, the researcher looks across studies to determine the magnitude of effects. It is worth following a systematic guideline such as PRISMA to establish a framework for the review (Moher et al., 2009).

The first step is to formulate the research question. Decide the keywords you will use to search for articles, the date from which you wish articles to be included, and the inclusion and exclusion criteria. Search the databases you have chosen for articles that meet your subject and eligibility criteria. It is also worth looking at reference lists from the articles you have selected to find other articles not so far identified.

Once the articles for inclusion have been identified they will need to be coded according to the variables chosen for the meta-analysis. Because these coding decisions are not always clear, two raters are often used to obtain some measure of reliability either by percent agreement or by a kappa coefficient. Enter the data extracted onto a database with relevant details of each study entered including, for example, type of intervention, follow-up periods, sample size, type of control group, and research design.

One of the problems in comparing a number of studies is that studies will report diverse outcomes according to the model they used. To determine effect sizes so that the meta-analysis is effective, a "common currency" of effects needs to be established in order for comparisons and aggregation to be made. Finney and Moyer (2010) suggest that the most common effect sizes used are *standardized mean difference, odds ratio,* and *correlation coefficient.* The standardized mean difference is "the difference between means on a continuous outcome variable for an intervention and a comparison condition, typically divided by the pooled standard deviation of the two groups." (Finney and Moyer, 2010, pp 321). By using standard deviations, one can measure by how many standard deviations, or what proportion of standard deviations, the intervention is performing better than the control group.

Another method of measuring effect size is by using the odds ratio. By calculating the probability of something changing divided by something not changing, a ratio may be obtained. An odds ratio of 1.00 would show that there was no difference between treatment and a control condition in which there were two possible outcomes.

The third method is the correlation coefficient, which can be used to express the relationship between a continuous intervention dimension (which is unusual in addiction studies) and the outcome (Finney & Moyer, 2010).

We have now established a method of calculating effect sizes, and, to find out whether there is indeed an effect and what that effect is, we must now aggregate them across the studies we have reviewed. This can be done with a *fixed-effects* or a *random-effects* approach. These two approaches deal with the study sampling errors, with the former assuming that the error in estimating the population effect size comes from random factors associated with subject-level sampling, whereas the latter assumes that there are study sampling errors in addition to subject-level sampling errors. A random-effects model is used more frequently because of a greater generalizability, although the fixed-effects model has a greater statistical power. Effects from larger sample sizes have less variance across studies and are therefore more precise. To test whether the overall effect size varies from zero, it is best to use specific statistical software designed to conduct meta-analyses (Finney & Moyer, 2010).

As with systematic reviews, a table should be presented detailing all the articles included in the study and describing all the relevant characteristics, including author, date of data collection, the main outcome findings, and methods of collecting the data. A forest plot that shows the range of findings for each study is also often included, detailing in comparison the range of effects in an intervention.

Issues with Meta-Analysis

There a number of issues that should be considered when conducting a meta-analysis. One may have to determine whether the effect sizes vary more than could be expected from subject-level sampling fluctuations in a fixed-effect model or, in a random-effect model, whether there are study-level random effects in addition to the subject-level sampling fluctuations. Are there additional factors that add variation in effect sizes explained by moderator variables? The moderator variables include different methods and participants across the studies and the interventions themselves. To test this, a homogeneity test can be used that will test for whether excess variation exists (Viechtbauer, 2007).

Another problem is publication bias. If the articles are selected carefully from peer-reviewed journals and conform to the criteria for inclusion, there is still the problem in that studies that show no positive or neutral results are often not published, either because the researchers do not submit for publication or because the papers are rejected for publication. Therefore, any articles that refute the research question may not be included in the databases searched and therefore the results may be skewed.

Selection of the articles needs to be done with great care. Only quantitative articles may be included—qualitative articles will not contribute a statistical outcome—and if the criteria are too strict, then the number of articles on which to base the analysis may be too small. On the other hand, if you the selection criteria are too wide, you may then include studies of poor quality that will affect the outcome of the meta-analysis. The other problem with selection of articles may be *agenda bias*, whereby the authors of the meta-analysis want to use the results to support a specific issue and may cherry pick the articles they include. Meta-analysis is complicated, and the analysis of the variance across articles is complex; therefore, it is always beneficial to get good statistical advice and to use an established statistical package for analyzing the data.

Conclusion and Final Advice

As previously mentioned, a good review article is hardly possible without a good literature search. The literature search has its own rules that generally apply to both original and review studies. A systematic review involves a literature search procedure guided by the principle of keeping an accurate and transparent record of the entire process! It is useful to create a summary Excel table where citations of studies will be recorded according to the selection criteria. It may seem like extra work at the beginning, but the author will come to appreciate this facility even before the first round of the peer-review process is over. Indeed, peer reviewers very easily notice any shortcomings we have tried to hide. It is therefore strongly recommended to draw up and enclose with the article a diagram in which you document the procedure for selecting the studies. This will help reviewers understand the approach and the results obtained, and, if any queries should arise, this evidence will make it easy to refute and explain any misgivings about the process or the results. For these purposes, it is also recommended to archive the documents in both printed and computerized versions; a physical file for hard copies and a separate electronic folder for computerized counterparts may be a useful option, with the latter providing the extra convenience of the "find" functionality.

To summarize, the ultimate goal when developing a review article is a systematic, straightforward, and transparent procedure. Both the reader and the editor must be clear about what the aims and methodology are, and all the results must be in line with the methods used. Although certain variations on standard procedures are possible, they always need to be explained and justified in discussion; otherwise you will most likely deal with them in the first round of the peer-review process. There are some specific approaches and tools for quality assessment of reviews (e.g., AMSTAR [Smith et al., 2011]; MOOSE [Stroup et al., 2000]) that can be relevant and very helpful in determining what is assessed and how to make the manuscript better.

Please visit the website of the International Society of Addiction Journal Editors (ISAJE) at www.isaje.net to access supplementary materials related to this chapter. Materials include additional reading, exercises, examples, PowerPoint presentations, videos, and e-learning lessons.

References

Atkins, D., Best, D., Briss, P. A., Eccles, M., Falck-Ytter, Y., Flottorp, S., . . ., Zaza, S., & GRADE Working Group. (2004). Grading quality of evidence and strength of recommendations. *BMJ, 328,* 1490.

Baumeister, R. F., & Leary, M. R. (1997). Writing narrative literature reviews. *Review of General Psychology, 1,* 311–320.

Bero, L. A., & Jadad, A. (1997). How consumers and policymakers can use systematic reviews for decision making. *Annals of Internal Medicine, 127,* 37–42.

Čablová, L., Pazderková, K., & Miovský, M. (2014). Parenting styles and alcohol use among children and adolescents: A systematic review. *Drugs: Education, Prevention, and Policy, 1,* 1–13.

Cornelson, L., McGowan, Y., Currie-Murphy, L., & Normand, C. (2014). Systematic review and meta-analysis of the economic impact of smoking bans in restaurants and bars. *Addiction, 109,* 720–728.

Culyer, A. J. (2014). *The dictionary of health economics* (3rd ed.). Cheltenham, England: Edward Elgar Publishing Limited.

Des Jarlais, D. C., Lyles, C., Crepaz, N., & TREND Group. (2004). Improving the reporting quality of nonrandomized evaluations of behavioral and public health interventions: The TREND statement. *American Journal of Public Health, 94,* 361–366.

Embleton, L., Mwangi, A., Vreeman, R., Ayuku, D., & Braitstein, P. (2013). The epidemiology of substance use among street children in resource-constrained settings: A systematic review and meta-analysis. *Addiction, 108,* 1722–1733.

Finney, J. W., & Moyer, A. (2010). Meta-analysis: Summarising findings on addiction intervention effects. In P. G. Miller, J. Strang, & P. M. Miller (Eds.), *Addiction research methods.* Oxford, England: Wiley-Blackwell.

Gabrhelík, R. (2013). Guidelines for Authors (updated January 2013). *Adiktologie.* Retrieved from http://www.adiktologie.cz/en/articles/detail/459/4021/GUIDELINES-FOR-AUTHORS.

Glantz, S. (2005, September). *Why you should not publish any tobacco funded research.* Presented at the annual ISAJE conference, San Francisco, CA.

Heidari S., Babor T. F., De Castro P., Tort S., & Curno M. (2016). Sex and Gender Equity in Research: Rationale for the SAGER guidelines and recommended use. *Research Integrity and Peer Review, 1:2.* DOI: https://doi.org/10.1186/s41073-016-0007-6

Higgins, J. P. T., & Green, S. (Eds.). (2008). *Cochrane handbook for systematic reviews of interventions* (Version 5.0.1 [updated September 2008]). The Cochrane Collaboration. Retrieved from www.cochrane-handbook.org.

Maisel, N. C., Blodgett, J. C., Wilbourne, P. L., Humphreys, K., & Finney, J. W. (2013). Meta-analysis of naltrexone and acamprosate for treating alcohol use disorders: When are these medications most helpful? *Addiction, 108,* 275–293.

Moher, M., Liberati, A., Tetzlaff, J., Altman, D. G., & The PRISMA Group. (2009). Preferred reporting items for systematic reviews and meta-analysis: The PRISMA statement. *Journal of Clinical Epidemiology, 62,* 1006–1012.

Pouget, E. R., Hagan, H., & Des Jarlais, D. (2012). Meta-analysis of hepatitis C seroconversion in relation to shared syringes and drug preparation equipment. *Addiction, 107,* 1057–1065.

Schulz, K. F., Altman, D. G., Moher, D., & CONSORT Group. (2010). CONSORT 2010 statement: Updated guidelines for reporting parallel group randomised trials. *PLoS Medicine, 7,* e1000251.

Smith, V., Devane, D., Begley, C. M., & Clarke, M. (2011). Methodology in conducting a systematic reviews of healthcare interventions. *BMC Medical Research Methodology, 11,* 15.

Society for the Study of Addiction (2015). Instructions for Authors. *Addiction.* Retrieved from http://www.addictionjournal.org/pages/authors (August 11, 2015).

Stroup, D. F., Berlin, J. A., Morton, S. C., Olkin, I., Williamson, G. D., Rennie, D., . . ., & Thacker, S. B. (2000). Meta-analysis of observational studies in epidemiology: A proposal for reporting. Meta-analysis of Observational Studies in Epidemiology (MOOSE) group. *JAMA, 283,* 2008–2012.

Viechtbauer, W. (2007). Hypothesis testing for population heterogeneity in meta-analysis. *British Journal of Mathematical and Statistical Psychology, 60,* 64–75.

von Elm, E., Altman, D. G., Egger, M., Pocock, S. J., Gøtzsche, P. C., Vandenbroucke, J. P., & STROBE Initiative. (2007). The Strengthening the Reporting of Observational Studies in Epidemiology (STROBE) statement: Guidelines for reporting observational studies. *PloS Medicine, 4,* e296.

CHAPTER 10

Use and Abuse of Citations

Robert West, Kerstin Stenius and Tom Kettunen

Introduction

Research output in the form of articles, books, and book chapters exists to be used by other researchers to inform subsequent research, influence policy decisions, and improve clinical practice. Authors need to consider how to make appropriate use of their previous publications and the work of others and how to ensure that their own work will be used appropriately.

A research article, book, policy document, or treatment manual should refer to other writing that is relevant to its message. Citation is the formal vehicle for doing this. It involves explicit reference to a piece of research output that, in principle, can be anything from an article in a journal to a website. Conventions applying to citation practice regulate the transmission of information, and citation conventions vary from one research field to another. The following text focuses primarily on what might be termed *cumulative research* in which the goal is to accumulate enduring knowledge and understanding.

There are two main types of citation (Box 10.1). In this chapter we use the term *referential citation* to refer to the situation in which a piece of research output (which may be empirical or conceptual) is being used for what it contributes to the field. The term *critical citation* is used when the citing piece points to what is considered a flaw in some research output.

The citation serves one or more essential functions: It enables the reader to examine the cited work to check the veracity of a statement that it is being used to support or the correctness of the use of a concept or interpretation of a process. When citing in support of a statement being made in one's own article, it also acknowledges the contribution made by the cited work. Both

How to cite this book chapter:
West, R, Stenius, K and Kettunen, T. 2017. Use and Abuse of Citations. In: Babor, T F, Stenius, K, Pates, R, Miovský, M, O'Reilly, J and Candon, P. (eds.) *Publishing Addiction Science: A Guide for the Perplexed*, Pp. 191–205. London: Ubiquity Press. DOI: https://doi.org/10.5334/bbd.j. License: CC-BY 4.0.

the *verification function* and the *acknowledgement function* are important. One may also use citations to document how a political debate, historical process, or specific concept has developed and has been defined. We can call this *the documentation function.*[1]

Regarding the verification function and the documentation function, the scope for intentional and unintentional distortion of research through unfounded assertions or misleading statements is enormous. In principle, every nonobvious factual claim should be supported in some way, either by citing direct evidence or by tracing a link through citations and/or inference to that evidence. Similarly every hypothesis, conceptual analysis, or statement of a theoretical position that is not advanced for the first time in a given article should trace a link to its source. Citations offer the readers an opportunity to determine for themselves whether the original source of a claim was justified and whether that claim is being accurately represented in the current piece.

Regarding the acknowledgement function, it is right and proper that researchers should receive credit for their work, and citation is the primary

Types of citations	Functions of citations
Referential citation: a work or part of a work is cited for what it contributes to the field	*Verification function:* the reader should be able to check the source for its accuracy and the accuracy with which it is reported
Critical citation: a work or part of a work is cited because it is believed to mislead the field	*Acknowledgement function:* the source is given credit for its contribution
	Documentation function: the source is identified as the object of the research in its own right

Box 10.1: Types and functions of citations.

means by which this is achieved. This is not merely a matter of etiquette: Employment and promotion of individual researchers are built on reputation, and citations play a crucial role in this. The institutions that employ researchers achieve kudos and in many cases funding on the basis of the reputations of their employees. Moreover, charities and government bodies that fund research must receive credit for the work they support. Their own income may well depend on it.

Deviations from Ideal Citation Practice

Citation practice often falls far short of the ideal (for a discussion, see Reyes, 2001). There are a number of sources one may use to find out about good practice in the use of citations in systematic reviews (e.g., Bannigan et al., 1997; Chalmers et al., 1993; Cook et al., 1995; Moher et al., 2009; Petticrew et al., 2008; Reeves et al., 2002; Stroup et al., 2000; Sutton et al., 1999; see also Chapter 9). Use of citations in less formal reviews, such as to introduce research reports, is subject to greater variability. The following paragraphs examine common deviations from ideal practice (see also Table 10.1).

Selective Citation through need for Conciseness

A legitimate reason to depart from ideal practice arises from the need for conciseness. Many times in a given field, a large number of studies may be cited in support of a given statement. In the absence of other constraints, the acknowledgement function might dictate that all relevant studies are cited. However, this would be impracticable. This raises the question of which article or articles to cite. There is a case for citing what we might call the *discovery article:* the first article to record the finding. However, this may be impossible to determine. Moreover, it may not represent the most robust support for the assertion in question. There is a case for citing a *review article* (an article that summarizes the research on a specific topic). This has the advantage of pointing the reader, at least indirectly, to a body of work rather than one or two studies that might be unrepresentative. The disadvantages are (a) the increased danger of misrepresentation because of hearsay and (b) failure to acknowledge the contribution of the original source.

A possible rule of thumb in determining policy relating to a specific finding is to aim to cite the discovery piece and no more than five other original sources that testify to the generality of the finding, unless there is an authoritative and noncontentious review that can be cited instead. When referring to a conceptual or theoretical exposition, the first major presentation of the current version should be used.

Selective Citation in Support of a Viewpoint

A common bias in reporting the literature is to select only (or primarily) studies that support a given hypothesis or idea (*viewpoint citation*). This is harder to avoid and to detect than one might imagine. If there were a well-defined body of literature that examined a particular hypothesis, and numerous high-quality studies conflicting with the hypothesis were ignored in a review, that would amount in the eyes of some to scientific misconduct. A reader who was not familiar with the area would be misled as much as if the author had fabricated data.

Less straightforward is the case where there are doubts about the methodological adequacy of conflicting studies. For example, studies that fail to detect the effect of an intervention may be small or involve inadequate implementation of the intervention. Unless one is explicitly attempting a comprehensive review in which there is the space to explore these issues, the citing author has to make a judgement about how far to go in ignoring weak studies. Given the realistic possibility that the citing author is not wholly disinterested in the matter, it is good practice to alert the reader to conflicting findings and make a brief comment about the weight that might be attached to these and why.

Even less straightforward is the case in which it is extremely difficult to determine what the corpus of findings on the topic is. This can happen for findings that typically do not form the main focus of articles. In the smoking literature, for example, it has been noted and is widely believed that depressed smokers are less likely to succeed in attempts to stop than are non-depressed smokers. There are certainly studies showing such an association (Covey, 1999; Glassman et al. 1990). However, often buried in reports of clinical trials and other studies are numerous reports of failures to find such an association, and indeed one meta-analysis has reported no association (Hitsman et al. 2003). There is no doubt that there are even more instances in which the association has been looked for and not found, with no subsequent report being made. At the very least, scientific prudence dictates that findings that are susceptible to this kind of phenomenon be cited with suitable caveats.

Selective Citation to Enhance Reputation

Self-citation or the citation of colleagues with a view to enhancing one's own or the colleague's reputation (*reputation citation*) is clearly unacceptable. It distorts science and the process of science and is personally damaging to individuals in less-powerful positions or to those who do not engage in that practice (see e.g. Fowler et al., 2007). One may debate how widespread this practice is, but there can be little doubt that self-serving bias runs at some level throughout the scientific literature (see e.g. Aksnes, 2003).

Self-citation can also apply to journals (articles in journals tending to cite articles from the same journal). This may arise for reasons other than reputation citation, some of which may be legitimate, but it can distort the literature. One study found significant differences in self-citation rates among journals of anesthesiology (Fassoulaki et al., 2000).

It may be thought that a bias of this kind would be easily detected and an appropriate correction could be applied. However, this is probably optimistic. It is not unreasonable that one's own name should feature prominently in a reference list given that one's research is presumably to some degree programmatic. A similar principle would hold true for one's close colleagues. It can be difficult therefore to tell when this bias is operating.

Selective Citation for Convenience

Using citations that are easy to find or that happen to have come to the attention of the author is not good practice but is probably very common. There may be many ways in which *convenience citation* can distort the literature. Insofar as more accessible articles may not represent the literature, use of convenience citations would create a biased impression. Searchable electronic databases, in principle, could mitigate the problem, but they can also lead to their own kind of distortion. It would be expected that they would favor English-language articles in journals indexed by the main databases. One would also expect more recent articles to gain preference because of the way that electronic databases sort the results of searches. Convenience citation would also be expected to favor the more popular journals. One might argue that this is no bad thing because it would be the better articles that would in general find their way into these journals. However, this is not necessarily so.

Selective Citation by Country of Origin

It goes without saying that a tendency to cite articles simply because they are from one's own country of origin is not good practice. Many researchers are under the impression that this occurs, however. Naturally, the greatest suspicion falls on the U.S. as the main producer of research output, and many non-U.S. researchers can probably recount cases where a U.S. author has cited predominantly or exclusively U.S. references, even when more appropriate ones from other countries exist. In fact, this bias has been found among both U.K. and U.S. researchers publishing in major medical journals (Campbell, 1990; Grange, 1999). Another study found that North American journals cite North American journals to a greater extent than did journals from other regions (Fassoulaki et al., 2000), but the opposite has also been found (Lancho Barrantes et al., 2012; Pasterkamp et al., 2007).

Citing Inaccessible Sources

It is quite common for authors to cite conference papers or their abstracts, submitted articles, in-house papers, or unpublished reports (the so-called *gray literature*). The problem with this kind of citation is that it does not fulfill the verification function of citation. Therefore, it is generally to be discouraged. There may be cases where it is the only option and important in fulfilling the acknowledgement or documentation role, but if this is not obvious, the use should be justified. If that citation is more than a few years old, the use becomes increasingly problematic. It is often reasonable to presume that if it is an article or abstract and the finding was robust, it would have found its way into the peer-reviewed literature.

It is becoming common to cite websites. This is reasonable but will pose increasing problems over time as websites move or become inaccessible. In general, for any statement intended to have lasting significance, this practice is best avoided until a system is devised for ensuring the longevity of web-based scientific literature. In policy analyses or descriptions of historical processes, though, references to sources such as websites and government documents may be a key part of the research process.

Citing Unevaluated Sources

When a citation is used to support a substantive statement, the implication is that the cited reference reports evidence in support of that statement. Inadequate though it is, peer review is the primary gatekeeper for this kind of report.

Convenience citation	selects citation material that is easy to find
Discovery article	the article that first puts forward a new concept
Gray literature	unpublished matter, such as conference presentations, submitted articles, and in-house papers and reports
Publication lag	the time between an article's acceptance by a journal and its publication
Reputation citation	cites a work or part of a work with a view to enhancing one's own reputation or that of a colleague
Review article	an article that summarizes the research on a specific topic
Viewpoint citation	cites a work or part of a work because it supports a given hypothesis or idea

Table 10.1: Terminology related to deviations from ideal citation practice.

However, it is commonplace for statements of fact to be supported by citations to book chapters, letters, conference presentations, abstracts, opinion pieces, and other material that has not been peer reviewed. Although in principle readers can track down the source and make their own evaluations, this is often impracticable. The only thing that comes close to a safeguard is that cited work has been through a peer-review process. Within the social sciences, though, even non–peer-reviewed books still remain a main source for new analytical concepts. In some cases, however, the review process for books is as rigorous as the peer-review process for journal articles.

Citing Without Reading

There is a great temptation to cite a work or part of a work on the strength of a report of what it says without going to the original source. Thus, if an article or a book chapter that we have access to makes a statement that is relevant to our work and cites another article in support of it, it is tempting to repeat the assertion and the citation without reading the original source material. This is clearly unacceptable because of the risk of misrepresentation. Equally, having identified an abstract of an article using an electronic database, an author may be tempted to cite the article without going to the full text. This is risky practice because one has not taken the opportunity to evaluate the research being cited by reading the methods and analyses used.

As a general principle, authors should not make reference to research output without having read and evaluated that output directly.

Overuse of Citations

Much of the earlier discussion concerned selective use of citations. Quite a common problem is the reverse: providing a long list of citations to support a single statement when fewer would be sufficient. If it is important that the work of the authors of all the various works be acknowledged or if the intention is to provide a comprehensive review, then a long list of citations is appropriate. Otherwise it can make an article unwieldy, and the rule of thumb of selective citation described earlier could be adopted.

Coercive Citation

During the peer-review process, editors can be tempted to help increase the standing of their journal by encouraging authors to add more citations to the journal, without specifying relevant articles or indicating where more references are needed. This practice is sometimes referred to as coercive self-citation.

Coercive citation is inappropriate as it undermines the integrity of academic publishing and it should be resisted by both authors and editors. Unfortunately, the practice is widespread and strategic. One study found that around 20% of academics in business disciplines, economics, sociology and psychology have experienced coercive citation practices (Wilhite & Fong, 2012). The study also found that editors soliciting superfluous citations are more likely to target manuscripts written by few authors, preferably by scholars of lower academic rank.

Getting Cited

All the above should suggest that the process of citation is subject to considerable bias, and, although there is a duty on researchers to minimize this, it is unlikely that bias will ever be eliminated. This being said, if one is writing an article that one believes is important, it would seem reasonable to try to ensure that it is drawn to the attention of its intended audience, and that means being cited. The choice of journal is obviously of relevance (see Chapter 3). And it may not be the most prestigious journal that offers the best chance but, rather, the best-quality *specialist* journal. The most prestigious journals tend to be generalist and, as such, may not be routinely read by many potential users of the research. Whatever outlet one uses for one's research, it can often be a good idea to take other steps to publicize the findings. Some researchers email or send copies of their articles to colleagues. One might post reference to them on listserves or publicize them on social media. With increasing use of Open Access, full text can often be made available on demand. Conference presentations and websites are also potentially useful sources of publicity.

Citation Indexes

We mentioned earlier that citations are often used as a marker of quality. There is a presumption that the more often an article is cited, in some sense the better it is. This extends to journals, for which the single most widely used measure of quality is the impact factor. The impact factor for a journal in a given year is calculated as the average number of citations in that year to articles in the preceding two years. Thus, if a journal published 50 articles in 2013 and 2014 and there were 100 citations to these articles in 2015, the journal's impact factor for 2015 would be 2.0. Citations of authors to their own work are included. Therefore, clearly the more prolific an author is and the more that authors cite their own work, the more useful those authors are to a journal wanting to maximize its impact factor.

Researchers are often judged by the citation counts of their articles and by the impact factors of the journals in which they publish. Funding decisions in many institutions are based in part on members of those institutions publishing in "high-impact" journals. Unfortunately there are many problems associated with using citation counts as a marker of quality and even more with using the impact factor (Hecht et al., 1998; Jones, 1999; Opthof, 1997; Seglen, 1997; Semenzato & Agostini, 2000). Some researchers have suggested that it may be possible to use citation counts and impact factor with appropriate caveats and corrections (Braun, 2003; Fassoulaki et al., 2002; Rostami-Hodjegan & Tucker, 2001), whereas others have argued that such use should be abandoned (Bloch & Walter, 2001; Ojasoo et al., 2002; Walter et al., 2003).

Regarding citation counts, the various biases in the use of citations discussed earlier should give an indication of the problem with using them as a marker of quality. In addition, it should be recalled that *critical citation* is quite commonplace. Therefore, an article might be cited precisely because it is weak or misleading. One article examined the association between peer ratings of quality and the numbers of citations between 1997 and 2000 to articles appearing in the journal *Addiction* in 1997 (West & McIlwaine, 2002). Although two independent reviewers agreed moderately in their ratings of the articles, the correlation between these ratings and the number of citations was almost zero. One factor that was correlated with citation count was the region of origin of the first author of the article: Articles from English speaking countries received more citations than those from continental Europe, which received more than those from the rest of the world. A larger analysis of citations to articles in emergency medicine revealed that the citation count of articles was predicted to some extent by the impact factor of the journal in which they appeared and to a more limited extent by quality of the articles (Callaham et al., 2002). A further study of citations to articles reporting randomized trials in hepatobiliary disease found a significant association with a positive outcome but no association with adjudged quality (Kjaergard & Gluud, 2002).

Apart from the biases already discussed, the fact that only a small proportion of predominantly U.S. journals are indexed in Web of Science would lead to a bias, particularly against non–English-speaking countries. One study reported that exclusion of core journals in emergency medicine had led citation counts in the field to remain low despite considerable expansion of the field (Gallagher & Barnaby, 1998). Another noted that the way to improve the impact factors of journals in dermatology was to increase the number of them indexed by Web of Science (Jemec, 2001). Another bias arises from researchers in some fields, such as biosciences, simply using more citations than researchers in other fields. This will disadvantage authors in low-citing fields, typically the social sciences. Another bias pertains to texts such as editorials, letters and book reviews not being included in the denominator of citable documents. When they are cited, this can distort the impact factors of small-volume journals. For example, journals publishing mostly "noncitable" book reviews can have surprisingly high impact factors (Jasco, 2009). There are a range of other factors that make citation counts potentially misleading as a marker of quality (Box 10.2).

- Articles are sometimes cited as criticism.
- Articles describing important original studies are often neglected in favor of reviews.
- There is a bias toward citing articles from one's own country or research group or articles that are easily accessible.
- Some fields of study generate more citations than others irrespective of how important the articles are, for example, fields with high levels of activity and mature fields.
- The importance and quality of a work or part of a work may relate to its policy or clinical implications rather than its use by other researchers.
- Other researchers may fail to grasp the importance of a work or part of a work.
- The citation indexes are biased toward U.S. and other English-language journals.

Box 10.2: Why citation counts are often misleading as a marker of quality.

Addressing some of these criticisms, the *Journal Citation Reports* introduced a number of augmentations in 2007, such as the five-year journal impact factor and Eigenfactor. The five-year impact factor score is similar in nature to the traditional two-year impact factor but deals with a five-year citation window, which can be more useful for research areas in which articles are published and cited at a slower pace. Eigenfactor is based on the structure of the scholarly citation network (based on incoming citations, weighting citations from highly ranked journals more heavily) and gives a numerical indicator of the overall contribution of a journal to the literature. Eigenfactor is influenced by the size of the journal (the more articles, the higher the score). Other journal-level metrics include an Article Influence Score and the SCImago Journal Rank.

The San Francisco Declaration on Research Assessment (DORA), published in May 2013, arose from concerns within the scientific community regarding how research output is evaluated, and how scientific literature is cited. It is signed by a broad coalition of researchers, editors, publishers, research societies, universities and funding agencies. The declaration includes a set of individual recommendations for parties involved in research assessment, as well as one general recommendation:

> Do not use journal-based metrics, such as Journal Impact Factors, as a surrogate measure of the quality of individual research articles, to assess an individual scientist's contributions, or in hiring, promotion, or funding decisions. (DORA, 2013)

DORA recommends that publishers use a variety of journal-based metrics to provide a more nuanced picture of how journals are performing. Another recommendation is to encourage a shift toward assessment based on the scientific content of an article, rather than the publication metrics of the journal (DORA, 2013). One way of promoting this shift is to provide article-level metrics, such as downloads, citation counts, and altmetrics. Altmetrics measure science dissemination more broadly than traditional research impact, looking at how articles are discussed in the news and social media, saved and bookmarked in reference management tools, and recommended in postpublication peer-review systems (such as F1000 rating) (Cheung, 2013; Leydesdorff, 2009). However, the usefulness of altmetrics is limited from a bibliometric perspective because they are difficult to standardize and some of the measures can be gamed (Priem, 2013).

Because the journal impact factor is badly suited for assessing the individual quality and quantity of scientific output by a researcher, a number of author-based bibliometric indicators have been developed. These include indices such as the h-index, hI-index, hm-index, i10-index, n-index, several m-indices, A-index, R-index, and the g-index. The multitude of indices reflects the difficulty in developing quantitative measures for assessing the quality of research (Fersht, 2009; Jasco, 2008; West et al., 2010a, 2010b).

Conclusions

Citations are the primary formal means by which scientific findings are communicated. In terms of full transmission of information, ideally citation practice would involve comprehensive and objective use of the whole corpus of relevant published literature. Clearly this is impracticable. However, it should still be possible to approximate this ideal by adopting a few guidelines. These recognize that citation serves the dual function of enabling verification of statements and acknowledging contributions.

In the case of formal reviews, the principles are well documented: The sources searched and the search rules should be clearly specified, as should the inclusion and exclusion criteria for articles. The sources should go beyond Web of Science databases and include searching reference lists of articles in the search domain. Regarding informal reviews, such as are used to introduce research reports, the following principles can be applied:

1. Support all nonobvious, substantive claims by citation or direct evidence.
2. Do not support statements of the obvious by citation.
3. If there is an authoritative review on a well-supported statement, this may be used in place of original articles.
4. When citing original articles, cite the discovery article together with a small number of other articles that illustrate the generality of the phenomenon.

5. Resist the propensity to do the following:
    a. prefer citations from your own country of origin unless the finding in question is country specific;
    b. prefer citations from yourself and colleagues;
    c. limit citations to those that support a contention, when in fact there are others that conflict with it;
    d. cite output that is readily retrievable if there are more appropriate references; and
    e. provide an unnecessarily large number of citations for a single statement.
6. Avoid citing inaccessible sources wherever possible.
7. When using citations in support of substantive statements, either use references that have been through some kind of peer-review process or provide an appropriate caveat.

Citation counts are widely used as an index of quality. Given that few if any researchers are able to follow all the above principles, together with the many other factors that influence the number of times a piece is cited, citation counts are a highly problematic index of quality. Journal impact factors are even more problematic. Authors should be aware of this and not be beguiled by their apparent objectivity. Ultimately, there appears at present to be no substitute for peer evaluation of research output, however flawed and subjective this might be.

Please visit the website of the International Society of Addiction Journal Editors (ISAJE) at www.isaje.net to access supplementary materials related to this chapter. Materials include additional reading, exercises, examples, PowerPoint presentations, videos, and e-learning lessons.

Note

[1] We are grateful to Klaus Mäkelä for this insight.

References

Aksnes, D. (2003). A macro study of self-citation. *Scientometrics, 56,* 235–246.

Bannigan, K., Droogan, J., & Entwistle, V. (1997). Systematic reviews: What do they involve? *Nursing Times, 93,* 52–53.

Bloch, S., & Walter, G. (2001). The Impact Factor: Time for change. *Australian and New Zealand Journal of Psychiatry, 35,* 563–568.

Braun, T. (2003). The reliability of total citation rankings. *Journal of Chemical Information and Modeling, 43,* 45–46.

Callaham, M., R. L. Wears, & Weber, E. (2002). Journal prestige, publication bias, and other characteristics associated with citation of published studies in peer-reviewed journals. *JAMA 287,* 2847–2850.

Campbell, F. M. (1990). National bias: A comparison of citation practices by health professionals. *Bulletin of the Medical Library Association, 78,* 376–382.

Chalmers, I., Enkin, M., & Keirse, M. J. (1993). Preparing and updating systematic reviews of randomized controlled trials of health care. *Milbank Quarterly, 71,* 411–437.

Cheung, M. K. (2013). Altmetrics: Too soon for use in assessment. *Nature, 494,* 176. DOI: https://doi.org/10.1038/494176d

Cook, D. J., Sackett, D. L., & Spitzer, W. O. (1995). Methodologic guidelines for systematic reviews of randomized control trials in health care from the Potsdam Consultation on Meta-Analysis. *Journal of Clinical Epidemiology, 48,* 167–171.

Covey, L. S. (1999). Tobacco cessation among patients with depression. *Primary Care, 26,* 691–706.

DORA. (2013). *San Francisco Declaration on Research Assessment: Putting science into the assessment of research.* Retrieved from http://www.ascb.org/dora-old/files/SFDeclarationFINAL.pdf.

Fassoulaki, A., Papilas, K., Paraskeva, A., & Patris, K. (2002). Impact factor bias and proposed adjustments for its determination. *Acta Anaesthesiologica Scandinavica, 46,* 902–5.

Fassoulaki, A., Paraskeva, A., Papilas, K., & Karabinis, G. (2000). Self-citations in six anaesthesia journals and their significance in determining the impact factor. *British Journal of Anaesthesia, 84,* 266–269.

Fersht, A. (2009). The most influential journals: Impact Factor and Eigenfactor. *PNAS, 69,* 6883–6884.

Fowler, J. H., & Aksnes, D. W. (2007). Does self-citation pay? *Scientometrics, 72,* 427–437.

Gallagher, E. J., & Barnaby, D. P. (1998). Evidence of methodologic bias in the derivation of the Science Citation Index impact factor [see comments]. *Annals of Emergency Medicine, 31,* 83–86.

Glassman, A. H., Helzer, J. E., Covey, L. S., Cottler, L. B., Stetner, F., Tipp, J. E., & Johnson, J. (1990). Smoking, smoking cessation, and major depression. *JAMA, 264,* 1546–1549.

Grange, R. I. (1999). National bias in citations in urology journals: Parochialism or availability? [see comments]. *BJU International, 84,* 601–603.

Hecht, F., Hecht, B. K., & Sandberg, A. A. (1998). The journal "impact factor": A misnamed, misleading, misused measure. *Cancer Genetics and Cytogenetics, 104,* 77–81.

Hitsman, B., Borrelli, B., McChargue, D. E., Spring, B., & Niaura, R. (2003). History of depression and smoking cessation outcome: A meta-analysis. *Journal of Consulting and Clinical Psychology, 71,* 657–663.

Jasco, P. (2008). The Pros and Cons of Computing the H-index Using Web of Science. *Online Information Review, 32,* 673–688.

Jasco, P. (2009). Five-year impact factor data in the Journal Citation Reports. *Online Information Review 33,* 603–614.

Jemec, G. B. (2001). Impact factors of dermatological journals for 1991–2000. *BMC Dermatology, 1,* 7.

Jones, A. W. (1999). The impact of Alcohol and Alcoholism among substance abuse journals. *Alcohol and Alcoholism, 34,* 25–34.

Kjaergard, L. L., & Gluud, C. (2002). Citation bias of hepato-biliary randomized clinical trials. *Journal of Clinical Epidemiology, 55,* 407–410.

Lancho Barrantes, B. S., Bote, G., Vicente, P., Rodríguez, Z. C., & de Moya Anegón, F. (2012). Citation flows in the zones of influence of scientific collaborations. *Journal of the American Society for Information Science and Technology, 63,* 481–489.

Leydesdorff, L. (2009). How are new citation-based journal indicators adding to the bibliometric toolbox? *Journal of the American Society for Information Science and Technology, 60,* 1327–1336.

Moher, D., Liberati, A., Tetzlaff, J, Altman, D. G., & The PRISMA Group. (2009). Preferred Reporting Items for Systematic Reviews and Meta-Analyses: The PRISMA Statement. *PLoS Med, 6,* e1000097.

Ojasoo, T., Maisonneuve, H., & Matillon, Y. (2002). [The impact factor of medical journals, a bibliometric indicator to be handled with care] [article in French]. *Presse Médicale, 31,* 775–781.

Opthof, T. (1997). Sense and nonsense about the impact factor. *Cardiovascular Research, 33,* 1–7.

Pasterkamp, G., Rotmans, J. I., de Klein, D. V. P., and Borst, C. (2007) Citation frequency: A biased measure of research impact significantly influenced by the geographical origin of research articles. *Scientometrics, 70,* 153–165.

Petticrew, M., & Roberts, H. (2008). *Systematic reviews in the social sciences: A practical guide.* John Wiley & Sons.

Priem, J. (2013). Beyond the paper. *Nature, 495,* 437–440.

Reeves, S., Koppel, I., Barr, H., Freeth, D., & Hammick, M. (2002). Twelve tips for undertaking a systematic review. *Medical Teacher, 24,* 358–363.

Reyes, H. (2001). [The references in articles published in biomedical journals] [article in Spanish]. *Revista Médica de Chile, 129,* 343–345.

Rostami-Hodjegan, A., & Tucker, G. T. (2001). Journal impact factors: A 'bioequivalence' issue? *British Journal of Clinical Pharmacology, 51,* 111–117.

Seglen, P. O. (1997). Why the impact factor of journals should not be used for evaluating research. *BMJ 314,* 498–502.

Semenzato, G., & Agostini, C. (2000). The impact factor: Deeds and misdeeds [Editorial]. *Sarcoidosis, Vasculitis, and Diffuse Lung Diseases, 17,* 22–26.

Stroup, D. F., Berlin, J. A., Morton, S. C., Olkin, I., Williamson, G. D., Rennie, D., . . . , & Thacker, S. B. (2000). Meta-analysis of observational studies in epidemiology: a proposal for reporting. *Jama, 283,* 2008–2012.

Sutton, A. J., Jones, D. R., Abrams, K. R., Sheldon, T. A., & Song, F. (1999). Systematic reviews and meta-analysis: A structured review of the methodological literature. *Journal of Health Services Research & Policy, 4,* 49–55.

Walter, G., Bloch, S., Hunt, G., & Fisher, K. (2003). Counting on citations: A flawed way to measure quality. *Medical Journal of Australia, 178,* 280–281.

West, J. D., Bergstrom, T. C., & Bergstrom, C. T. (2010a). Big Macs and Eigenfactor scores: Don't let correlation coefficients fool you. *Journal of the American Society for Information Science & Technology, 61,* 1800–1807.

West, J. D., Bergstrom, T. C., & Bergstrom, C. T. (2010b). The Eigenfactor metrics: A network approach to assessing scholarly journals. *College & Research Libraries, 71,* 236–244.

West, R., & McIlwaine, A. (2002). What do citation counts count for in the field of addiction? An empirical evaluation of citation counts and their link with peer ratings of quality. *Addiction, 97,* 501–504.

Wilhite, A. W., & Fong, E. A. (2012). Coercive citation in academic publishing. *Science, 335,* 542–543.

CHAPTER 11

Coin of the Realm: Practical Procedures for Determining Authorship

Thomas F. Babor, Dominique Morisano and
Jonathan Noel

Like a coin, authorship has two sides: credit and responsibility. One receives professional credit from his/her publications and takes responsibility for their contents.

Biagioli et al. (1999, p. 2)

Introduction

Authorship credit is conceivably the most important and least understood area of professional life for members of the scientific community. Because promotion, prestige, and productivity are judged largely by publication activity, authorship credit has become the "coin of the realm" in the scientific marketplace (Wilcox, 1998). The two sides of this coin are credit and accountability. The assignment of individual credit to a publication implies certain ethical and scientific imperatives that are of tremendous importance to the scientific enterprise (Rennie & Flanagin, 1994). These imperatives include the certification of public responsibility for the truth of a publication and the equitable assignment of credit to those who have contributed in a substantive way to its contents.

The need for clear and consistent procedures for the determination of authorship credits comes from two considerations. First, many journals are now demanding that articles be prepared in a way that is consistent with the

How to cite this book chapter:
Babor, T F, Morisano, D and Noel, J. 2017. Coin of the Realm: Practical Procedures for Determining Authorship. In: Babor, T F, Stenius, K, Pates, R, Miovský, M, O'Reilly, J and Candon, P. (eds.) *Publishing Addiction Science: A Guide for the Perplexed*, Pp. 207–227. London: Ubiquity Press. DOI: https://doi.org/10.5334/bbd.k. License: CC-BY 4.0.

principles of responsible authorship. Second, a clear consensus about the conditions governing authorship decisions would make the work of individual authors much easier.

Numerous professional organizations (e.g., American Psychological Association, 2010), expert panels (International Committee of Medical Journal Editors, 1991, 2003, 2013), and individual commentators (Rennie et al., 1997) have developed policies and procedures dealing with individual, group, and corporate authorship. In this chapter, we review some of these guidelines from both the practical and ethical perspectives, in an attempt to develop workable procedures that authors can follow during the course of preparing and publishing a scientific article. In addition, we consider authorship problems that sometimes arise in the course of a publication cycle.

Authorship problems seem to be occurring with increasing frequency (Wilcox, 1998). Of 785 authors abstracted from 121 articles published in *The Lancet*, 44% did not meet the most lenient guidelines for authorship and 60% of the most common contributor's activities overlapped with those on acknowledgement lists (Yank & Rennie, 1999). Among Cochrane Reviews, 39% of publications had evidence of honorary authors, and 9% had evidence of ghost authors (Mowatt et al., 2002). An analysis of ghost and honorary authorship among articles published within six leading medical journals (e.g., *JAMA*, *The Lancet*) in 2008 found that, although there appeared to have been a decrease in ghost authorship, specifically over the previous decade, the prevalence of articles with honorary and/or ghost authorship was still 21% (Wislar et al., 2011). Within 10 top peer-reviewed nursing journals, an even greater number (42%) of articles published in a two-year period contained honorary authors, and 27.6% had ghost authors (Kennedy et al., 2014). Undeserved authorships; failure to credit collaborating authors; relaxed policies for students, research assistants, and postdoctoral fellows; and an excessive number of co-authors are all serious problems. Some journals have gone so far as to limit the number of authors who can be listed on a submission (e.g., *The American Journal of Public Health* lists the cap as six).

The pervasiveness of ethical issues in authorship is suggested by the extent to which scientific readers can be amused by the satirical humor epitomized in the "Ode to multi-authorship" quoted in Box 11.1.

All cases complete, the study was over
the data were entered, lost once, and recovered.
Results were greeted with considerable glee
p value (two-tailed) equalling 0·0493.
The severity of illness, oh what a discovery,
was inversely proportional to the chance of recovery.
When the paper's first draft had only begun

the wannabe authors lined up one by one.
To jockey for their eternal positions
(for who would be first, second, and third)
and whom "et aled" in all further citations.
Each centre had seniors, each senior ten bees,
the bees had technicians and nurses to please.
The list it grew longer and longer each day,
as new authors appeared to enter the fray.
Each fought with such fury to stake his or her place
being just a "participant" would be a disgrace.
For the appendix is piled with hundreds of others
and seen by no one but spouses and mothers.
If to "publish or perish" is how academics are bred
then to miss the masthead is near to be dead.
As the number of authors continued to grow
they outnumbered the patients by two to one or so.
While PIs faxed memos to company headquarters
the bees and the nurses took care of the orders.
They'd signed up the patients, and followed them weekly
heard their complaints, and kept casebooks so neatly.
There were seniors from centres that enrolled two or three
who threatened "foul play" if not on the marquee.
But the juniors and helpers who worked into the night
were simply "acknowledged" or left off outright.
"Calm down" cried the seniors to the quivering drones
there's place for you all on the RPU clones.
When the paper was finished and sent for review
six authors didn't know that the study was through.
Oh the work was so hard, and the fights oh so bitter
for the glory of publishing and grabbing the glitter.
Imagine the wars when in six months or better
The Editor's response, "please make it a letter".

RPU=repeating publishable unit; PI=principal investigator

Reprinted from *The Lancet*, 348, HW Horowitz, NH Fiebach, SM Levitz, J Seibel, EH Smail, EE Telzak, GP Wormser, RB Nadelman, M Montecalvo, J Nowakowski, and J Raffall, "Ode to multiauthorship: A multicentre, prospective random poem, 1746, 1996, with permission from Elsevier.

Box 11.1: Ode to multiauthorship: A multicentre, prospective random poem.

Conventions in Assigning Order of Authorship

One of the difficulties in determining the criteria for authorship comes from the different traditions and practices that have been used to distribute authorship credits. Table 11.1 provides definitions of common authorship terms and ethical issues, some of which are also discussed in Chapters 5 and 14.

Authors are sometimes listed in alphabetical order to avoid controversy about the relative contributions of different authors, especially when the contributions have been fairly equal. A related convention is to list authors in reverse alphabetical order, presumably to avoid the preference given to persons whose surname begins with a letter that appears early in the alphabet. Another convention is to list the laboratory director, center director, or other prominent person last. As noted in other parts of this chapter, this convention is not ethical unless that individual has made a substantial contribution to the publication and is not being listed merely to flatter the powerful or to add to the prestige value of the authorship list. This convention can also cause confusion when comparing contributions across fields. For instance, a last author might be presumed by some professionals to have contributed the least to an article and by others to have backed the entire project.

The convention followed most frequently in the addiction field is to list authors according to their relative contributions, with the first author assumed to be responsible for writing the article, corresponding with the journal editor, and making the most substantive contributions. The first author in such a system is sometimes called the corresponding author. In some cases a senior researcher who is not the first author is designated as corresponding author to facilitate the progress of the manuscript through the peer-review process. This practice is not acceptable if the main purpose is to take advantage of this researcher's influence and prestige, rather than to reflect actual contributions to the manuscript.

Although the convention is assumed to be based on the equitable distribution of authorship credits, the relative ordering of authors is often dependent on the first author's subjective judgment of others' contributions. In the absence of conducting an inventory of contributions, effort, and follow through, it is likely that some contributors will receive more credit than they deserve, and others less, solely because of the ambiguity and arbitrariness of the process.

With the growth of multicenter clinical trials and other "big-science" collaborative projects, corporate authorship has also increased. This convention lists a team name as the author, with a footnote or acknowledgement describing the contributors and the corresponding author. One reason for this convention is to make citations and referencing more efficient in cases where there are large numbers of contributors. Corporate authorship might also help to avoid the difficulties associated with determining who contributed what to a

Coercion authorship
is a gift authorship that is demanded rather than voluntarily awarded.
Contributorship
consists of listing the contributions of each person involved in the project, avoiding the attribution of authorship entirely.
Corporate authorship
lists the name of a project as author, along with a separate acknowledgement describing the contributors and the corresponding author (as an alternative to long author lists in multi-authored reports).
Corresponding author
is often the first author listed on an article, assumed to be the main researcher and writer of the article and the person responsible for corresponding with the journal editor. In some cases the corresponding author is not listed first when the writing and corresponding functions are divided.
Ghost authorship
is the failure to include as co-author of a work a person who satisfies the criteria for authorship (e.g., a science writer employed by a drug company).
Gift authorship
awards authorship credit because of a person's power or prestige rather than for substantial contribution to the work.
Group authorship
See "Corporate authorship."
Guarantor
is the person who takes responsibility for the contents and integrity of the work as a whole.
Honorary authorship
See "gift authorship."
Mutual-admiration authorship
occurs when two or more researchers agree to list each others' names on their own articles despite the others' minimal involvement.
Mutual-support authorship
See "mutual-admiration authorship."
Pressured authorship
See "Coercion authorship."
Surprise authorship
occurs when a researcher finds out after publication that his or her name appears on an article.

Table 11.1: Forms of authorship.

multi-authored article, and how much credit each author should receive. Some journals require contributors to formally name at least one person in the masthead, however (e.g., Alexander Bloggins for the Addiction Research Group).

When participating in multidisciplinary or international collaborations, differing authorship conventions must also be taken into account, as authorship criteria and authorship order can have significantly different connotations in different disciplines (Anderson et al., 2011). As noted previously, in some disciplines, the last author may indicate the person who contributed the least effort, whereas in others it might signify the senior author or laboratory head.

Because of the problems associated with determining who merits authorship credit, one editor (Smith, 1997) proposed the concept of contributorship. This involves listing the contributions of each person involved in the project, and avoiding the attribution of authorship entirely. Although this convention has not been adopted by any journal in its pure form (probably because the problems it causes with referencing), some journals, such as the *American Journal of Public Health,* request that all authors list their contributions when an article is submitted and publish a summary as a footnote or acknowledgement (*American Journal of Public Health* Instructions for Authors at ajph.aphapublications. org/page/authors.html).

In summary, a variety of conventions have been used to arrange the names of individual contributors in multi-authored articles. Some conventions are used more than others, with the main-author-first convention used most often. Other conventions (e.g., group authorship) tend to be used in special situations as the case demands. The purpose of these conventions, particularly more recent variants, is to assure that proper credit is assigned so that individual responsibility for a publication can be inferred by the reader.

Publication Policies and Publication Misconduct

Over the past 25 years, journal editors, research administrators, and funding agencies have devoted increasing attention to the ethical and practical issues of scientific authorship. Concern about authorship has been heightened by a number of events and situations that have at times compromised, and at other times embarrassed, the entire scientific enterprise (Box 11.2 and Box 11.3).

The most flagrant examples involve scientific misconduct. In a number of well-publicized cases (Broad & Wade, 1982), investigators have published scientific articles that have been retracted because the data were fraudulent or the contents plagiarized from other sources. What is remarkable about many of these cases is that, in addition to the person directly involved in scientific misconduct (e.g., John Darsee, who was the lead author on numerous fraudulent articles; Relman, 1983), there have typically been a number of co-authors who apparently had no idea that the senior author was fabricating data or copying others' ideas. This implies that in some cases co-authors are not in a position

In 1983 and 1986, the International Advertising Association published pro-tobacco reports on tobacco advertising bans and smoking prevalence, with the work credited to Dr. J. J. Boddewyn of Baruch College, The City University of New York (Davis, 2008). Supporters of the tobacco industry enthusiastically touted the reports, but a later review of publicly available tobacco industry documents paints a different picture. Not only were the reports ghost written by Paul Bingham, then an employee of British American Tobacco, but Dr. Boddewyn was also a paid consultant of the tobacco industry, and the research itself was highly flawed. The relationship between Mr. Bingham, British American Tobacco, Dr. Boddewyn, and the International Advertising Association was not disclosed in the reports or in later hearings in front of the U.S. Congress.

Box 11.2: Ghost authorship by the tobacco industry.

In the journal *Science,* Dr. Gerald P. Schatten was listed as a co-corresponding author and senior author of an article on a high-efficiency method for generating stem cells (University of Pittsburg, 2006). Soon after publication, allegations of scientific misconduct, including scientific fraud and data manipulation, on the part of Dr. Woo Suk Hwang, the lead author, were made public and ultimately the article was retracted. Although Dr. Schatten was absolved from participating in any misconduct, he was culpable for research misbehavior. Dr. Schatten wrote much of the article but did not verify the authenticity of the raw data and did not critically examine discrepancies that occurred through the drafting process. An investigative board ruled that Dr. Schatten assumed senior authorship to enhance his scientific reputation, improve opportunities for funding, and obtain financial benefit. The board also ruled that only a few of the 25 authors listed had actually read the article before submission.

Box 11.3: Gift authorship of a retracted article.

to take public responsibility for the contents of a scientific report, which is now considered to be one of the main criteria for authorship credit. In reality, there is a significant amount of basic trust across a number of domains that authors must invest in each other when collaborating on a publication, no matter what

their authorship position. Basic domains include honesty regarding the originality of the origins of any writing contributions, open disclosure about any
conflicts of interest (e.g., financial investment in a business that is dependent on
research outcomes, personal relationships with potential reviewers), and being
thorough and ethical in any data entry and statistical analyses. With the rise in
publication pressures that authors face at their own institutions and funding
agencies (e.g., having to produce a minimum number of publications per year
to stay employed), it is important to address a range of ethical concerns in publishing. In its updated statement on authorship standards for submissions to
biomedical journals, the International Committee of Medical Journal Editors
(2013) indicates that authors should be able to identify the specific parts of an
article that the other co-authors have been responsible for.

Extreme cases aside, the abuse of scientific authorship has been suspected in
an even greater number of cases where the scientific misconduct is much more
subtle. Examples include the addition of authors to curry favor, conferring co-
authorship by virtue of status or power, rewarding students or junior faculty
with co-authorship to advance their careers, and adding a prominent name to
a list of co-authors to receive a more sympathetic editorial review. Related to
these problems and to the ever-growing importance of "research productivity"
are disturbing trends toward the proliferation of authorship credits attached to
publications, a growth in the number of mediocre quality publications ("paper
inflation"), and the multiplication of reports using the "least publishable unit"
to maximize the output from a single study (Lafollette, 1992).

In part to prevent these kinds of problems, many journal editors and other
individuals in scientific publishing have promoted policies designed both to
detect misconduct and prevent the more blatant forms of authorship abuse.
These policies include publishing detailed descriptions of the criteria for scientific authorship, requiring that all authors sign a statement of authorship
responsibility, putting limits on the number of authors listed on the masthead,
and requesting that co-authors provide a written explanation of their individual contributions to a publication.

How does all of this apply to individual authors? Even if most authors in
the addiction field have never encountered an instance of data fabrication or
plagiarism, they are likely to encounter the more subtle forms of irresponsible authorship and publication misconduct, such as gift authorship and ghost
authorship (Flanagin et al., 1998). Honorary or gift authorship consists of
awarding authorship credit because of the person's power and prestige or as
"payment" for another kind of contribution rather than for time, effort, and
substantive contributions to the work. An extreme example of this is sur-
prise authorship, where a researcher finds out that his or her name appears
on an article only after publication (Anderson et al., 2011). When someone
demands (and receives) an honorary authorship, it is sometimes called a coercion authorship or pressured authorship (Claxton, 2005; Freeser, 2008). Closely
related to gift authorship is mutual-admiration or mutual-support authorship,

in which two or more researchers agree to list each other as authors despite little involvement in each other's articles, usually as a means to expand their individual publication histories (Claxton, 2005). Ghost authorship refers to the failure to include as co-authors those who satisfy the criteria for authorship (Sheikh, 2000). This happens most often in the publication of pharmaceutical company trials in which an industry-paid scientific writer drafts the article but is not listed as a co-author to avoid the perception of conflict of interest. It also occurs with funded students and research assistants (Newman & Jones, 2006) who might contribute substantively to a publication but do not receive credit because the contribution is considered "part of the job."

In the remainder of this chapter, we review guidelines that have been developed to deal with publication misconduct and then some practical steps that can be taken by individuals, project teams, centers, departments, and professional organizations to ensure responsible authorship.

Formal Guidelines

To develop a more coherent, equitable, and ethical set of guidelines for addiction journals, various policies have been proposed in the scientific literature. These policies include the guidelines recommended by the Council of Science Editors (Biagioli et al., 1999), the Sigma Xi standards for responsible authorship (Jackson & Prados, 1983), the statement of the International Committee of Medical Journal Editors (2013), and a variety of proposals from individual commentators (e.g., Broad & Wade, 1982; Fine & Kurdek, 1993; Newman & Jones, 2006). Box 11.4 describes the general guidelines developed by the American Psychological Association (2010). These have been the subject of a considerable amount of interpretation and discussion in the psychological literature, and some attempts have been made to develop operational definitions of the specific criteria.

Winston (1985) developed a system in which points are assigned for various professional contributions to a scholarly publication, with research design and report writing earning the most points. A certain number of points must be earned to qualify for authorship credit, and the individual with the highest number is granted first authorship.

One of the most cited sources on authorship is the 1985 consensus statement of the International Committee of Medical Journal Editors (1985). The statement indicated that only those in a position to take public responsibility for the work could claim authorship. Although this definition would preclude gift authorship and help to minimize ghost authorship, there were still problems with the definition of a "substantial" contribution (see Yank & Rennie, 1999) especially in situations in which collaborating investigators band together on a project to take advantage of expertise that is unlikely to be concentrated in one individual. These problems were corrected in a 2003 revision to this statement

Psychologists take responsibility and credit, including authorship credit, only for work they have actually performed or to which they have substantially contributed. Principal authorship and other publication credits accurately reflect the relative scientific or professional contributions of the individuals involved, regardless of their relative status. Mere possession of an institutional position, such as department chair, does not justify authorship credit. Minor contributions to the research or to the writing for publications are acknowledged appropriately, such as in footnotes or in an introductory statement. Except under exceptional circumstances, a student is listed as principal author on any multi-authored article that is substantially based on the student's doctoral dissertation. Faculty advisors discuss publication credit with students as early as feasible and throughout the research and publication process as appropriate.

Box 11.4: Authorship guidelines proposed by the american psychological association.
Source: Section 8.12, American Psychological Association (2010).

and further revised in 2013 (see www.icmje.org). The International Committee of Medical Journal Editors now indicates that each author should meet the following criteria: (a) substantial contributions to the conception or design of the work or the acquisition, analysis, or interpretation of the data; (b) drafting the work or revising it critically for important intellectual content; (c) approval of the final version to be published; and (d) agreement to be accountable for all aspects of the work in ensuring the questions related to the accuracy or integrity of any part of the work are appropriately investigated and resolved. In addition, the International Committee of Medical Journal Editors recommends that an author should have confidence in the contributions of their co-authors and be able to identify which parts of the work he or she was responsible for. Additional changes were made by the International Committee of Medical Journal Editors to deal with contributors who do not meet authorship criteria, such as people who provide general supervision or administrative support for a research group, technical help, writing assistance, language editing, or proofreading. These individuals and their contributions should be listed in an acknowledgements section. To the extent that a listing of such persons could be interpreted as an endorsement of the data or conclusions, the International Committee of Medical Journal Editors concluded that all persons listed must provide written permission to be acknowledged.

Practical Steps to Determine Authorship

The foregoing discussion of conventions, problems, and policies suggests that authorship of an article is foremost a social process that requires a considerable amount of discussion, negotiation, and influence. If there is a general perception that the procedures for attributing authorship credits are inadequate and ineffective (see Yank & Rennie, 1999), then it may be because the social nature of authorship has not been taken into account in the design of policies and procedures for responsible authorship. Most guidelines focus on individual accountability in relation to abstract ethical principles, with bureaucratic controls and punitive sanctions emphasized instead of practical guidance about what to do at the level of the group where real influence and control are concentrated. In this section, we describe a model process to demonstrate how many of the helpful suggestions provided in the literature on scientific authorship can be implemented in a practical, systematic, and open way. The process is based on the assumption that, because the writing of a multi-authored article is a social process, the responsibility, accountability, and equitable distribution of credit reside in the group of individuals most responsible for conducting the research and writing the article. This process can easily be implemented by an external agency or even within an institution, department, or research center. It needs to be conducted in an open, democratic, and ethical way so that all collaborating investigators agree to accept the basic values of scientific integrity.

As in any group process, one or more individuals need to take a leadership role. There is general agreement in the scientific community that the person most closely associated with the project should take responsibility for drafting the article and being first author. Exceptions to this rule are possible, such as when the investigator who conceived and directed a project cedes responsibility to a junior investigator who made special contributions and who is capable of carrying the written report to a successful conclusion. A crucial skill that should be taken into account in the choice of one or more leaders for a scientific publication is familiarity with the authorship issues described in this chapter. If the person has had no formal training in research ethics, the articles cited in the reference section of this chapter should be reviewed, giving special attention to several key sources (e.g., Fine & Kurdek, 1993; International Committee of Medical Journal Editors, 2013).

To avoid conflict, misunderstandings, and publication misconduct, both the lead author and the group should follow generally accepted procedures that are characterized by openness and transparency and should decide as early as possible who will be listed as an author, the order of authorship, and the other contributors to mention in the acknowledgments (American Psychological Association, 2010). In the following paragraphs, we provide an outline for a model that can be modified to fit the needs of a project team.

The model requires the completion of specific tasks at each of three stages in the publication process. As described below, periodic discussions about authorship and accountability should be conducted at the planning stage, the drafting stage, and the finalization stage of a publication. According to Lafollette (1992), "The issue is absolutely clear. Who did what and how much? Answering those questions early on—and continuing to ask them as projects change—can help to prevent disputes or embarrassment later" (p. 107).

Planning Stage

The planning stage of the publication process begins when a scientific investigation or other project (e.g., a review article) has advanced to the point where it is likely that a scientific article is appropriate or warranted. This decision is usually made by the project leader, who either takes direct responsibility for the direction of the publication or designates one or more individuals to initiate the publication planning process. The following tasks and activities are suggested.

- One or more senior members of the research or writing team take responsibility for developing an outline of the article, a timetable for the completion of the article, and a list of potential co-authors, based on actual contributions to date and expected contributions in the future. The outline is distributed to all prospective authors, with the understanding that authorship will depend on substantive contributions, as well as effort and follow through, as described in relevant policies and publications (including this chapter).
- Plans are made for a periodic reassessment of the research team's contributions throughout the planning, drafting, and finalization stages. If it is found that previous expectations are not being met, then assignment of authorship credit may be modified, based on actual contributions at the time of publication completion.
- Relevant policies and publications (including copies of this chapter) are distributed to prospective authors along with the outline.
- A meeting is called to discuss the proposed publication and the distribution of responsibilities for its completion. Assignments are made for data analysis and writing sections of the first draft. A timeline of key tasks is distributed and discussed.

Drafting Stage

After the first draft of an article is completed or as relevant sections are finished, the drafting author or authors circulate the article for comments. At this stage, potential authors must be reminded not only about their rights to possible authorship but also about their responsibilities.

A crucial task at this stage is to identify who qualifies for formal authorship credit according to generally accepted criteria for responsible authorship. One way to accomplish this task is to ask all potential contributing authors (including the lead author) to describe their contributions to the project. Box 11.5 provides a checklist of contributions that prospective authors could be asked to complete by the lead author in order to determine eligibility for authorship at this stage. Although this one was designed for original research reports and may not apply to all publication types (e.g., reviews), similar disclosure checklists have been found to be useful for determining authorship credit (Yank & Rennie, 1999).

Once the checklists are completed, the lead author could call a meeting to discuss authorship and other matters related to the proposed publication. At the meeting, each person is asked to describe his or her contributions to the project to date. In such a setting, individuals often reveal contributions that others were not aware of and, in other cases, describe activities that might not be considered substantial in comparison with those of others. At this time, it is important to discuss generally accepted criteria for authorship, such as those listed in Box 11.5, to make sure that everyone agrees on the standards for determining who should be listed on the article and in what order the names should be arranged. To provide authority to the process, it could be advantageous to mention that most journals now require a similar process of asking authors to sign a statement attesting that they have met minimal criteria for authorship, and some journals (e.g., *The Lancet, BMJ, American Journal of Public Health*) require authors to describe their individual contributions, the text of which is published along with the article.

One of the most difficult decisions in the assignment of authorship credit is the distinction between major (or substantial) and minor contributions. A major contribution usually involves the independent development or interpretation of ideas that are crucial to the advancement of a scientific study or a scholarly article. It may also involve the use of special skills to perform a complex task without which the project could not have been done, such as the application of a sophisticated statistical technique. The emphasis in these definitions is more on quality than quantity. All persons making major contributions should receive authorship credit, provided that they also participate in the writing of the article and any revisions required by the editor. Such individuals should also be capable of taking public responsibility for both general and specific aspects of the publication, recognizing that opinions differ as to what this means. Although the checklist provided in Box 11.5 was compiled from a variety of sources, we borrowed heavily from Yank and Rennie (1999), who distinguished between "major" and "partial" contributions. In a content analysis of articles in which authors provided a description of their roles in the publication process, they also report the 10 most common author contributions. A major contribution meant that the contributor fulfilled a majority of the activities for a given category (examples below). A partial or minor contribution referred to a more limited role, presumably in terms of time, effort, or substance.

Instructions: Use the checklist to describe your contributions to the project to date. Under each item you have checked, describe the nature of your contribution, the amount of effort you put into it (e.g., hours, days, months), and whether your contribution fulfilled all of the requirements for that task or some of the requirements (e.g., in collaboration with others, you wrote part of the article or you collected part of the data).

- Were responsible for conception of the project (planning meetings, drafting of research proposal, etc.)
- Reviewed the literature
- Obtained funding or other resources
- Assembled the project team
- Coordinated study (5) by assigning responsibilities and tasks
- Trained of personnel
- Supervised personnel
- Obtained human (or animal) subjects approvals
- Designed the methodology or experimental design (2)
- Advised on design or analysis (9)
- Wrote the research protocol
- Collected data (4), including follow-up data
- Performed clinical analysis or management (6)
- Performed randomization or matching
- Performed statistical analysis of data (8)
- Interpreted the data (3)
- Performed economic analysis of data
- Managed data (10)
- Provided technical services (coding questionnaires, laboratory analyses (7), etc.)
- Provided or recruited patients
- Provided materials or facilities
- Presented and defended findings in a public forum
- Wrote draft of article
- Wrote final version of article (1)
- Submitted report for publication
- Responded to reviewers' comments
- Were responsible for other activity or service (describe)

Box 11.5: Checklist for conducting an inventory of major and minor contributions to a scientific article.

Note: The numbers in parentheses refer to the top-10 overall categories of contribution identified by Yank and Rennie (1999) in a content analysis of articles according to the most frequently mentioned contributions to authorship.

Examples of major contributions that fulfilled Yank and Rennie's (1999) "lenient" interpretation of the International Committee of Medical Journal Editors (1991) authorship criteria were (a) conception of the idea for the study or article, (b) design of the study, (c) statistical analysis or interpretation of data, (d) laboratory analysis, (e) management or analysis of clinical aspects, and (f) performance of field work or epidemiology. Anyone who wrote or revised the article (even sections) fulfilled the second part of the criteria (i.e., drafted the article or revised it critically for important intellectual content).

In considering the relative importance of major contributions, we believe two additional factors should be taken into account by the project leader and team: effort and follow through. Effort pertains to the amount of time spent on the particular contribution. Follow through involves active participation at various stages throughout the project. For example, if a person has participated in a study in a minor way or has made a major contribution that involves minimal effort (e.g., the development of an idea for the study or a novel hypothesis) and/or follow through, this does not necessarily entitle the individual to authorship if other persons have made greater contributions with respect to effort and follow through.

Nonsubstantive considerations should not determine the order of authorship or whether to include an individual as an author. Examples of nonsubstantive factors include rank or status, need for publication credits to justify advancement, involvement in the project as a consequence of routine duties for which the individual is paid (e.g., collecting laboratory samples), or ability to provide access to study participants. The person who is named as the principal investigator of a project or a grant for administrative reasons might not even qualify for authorship under these circumstances if she or he has had no role in the design and conduct of a particular project (e.g., the secondary analysis of data collected for another purpose).

Members of a research team also need to recognize that, in general, individuals will be expected to contribute to projects in a collegial fashion without necessarily receiving credit in all project publications. And, as noted in Chapter 5, the group may want to give consideration to the special situation of students and postdoctoral fellows where different standards for a contribution may apply.

Taking all of the above information into account, it should not be difficult in most cases to reach consensus about who qualifies for authorship and what the most equitable relative ranking of contributions should be. When contributions are discussed in an open forum in relation to generally accepted criteria and ethical principles, secondary (nonsubstantive) considerations tend to be difficult to defend, especially when there is a written record of each individual's perceived contributions. If there are discrepancies between what an individual perceives to be his or her contributions and the perceptions of others, these differences often can be resolved through open discussion.

Finalization Stage

Before an article is formally submitted to a journal, a corresponding author needs to be designated. This person is usually the first author, but sometimes it is also the senior project leader in cases in which the first author is inexperienced with publication submission. A prominent or senior co-author should never be designated as corresponding author solely to influence the review process. If there is general agreement about the authorship order throughout the writing process, this order can be reviewed again at the final stage to determine whether preparation and revision altered the relative order of contributions enough to require changes.

Authorship Disputes

If attempts to resolve authorship status before writing or publishing a manuscript are unsuccessful, four processes for authorship dispute resolution have been proposed: direct dialogue, mediation, peer panel, and a binding decision (National Institutes of Health, 2010). Direct dialogue requires the parties in a dispute to discuss their differences with each other in order to reach an agreeable solution. If direct dialogue is unsuccessful, they may enter mediation, which uses a neutral, third-party mediator to assist in finding a resolution. Parties in dispute may also present their perspectives on authorship to a three-person peer panel and agree to abide by the panel's decision. If the dispute remains unresolved, then a scientific director or person in a similar position may make a binding decision. Although these processes have been created by a U.S. institution, they are applicable to any research environment and can be modified to best suit the authors' circumstances.

Conclusion

Intellectual honesty is a fundamental ingredient of scientific integrity, and this extends to the need for complete accuracy and transparency in representing contributions to research reports and other scientific writing. The contributions of colleagues and collaborators need to be recognized in all scientific publications, but authorship must be assumed or awarded only on the basis of substantive contributions to an article and the ability of its authors to take public responsibility for its contents or, at least, for major parts of the contents. Decisions regarding authorship should be seen as part of a process that begins with the development of a publication plan and ends with the final revision of an accepted article. In between, it is best to have all potential contributors to a publication participate in an open process of stating their perceived contributions to a given project in the context of generally accepted criteria for authorship. Such a process is likely to manage expectations and prevent publication misconduct as well as misunderstandings and conflicts. To the extent that authorship credit continues to be seen as the coin of the realm in addiction science, both sides of the coin (credit and responsibility) need to be valued.

Authorship Credit Exercise

Appendix A contains two case studies that describe sensitive and possibly contentious authorship credit scenarios. For each case, answer the questions at the end and then discuss your answers with colleagues or a mentor in order to apply the principles described in this chapter. Also review Chapters 5, 14, and 15 for additional information about resolving ethical dilemmas in relation to authorship.

Acknowledgements

The authors thank Ian Stolerman for his helpful comments and suggestions.

Please visit the website of the International Society of Addiction Journal Editors (ISAJE) at www.isaje.net to access supplementary materials related to this chapter. Materials include additional reading, exercises, examples, PowerPoint presentations, videos, and e-learning lessons.

References and Additional Reading

American Psychological Association Ethics Committee. (2010). *Ethical principles of psychologists and code of conduct including 2010 amendments*. Retrieved from http://www.apa.org/ethics/code/principles.pdf.

Anderson, M., Kot, F. C., Shaw, M. A., Lepkowski, C. C., & De Vries, R. G. (2011). Authorship diplomacy: Cross-national differences complicate allocation of credit and responsibility. *American Scientist, 99*, 204–207. DOI: https://doi.org/10.1511/2011.90.204

Biagioli, M., Crane, J., Derish, P., Gruber, M., Rennie, D., & Horton, R. (1999). *CSE Task force on authorship draft white paper*. Retrieved from http://www. councilscienceeditors.org/resource-library/editorial-policies/cse-policies/ retreat-and-task-force-papers/authorship-task-force/cse-task-force-on-authorship/.

Broad, W., & Wade, N. (1982). *Betrayers of the truth: Fraud and deceit in the halls of science*. New York, NY: Simon & Schuster.

Claxton, L. D. (2005). Scientific authorship. Part 2: History, recurring issues, practices, and guidelines. *Mutation Research, 589,* 31–45. DOI: https://doi.org/10.1016/j.mrrev.2004.07.003

Davis, R. M. (2008). British American Tobacco ghost-wrote reports on tobacco advertising bans by the International Advertising Association and J J Boddewyn. *Tobacco Control, 17,* 211–214. DOI: https://doi.org/10.1136/tc.2008.025148

Feeser, V. R., & Simon, J. R. (2008). The ethical assignment of authorship in scientific publications: Issues and guidelines. *Academic Emergency Medicine, 15,* 963–969. DOI: https://doi.org/10.1111/j.1553-2712.2008.00239.x

Fine, M. A., & Kurdek, L. A. (1993). Reflections on determining authorship credit and authorship order on faculty-student collaborations. *American Psychologist, 48,* 1141–1147. DOI: https://doi.org/10.1037/0003-066X.48.11.1141

Flanagin, A., Carey, L. A., Fontanarosa, P. B., Phillips, S. G., Pace, B. P., Lundberg, G. D., & Rennie, D. (1998). Prevalence of articles with honorary authors and ghost authors in peer-reviewed medical journals. *JAMA, 280,* 222–224. DOI: https://doi.org/10.1001/jama.280.3.222

Horowitz, H. W., Fiebach, N.H., Levitz, S. M., Seibel, J., Smail, E. H., Telzak, E. E., . . . Raffalli, J. (1996). Ode to multiauthorship: A multicentre, prospective random poem [Letter to the Editor]. *The Lancet, 348,* 1746. DOI: https://doi.org/10.1016/S0140-6736(05)65883-7

International Committee of Medical Journal Editors. (1985). Guidelines on authorship. *British Medical Journal, 291,* 722. DOI: https://doi.org/10.1136/bmj.291.6497.722

International Committee of Medical Journal Editors. (1991). Uniform requirements for manuscripts submitted to biomedical journals. *New England Journal of Medicine, 324,* 424–428. DOI: https://doi.org/10.1056/NEJM199102073240624

International Committee of Medical Journal Editors. (2003). *Uniform requirements for manuscripts submitted to biomedical journals: Writing and editing for biomedical publication.* Retrieved from http://www.icmje.org/about-icmje/faqs/icmje-recommendations/.

International Committee of Medical Journal Editors. (2013). *Recommendations for the conduct, reporting, editing, and publication of scholarly work in medical journals.* Retrieved from http://www.icmje.org/icmje-recommendations.pdf.

Jackson, C. I., & Prados, J. W. (1983). Honor in science. *American Scientist, 71,* 462–464.

Kennedy, M. S., Barnsteiner, J., & Daly, J. (2014). Honorary and ghost authorship in nursing publications. *Journal of Nursing Scholarship, 46,* 416–422. DOI: https://doi.org/10.1111/jnu.12093

LaFollette, M. C. (1992). *Stealing into print: Fraud, plagiarism, and misconduct in scientific publishing.* Berkeley, CA: University of California Press.

Mowatt, G., Shirran, L., Grimshaw, J. M., Rennie, D., Flanagin, A., Yank, V.,
. . ., Bero, L. A. (2002). Prevalence of honorary and ghost authorship in
Cochrane reviews. *JAMA, 287,* 2769–2771. DOI: https://doi.org/10.1001/
jama.287.21.2769

National Institutes of Health. (2010). *Processes for authorship dispute resolution.* Retrieved from https://oir.nih.gov/sourcebook/ethical-conduct/
responsible-conduct-research-training/processes-authorship-dispute-
resolution.

Newman, A., & Jones, R. (2006). Authorship of research papers: Ethical and
professional issues for short-term researchers. *Journal of Medical Ethics, 32,*
420–423. DOI: https://doi.org/10.1136/jme.2005.012757

Relman, A. S. (1983). Lessons from the Darsee affair. [Editorial]. *New England Journal of Medicine, 308,* 1415–1417. DOI: https://doi.org/10.1056/
NEJM198306093082311

Rennie, D., & Flanagin, A. (1994). The Second International Congress on
Peer Review in Biomedical Publication. *JAMA, 272,* 91. DOI: https://doi.
org/10.1001/jama.261.5.749

Rennie, D., Yank, V., & Emanuel, L. (1997). When authorship fails: A proposal
to make contributors accountable. *JAMA, 278,* 579–585. DOI: https://doi.
org/10.1001/jama.1997.03550070071041

Sheikh, A. (2000). Publication ethics and the research assessment exercise:
Reflections on the troubled question of authorship. *Journal of Medical Ethics,
26,* 422–426. DOI: https://doi.org/10.1136/jme.26.6.422

Smith, R. (1997). Authorship: Time for a paradigm shift? [Editorial]. *BMJ, 314,*
992. DOI: https://doi.org/10.1136/bmj.314.7086.992

University of Pittsburgh. (2006). *University of Pittsburgh summary investigative report on allegations of possible scientific misconduct on the part of
Gerald P. Schatten, Ph.D.* Retrieved from https://ecommons.cornell.edu/
bitstream/handle/1813/11589/Gerald_Schatten_Final_Report_2.08.
pdf?sequence=1&isAllowed=y.

Wilcox, L. J. (1998). Authorship: The coin of the realm, the source of
complaints. *JAMA, 280,* 216–217. DOI: https://doi.org/10.1001/
jama.280.3.216

Winston, R. B. (1985). A suggested procedure for determining order of authorship in research publications. *Journal of Counseling & Development, 63,*
515–518. DOI: https://doi.org/10.1002/j.1556-6676.1985.tb02749.x

Wislar, J. S., Flanagin, A., Fontanarosa, P. B., & Deangelis, C. D. (2011). Honorary and ghost authorship in high impact biomedical journals: a cross sec-
tional survey. *BMJ, 343,* d6128. DOI: https://doi.org/10.1136/bmj.d6128

Yank, V., & Rennie, D. (1999). Disclosure of researcher contributions: A study
of original research articles in *The Lancet. Annals of Internal Medicine,
130,* 661–670. DOI: https://doi.org/10.7326/0003-4819-130-8-199904200-
00013

Appendix A: Authorship Credit Scenarios

Multicentered Trial with Multiple Investigators

Dr. Joe Camel is an assistant professor at Small State University where he is the principal investigator of a large, multicenter trial to determine the effectiveness of a new nicotine inhaler at reducing cigarette use and nicotine cravings. The main findings of the study were positive and have already been published in the Journal of Reputable Results. *To maximize use of the data collected, Dr. Camel has made the raw data available to each of his colleagues for secondary analyses. It was agreed on by the group that a brief outline of the analyses to be performed and a list of potential co-authors should be prepared by those requesting to use the data to ensure there are no duplicate analyses. The group also agreed to prepare comments and critiques in response to data requests.*

Dr. Muck E. Muck, a professor at Ivy League University, informs the Small State group that his team would like to perform an analysis on the effect of alcohol use in nicotine-cessation therapy. In response, Dr. Camel insists on being listed as the last and corresponding author even though he will not contribute to the data analysis, interpretation of the results, or manuscript preparation. Dr. Camel tells Dr. Muck that, as principal investigator of the trial, he has the right to be listed as an author on all related publications, and because he made the data freely available to Dr. Muck, he will not supply the data unless he does so.

Discussion Questions

1. What are the ethical implications and whose interests are involved?
2. What should Dr. Muck do about the manuscript and the request to add Dr. Camel as a co-author?
3. What should have been discussed among the collaborators before the raw data was made available?

Junior Investigators Sharing Authorship on Each Other's Articles

Dr. Allen Quidproquo and Dr. Miriam Scratchmyback are the only postdoctoral fellows at the National Center for Addiction Science. They have both been working to publish their dissertation results. Dr. Quidproquo's research focuses on the association of genes with initiation of substance use, whereas Dr. Scratchmyback researches the role of visual cues in treatment and relapse. The two fellows agree that their research has little in common and rarely discuss research topics in the office. But, being the only postdoctoral fellows at their center, they often share meals together, talk about their nonacademic lives, and have quickly become friends.

During one meal, Dr. Quidproquo talks about the pressure he is under to publish as often as possible. He can only stretch his data so far and has only a handful of publications to his credit. Dr. Scratchmyback has already been included as an author on more than a dozen publications. Therefore, Dr. Quidproroquo asks Dr. Scratchmyback if he could be a co-author on her publications to bump up his publication numbers, and, in return, he will list Dr. Scratchmyback as a co-author on all of his publications. Dr. Quidproquo reasons that this arrangement would effectively double the amount of publications on his list and substantially add to Dr. Scratchmyback's list as well. He reasons this would better position them for future funding opportunities, faculty positions, and other research awards.

Discussion Questions

1. How should Dr. Scratchmyback respond to her friend's request?
2. What can Dr. Scratchmyback do to maintain her own scientific integrity and/or prevent his colleague from committing scientific misconduct?
3. To what extent does either fellow stand to gain or lose from this arrangement?

CHAPTER 12

Preparing Manuscripts and Responding to Reviewers' Reports: Inside the Editorial Black Box

Ian Stolerman and Richard Pates

Introduction

This chapter describes how the peer-review process works and presents suggestions to authors of manuscripts. It is based on the experiences of scientists and
clinicians who have many years of experience as editors of prominent addiction journals. The task of the editor is to publish manuscripts appropriate for
the journal and to assist would-be authors in the production of suitable material. Many of the problems facing authors writing for scholarly peer-reviewed
journals in the addiction field are similar to those in other fields. Therefore, it
is recommended that readers consult one or more of the full-length books that
have already been published in this general area. Hundreds of books have been
published, as can be seen by searching Amazon.com or PubMed for "scientific
writing." For example, a search of Amazon.com on December 1, 2014, produced 16,904 results, many of which were relevant. A short list of recent books
is provided at the end of this chapter in Appendix A.

Nearly all academic journals now work exclusively with computerized systems
that allow for submitting manuscripts, sending articles to reviewers, responding
to the reviewers' comments, and making a decision on the manuscript (e.g.,
accept/minor changes/major changes/reject). The advantages of these systems
are increased efficiency for the editorial staff and an easier submission role for
the author. It also makes it easier to keep track of manuscripts. As a rule, nearly
all the communications to and from the journal are now done electronically.

How to cite this book chapter:
Stolerman, I and Pates, R. 2017. Preparing Manuscripts and Responding to Reviewers'
    Reports: Inside the Editorial Black Box. In: Babor, T F, Stenius, K, Pates, R,
    Miovský, M, O'Reilly, J and Candon, P. (eds.) *Publishing Addiction Science: A
    Guide for the Perplexed*, Pp. 229–244. London: Ubiquity Press. DOI: https://doi.
    org/10.5334/bbd.l. License: CC-BY 4.0.

Triage: the First Selection

The author's quest to find a suitable publication outlet ends with a letter stating, "I am pleased to inform you that your manuscript is acceptable for publication. . . ." But the first step is to get the manuscript into the peer-review system. Yet having your article peer reviewed is not the inevitable consequence of submitting to a peer-reviewed journal. Some journals state formally that they operate a system of "triage," whereby the editor or his or her assistants decide which submitted articles will be entered into the peer-review process. In practice, it is likely that all journals have such a system to protect the profile of the journal and to avoid bothering authors and peer reviewers with a long and laborious evaluation process when it is easy to predict a negative result (see Box 12.1). Thus, if something is received that clearly has no hope of acceptance, it may be rejected without review. Here, the difference between journals is quantitative rather than qualitative: In the journals of highest impact in science and medicine generally, including addiction research, it may be that more than half of the submissions are rejected at this stage. Some addiction journals will, however, accept almost all articles, or reject only 20% or 25%. For some information on acceptance rates of addiction journals, see Chapter 3 and its appendix.

There are some aspects of manuscript preparation that are so easy that everyone should get them right. To ensure your manuscript has the best chance of passing through triage, make sure you do all these things as set out in the instructions to authors that every journal provides. Follow all advice and recommendations exactly, format your submission precisely as requested, make sure that all sections are complete, and be sure that no tables, figures, or figure legends are missing. Check the reference list to ensure that all cited references are in it, and no others. Check the accuracy of each citation. Look in the journal to see exactly how references are styled. Then check them over again, after you have made the corrections, until no more errors can be found. This sort of work is tedious but does not need expensive resources, profound knowledge of the subject, or outstanding intellectual ability. If the editor sees at a glance that you do not even get these straightforward, mechanical things right, he or she may well develop a jaundiced view about your capability to deal with more complex matters. Try to look at your own manuscript as an editor might. If you do not bother to do the easier things required of an author, the editor might reasonably conclude that you will not be able to do complex revisions either, and you may not be given the opportunity to revise and resubmit.

Communication with more Experienced Writers

Would-be authors may seek the advice of more experienced colleagues at almost any stage of the publication process. When planning a publication, discussion

- The submission is outside the scope of the journal (e.g., it is about a misused substance but it is not relevant in any discernible way to misuse of or dependence on it).
- The manuscript type is not appropriate (e.g., a case report is submitted to a journal that does not publish case reports).
- It contains clear ethical problems such as apparent violation of current generally accepted standards for the treatment of human or animal subjects.
- The article is poorly organized.
- The report is purely descriptive, has no hypotheses, or reaches no conclusions.
- There are major methodological weaknesses.
- The article appears to offer nothing new.
- Instructions to authors are flagrantly ignored in some way not mentioned here.

Box 12.1: Reasons for rejection by triage.
Note: The editor has a duty to reviewers, as well as to authors, and tries not to waste reviewers' time by requesting evaluations of work that has no chance of acceptance for one or more of the reasons above.

with colleagues after presenting the work at a seminar in the home institution may yield some tips as to the type of journal that may be interested in the study. Subsequently, during preparation of the manuscript, it may be appropriate to seek the advice of local colleagues on technical aspects, such as statistical analyses. When a manuscript exists in a complete form, it is often immensely helpful to ask at least one person to read it and make comments and suggestions. People are very often willing to help if authors make clear that they value an expert opinion on aspects such as coverage of the literature, clarity, style, language, and validity of conclusions. If there is no person in the author's own institution, it is possible to approach outsiders and ask if they would be willing to comment. Both people whom you know personally and others who have published in the area are likely to feel flattered and pleased that you value their opinion and may well provide advice. The manuscript that cannot be improved has yet to be written, and even experienced authors often seek the opinions of colleagues because, after working on a manuscript for months through revision after revision, authors may find it difficult to spot the little problems that spring to the attention of a new reader.

Writing in a Foreign Language

It is an unavoidable fact that many authors have to write in a language other than their own, and conveying complex scientific ideas with clarity and precision can be a difficult task even in one's native language. Authors may therefore find it worthwhile to seek the assistance of colleagues with more experience in writing in the chosen language and, if possible, enlist a native speaker of the language to correct the manuscript. If that is not possible, it may be necessary to obtain the assistance of a professional translator to suggest corrections. Journals provide varying amounts of assistance in the correction of errors after accepting a manuscript for publication, but they cannot do anything to assist reviewers of poorly written manuscripts. A fuller consideration of language issues and language editing services may be found in Chapter 4.

The Peer-Review Process: Selection of Reviewers

The next step for the editor is selection of reviewers who will advise him or her of the strengths and weaknesses of the work and recommend whether or not it should be accepted, reconsidered after revision, or rejected outright. To improve the manuscript, reviewers are also expected to make constructive suggestions in a report that can be sent to the authors. The criteria used for selecting reviewers are diverse, and probably few if any journals have tightly defined procedures. Box 12.2 shows the main criteria used by editors to identify reviewers. The number of reviewers for each article varies within and between journals, but most commonly there are two. The editors of some journals may work with only one reviewer, but this seems to be increasingly rare. Occasionally, three or more reviewers are used, depending on the journal and the editor's perception of the complexity and significance of the work. For example, multidisciplinary manuscripts may require more than two reviewers to ensure sufficient expertise. Similarly, if a study seems likely to have a major practical impact, for example on policy or treatment, the editor may wish to be especially certain that it is assessed thoroughly. If the two reviewers initially selected disagree about the article, an editor may seek additional advice from a third person to reach a decision.

For all reports, regardless of whether they are quantitative or qualitative, each journal has its own set of instructions for reviewers; journals differ with respect to the attributes of their "ideal" manuscripts. There will sometimes be a requirement for reviewers to complete a questionnaire as part of the review, with ratings of the manuscript according to criteria such as importance and likely impact on the field, as well as technical competence. The reviewers are usually also asked to make a recommendation on the fate of the manuscript and to justify it in confidential comments to the editor. Finally, reviewers are in all cases expected to produce a report that the editor will forward to the authors. The main purposes of this report are (a) to make suggestions enabling the author to improve the manuscript and (b) to list criticisms that the reviewer believes need to be addressed if the report is to be published. The report to the authors should not include specific recommendations for acceptance or rejection of the manuscript because that decision is the editor's. Chapter 13 provides further advice on how to become a competent reviewer.

Reviewers are asked to act according to ethical guidelines that are presented and discussed elsewhere in this book (see Chapters 14 and 15). The task of the editor is to reconcile sometimes conflicting reports from different reviewers and to make a personal judgment based on a variety of other considerations. The task is made more difficult if the reviews contain conflicting recommendations for publication.

- Recognized expertise in the specific field of the manuscript as noted in the journal's database of previous reviewers and authors.
- Previous invitations to the reviewer that have resulted in thorough, well-written, polite reviews submitted in a timely manner.
- Record of recent publications in the field as determined by searches of databases such as MEDLINE and PsycINFO.

The following individuals are typically excluded from consideration as reviewers:

- Persons who are known to have a very close connection to the authors or to have a conflict of interest with the authors will be avoided.
- People who are currently reviewing another manuscript for the same journal or who have reviewed one within a set period (e.g., three months) will be avoided.
- Those who work is excessively praised or criticized in the manuscript to be assessed are avoided.

Box 12.2: Some criteria editors use to identify reviewers for a particular manuscript.
Note: Some journals ask authors to suggest reviewers or to name persons they do not wish to have as reviewers. How to use these suggestions is the editor's decision. Different bulleted points from those above will be used in combination to reach a decision on whom to invite, and there will inevitably be appreciable variations between journals with respect to the use of these different methods of selection.

Criteria for Evaluation of Manuscripts

If the journal has published its instructions to reviewers or put them on a website, these instructions will give you an idea about the features at which both editors and reviewers will look. Many journals probably look for the same desirable features of highly rated studies.

If a study is quantitative, the criteria include the use of a sufficiently large and suitably representative sample of the population under study, the presence of a high response rate among invited participants, the use of valid measures, the absence of procedural biases, minimal confounding of one independent variable with another, and the use of appropriate controls. Similarly, reviewers will look for as full a description of the methods as available space allows, with reference to earlier publications that provide more detail and establish the validity of the methods and measuring instruments (where applicable). Results must be described in a clear and logical sequence, with all necessary information presented. No more detail should be provided than can be covered in the discussion section. The discussion should bring out the importance of all the main findings and indicate how the work advances the state of knowledge and understanding in the relevant subfield. In addition, alternative interpretations of the data may be given, thus acknowledging limitations of the study. Reviewers pay attention to all the preceding points—and to many others.

In quantitative research, the data analysis section is prone to several problems. These include the following:

- failing to deal adequately with confounding variables;
- claiming to have shown something without performing a (statistical) test that supports it directly and unequivocally;
- failing to control for multiple comparisons; and
- drawing inappropriate conclusions from non-significant associations or differences: we probably all realize that lack of significance means only that we have failed to find an effect and does not prove that no effect exists, but we don't always remember this in our enthusiasm to explain how our results fail to support the ideas of a scientist whose theory we dislike.

Authors developing reports of randomized controlled trials may wish to follow the CONSORT (Consolidated Standards of Reporting Trials) checklist, which includes 22 items considered essential to judge the reliability or relevance of the findings (presented in Appendix B to this chapter in slightly abbreviated form).

The criteria for the evaluation of *qualitative* reports vary depending on the type of data and methods of analysis (e.g., participant observation and ethnography, qualitative interviews, content analysis, textual analysis, discourse analysis, ethnography and conversation analysis). Chapter 8 provides more information about how to write and publish articles using qualitative methods. Most types of qualitative reports should do the following.

- Give clear criteria for the selection of data or subjects. Position the material carefully in the social and cultural space. For example, different genres of fiction represent different segments of the culture.
- Present a detailed account of where and when data were collected or which existing data sets were used. In studies based on fieldwork, describe the relation between the fieldworkers and the subjects and discuss the possible influence of the data collection on the phenomenon under study. Keep careful records of the data so that they can be provided for independent examination if necessary.
- Clearly state how the analysis was done, with an indication of whether reliability was assessed—for instance by replicating the analysis.
- Describe any themes, concepts, and categories derived from the data. Divide the interpretation process into short steps, specifying rules of classification and interpretation.
- Outline steps taken to guard against selectivity in the use of data; discuss exceptions and deviant cases. Ideally, the reader should be able to apply the same classifications, take the same analytic steps, and reach the same kind of results with another data set.
- Present data systematically so that quotations, field notes, and so on are easily identifiable.
- Offer enough primary evidence to show a relation between evidence and conclusions, but avoid the presentation of too many illustrations; the focus should be on the most representative examples.

Common Problems with Manuscripts

All parts of a manuscript are open to criticism from the title onward. The first requirement for gaining the confidence of an editor or a reviewer is to describe the findings objectively and in a sober style without the use of hyperbolic language. If your data are good, they will speak for themselves. It is always better for the reader to find that the results themselves are stronger than you claim.

Every "data not shown" statement may raise reviewers' suspicions that the authors are trying to hide something. If there really is not enough space to show important data graphically or in tabular form, then give some examples of the more important of such results in the text (with means, standard errors, or other indicators of variance and numbers of subjects, if it is a quantitative study).

The discussion section is the most difficult part of a manuscript to write, and it often shows. Sometimes the opening paragraph is only a summary of the results, which is not satisfactory. One approach is to decide which are the main new findings, mention only them, and summarize two or three important conclusions that follow from them. It is also common to find that the discussion does not focus on the aims as stated in the introduction and sometimes discusses issues on which no background was given. Such a failure to place findings in the context of previous knowledge means that the case for publication is not made. Reviewers and editors want to know what is new, what is confirmatory, and what fails to confirm previous findings. Instead, authors may attempt to extract too much from their data by trying to address too many different issues. The effect of this error is to dilute strong conclusions with weakly supported ones, giving an overall unfavorable impression leading to rejection.

The discussion should also consider alternative interpretations of the study and acknowledge major limitations. These may arise from methodological weaknesses or unexpected findings that could not be pursued to a firm conclusion because of practical limitations, such as the project period coming to an end or a financial constraint (these nonscientific reasons do not need to be stated). If the reviewers discover these weaknesses, they will consider themselves smart and are likely to make sure you know it; if you show that you are aware of the limitations and understand the implications, they will perceive you as smart and honest, which counts for a lot.

Do not waste time and space discussing "trends" that are not statistically significant; if the effect is not there, its implications do not need discussing. Remember that there are more than enough "significant" effects that do not replicate and there is no need to create new myths. If you believe that a real and important difference was undetected because of a lack of statistical power, the study needs to be repeated; that may be a factor to discuss.

Somewhat different problems are associated with reviews and theoretical articles. If a review claims to be comprehensive, it should state the way the literature search was carried out and define the criteria used for including articles (see Chapter 9). Articles that do not claim to be integrative reviews but rather argue the case for a particular theoretical viewpoint or set of ideas are often less comprehensive. In such instances, authors often cite publications that support their own position in a rather uncritical manner, and they may refer to few or no articles that oppose it. Editors may then firmly but politely ask the author to state the assumptions made and ensure that the article clearly indicates any controversial issues. Alternatively, where the intention is to let a distinguished writer express a personal view based on his or her selective citation of the literature, it should be made clear that a case is being made for a theory and that a balanced assessment of the state of the field is not being attempted.

Finally, remember you are writing a scholarly article and not running a campaign! Do not enter into politics and polemics. For example, if your main finding is that a widely used intervention is less favorable than another that lacks some sort of official approval for general use in your country, make the case for its relative merits and, if appropriate, argue for a policy change. But do not abuse the politicians and do not keep repeating the argument in more and more florid and emphatic language. Political battles are not won in the pages of academic journals.

The Editor's Decision

The much-anticipated response from the editor finally arrives, together with the statements from the reviewers. The editor will often need to reach a decision based on the balance between innovation and quality of work. The perfect manuscript would have important new ideas with far-reaching importance backed up by sound data obtained by means of thoroughly validated methods. In reality, such manuscripts are seen only rarely, if ever, and the editor and the reviewers have to make judgments. If the approach to a problem is highly novel or the study is a potential stimulus for further valuable work, a manuscript may be accepted with data that are less than wholly convincing. On the other hand, if there is not very much that is really new but the study is the first one to address a particular methodological weakness of previous work, then clear data of high quality will probably be essential.

The reviewers' reports and recommendations inevitably influence the editor's decision, but they are not the sole determining factors. Editors may study a manuscript in varying amounts of detail and may have concerns that are not reflected in reviewers' reports. These concerns may relate to any of the range of issues that the reviewers also address but may especially relate to the appropriateness of the subject matter for the journal, whether there are any ethical problems, and whether the importance of the work is sufficient to justify publication in their journal rather than in a publication of lesser status that may be struggling to fill its pages. Studies may be technically competent and presented well but may be unimportant because they merely confirm well-known facts or focus on apparently trivial issues. When the reports of reviewers are in agreement with each other, the editor will most frequently accept the recommendations made. It is a brave editor indeed who overturns the opinions of two independent experts—reviewers may soon stop assisting an editor who consistently ignores the advice given. When reviewers disagree, the editor may seek to sort out the matter by studying the manuscript and coming down in support of one or other reviewer; this is the ideal method if the editor can reach a clear view because he or she can reach a decision quickly without wasting another expert's valuable time. However, sometimes reviewers reach opposing conclusions on the basis of equally well-argued cases, and then the editor may feel it is essential to obtain advice from a third person. This is especially likely to occur if the work is outside the editor's main area of expertise.

When a third reviewer reaches a definite view supporting one or the other of the earlier reviewers, then the way forward is clear; but this does not always happen. If the first reviewer supports publication strongly and the second reviewer recommends rejection, the third reviewer quite often says the manuscript is weak but may reach publication standard after major revision; in such cases the contribution of the third reviewer may swing the decision one way or the other depending on the journal's needs at the time. If the journal is trying to raise the standard of published items, such marginal manuscripts will probably be rejected, whereas if the study is in a field that is under-represented in the journal, the editor may wish to include it. An editor may also seek to publish the article in a shorter form that reflects its lesser merits.

The abusive reviewer is a particular annoyance to editors. The most commonly identified, although happily quite rare, form of abuse occurs when a reviewer attempts a review of the author instead of the manuscript. It is one thing, and perfectly acceptable, to state that an argument is constructed poorly and is unconvincing, that it is presented badly, or that it does not take account of previous knowledge; it is quite another thing to assert that the author is stupid, careless, or ignorant. Editors have a duty to alter or remove such inappropriate remarks from a reviewer's report so that unnecessary distress is not caused and the author will be encouraged to improve the manuscript. If the reviewer is young and inexperienced, the editor may also explain the problem with the report, whereas a senior person will more likely not be invited to review again. Additional advice for inexperienced reviewers may be found in Chapter 13 of this book.

A particularly difficult situation arises if the review process generates suspicion that the author has engaged in scientific misconduct or another form of unethical behavior. Such misconduct may be either minor or major in nature, and the editor typically has available a range of sanctions to apply. These may include refusing to consider further work from the author, reporting the matter to the author's institution or employer, and publishing a statement in the journal to alert the scientific community to the issue. The availability of a code of practice by which editors can abide in such circumstances is very helpful (see The Farmington Consensus for the ethical practice guidelines developed by the International Society of Addiction Journal Editors). Editors are also wary of trying to resolve contentious ethical issues; they often do not have the resources to conduct a full investigation. Equally important, they cannot simply brush the matter aside by refusing to publish suspect material but must take reasonable steps to ensure that appropriate action is taken. These and other related issues are discussed in greater detail in Chapters 14 and 15, which deal with ethical considerations in scientific publishing.

Responding to Reviewers' Reports: General Rules of Conduct

Authors who regularly achieve immediate acceptance of a manuscript as submitted are rare indeed. Revisions are almost always required before acceptance, and in many cases a final decision cannot be reached until the revised version has been assessed. Therefore, the way in which authors respond to the reports of reviewers and to the editor can have a major influence on the outcome. If editors invite resubmission, they expect to receive the manuscript back again.

An invitation to resubmit is not a half-hearted and cowardly way of saying the work is unpublishable but rather an implicit suggestion that the editor
remains interested in the article and that it is likely to be accepted if the author
is responsive to the questions and recommendations of the reviewers. In such
cases, it is nearly always worth resubmitting unless there is some clear and unavoidable requirement with which you cannot possibly comply. The remainder
of this section offers guidance for authors on how to navigate through this maze
successfully.

The overriding aim of the response is to engender trust among editors and
reviewers. Authors should never claim to have made changes that in fact they
have not done. If the cover letter says all requested changes have been made
and an editor or reviewer checks two or three points at random and finds nothing much has changed, he or she may reject the manuscript without looking
carefully at the rest of it. If you have made major changes by rewriting whole
sections of the manuscript, state that is the case and identify the sections. Alternatively, if just a few words needed to be inserted or deleted, make clear which
words were changed so that reviewers can see what has been done. If you were
asked to shorten something, you should almost always do so and perhaps state
by how much (i.e., by how many words or pages). Do not try to fool the editor
by printing the new version in smaller type or by other stylistic changes. Be
polite, even if you feel that the reviewers have not understood your intentions.
When you have been through all the points of criticism, you should have an
idea of the changes you think are appropriate. Will they be enough?

If after reading the reports you have concluded that none of the recommendations is worth accepting and you do not want to make any changes, it is common sense to take a break from the job and look at it again on another day! It
is simply not realistic to expect editors and reviewers to accept that none of the
changes they request and the criticisms they make is well founded. Reviewers
spend anything from an hour to a full day preparing their reports. If you dis-
miss this effort out of hand, you will get nothing published. You must therefore
aim to make changes to deal with as many as possible of the points raised and,
preferably, with a clear majority of them.

Occasionally authors may feel that an editor's decision to reject their manuscript was unnecessary because the criticisms made could be answered through
revisions. In such cases, in which there was no other clear reason given for
rejection, authors may wish to seek approval to resubmit. For example, there
may be no criticism of the conduct of the study or analysis of the data, but
the reviewer may feel that the interpretation is so seriously flawed that the
conclusions are not supported by the data. The manuscript might therefore be
publishable if the authors are willing to revise their conclusions. Resubmission
after rejection should be preceded by a carefully considered letter to the editor
explaining why you believe that you can deal with the criticisms made. The
editor will then decide whether to alter the previous negative decision and may
agree to consider a revised version. Seek approval before resubmitting because

if a rejected manuscript is resubmitted without prior agreement, it is very likely that the editor will refuse to consider it. Some journals have stated their appeals procedure whereas others deal with appeals on an ad hoc basis.

The Cover Letter: Make Life Easier for the Editor

Once you complete the changes to the manuscript, write a detailed reply to the reviewers. It is worth spending a significant amount of time getting your reply to reviewers as near to perfect as you can. Sometimes constructing the letter takes as long as revising the article, but it will not take as long as botching the job and then being obliged to reformat the manuscript for another journal to start the whole process over again. Nevertheless, it is best to keep this reply as short as possible. Typical successful replies will be in the range of one to three single-spaced pages. If the reviewer makes a point in just three lines and you need a page to rebut it, it is likely you have not gotten straight to the heart of the matter, and your reply will probably not be convincing. It is best to write the minimum needed to refute the criticism.

If the reviewer cannot understand a point that you made in the manuscript, it may be because he or she is lacking in intellectual capacity (as we often think when we encounter such comments on our own work). However, if one person does not follow what you have written, the same may apply to others. Reviewers are all published research workers and are often the very people whom you might hope would read your article; if a reviewer cannot understand your point, try to analyze your text to see how the misunderstanding may have arisen. Then make changes to ensure it will not happen again.

Do remember that if a reviewer asks a question, other readers may want to know the answer to it too. The answer should therefore usually be contained in the revised manuscript and not in the cover letter. Save the reviewers' time and they will love you; do not answer a question in the letter and then refer reviewers to a section in the manuscript that they have to read over and over to check if it is really there!

If possible, reply in numbered sections that correspond with the reviewers' numbered points. Explain the revisions you made to deal with most of the criticisms, and also explain why you did not deal with the rest. Describe briefly each change you made, referring to the relevant page or paragraph in the revised manuscript. Try not to respond in a combative, overly assertive style. If there are major and important changes recommended that you are sure are wrong, then present a concise, logically argued rebuttal. If there are minor changes requested that you feel do not really improve matters, do them anyway because it helps a lot if you can truthfully claim to have dealt with the majority of points. At all stages, remember that although reviewers and editors may appear to be distant, self-opinionated, and arrogant, they are also human beings with their own feelings, emotions, and problems. If you want acceptances, make life easy for them by writing clearly, and do not antagonize them with criticisms or gratuitous insults, however unwise and misguided you think the reviewers may be. It is also worth making the changes via a tracking system in a different color (such as red) on the manuscript to show clearly where changes, additions, and deletions have been made.

It is sensible to maximize and stress the points that you agree with that the reviewers wrote and to acknowledge their contribution when they have made suggestions that improve the manuscript. Do not build minor disagreements into major issues. You probably need to make only minor changes to accommodate them and then mention the changes in the cover letter and should not waste time arguing and or risk offending the reviewer in the process. However, it is not necessary or appropriate to minimize disagreements to the point of dishonesty; they should be dealt with by logical rebuttal in the cover letter and, sometimes, by acknowledging and discussing the point in the manuscript.

Perhaps the most difficult case occurs when you feel that a reviewer shows a bias towards a theoretical approach that differs from yours, and therefore undervalues the work. Here you can explain in the cover letter that there are different approaches to the problem (state what these are), that yours is equally valid, that there is a genuine difference of opinion and that you have a different but scientifically legitimate point of view. However, this strategy is probably unwise unless you have a strong case and there is no other way to deal with the issue. In the end, the editor will have to decide and what one person perceives as objective and unbiased looks very different from another viewpoint. At the end of the day, the editor wants to have articles to publish. The number of acceptances rather than of rejections is therefore the mark of success and of an editor's job well done. Authors, editors, reviewers, and publishers must all work together to ensure the production of a journal of high quality that achieves its intended objectives.

Acknowledgements

We thank Klaus Mäkelä and Kerstin Stenius for assistance with drafting the material about qualitative research.

Please visit the website of the International Society of Addiction Journal Editors (ISAJE) at www.isaje.net to access supplementary materials related to this chapter. Materials include additional reading, exercises, examples, PowerPoint presentations, videos, and e-learning lessons.

Appendix A: General Publications on Scientific and Medical Publishing

This is a very short selection from the huge number of publications. Many additional works may be found by searching biomedical databases such as PubMed or on-line booksellers.

Albert, T. (2000). *The A-Z of Medical Writing*. London, England: BMJ Books.

American Psychological Association. (2010). *Publication Manual of the American Psychological Association* (6th ed). Washington, DC: Author.

British Medical Association. http://bma.org.uk/about-the-bma/bma-library/library-guide/reference-styles This is a useful website established by the British Medical Association, giving general guidance on resources for people publishing in the biomedical field.

Hofmann, A. K. (2013). *Scientific Writing and Communication: Papers, Proposals, and Presentations*. New York, NY: Oxford University Press.

Huth, E. J. (1990). *How to Write and Publish Papers in the Medical Sciences* (2nd ed). New York, NY: Williams & Wilkins.

Iverson, C. (Ed.). (1998). *American Medical Association Manual of Style: A Guide for Authors and Editors* (9th ed.). New York, NY: Williams & Wilkins.

Katz, M. J. (2009). *From Research to Manuscript: A Guide to Scientific Writing* (2nd ed.). New York, NY: Springer.

McInerney, D. M. (2002). *Publishing Your Psychology Research: A Guide to Writing for Journals in Psychology and Related Fields*. Thousand Oaks, CA: Sage.

Moher, D., Schulz, K. F., Altman, D. G., & for the CONSORT Group. (2001). The CONSORT statement: Revised recommendations for improving the quality of reports of parallel-group randomised trials. *The Lancet, 357*, 1191–1194.

Peat, J., Elliott, E., Baur, L., & Keena, V. (2002). *Scientific Writing: Easy When You Know How*. London, England: BMJ Books.

Richardson, P. (Ed.). (2002). *A Guide to Medical Publishing and Writing: Your Questions Answered*. London, England: Quay Books.

Strunk, W., Jr., & White, E. B. (1999). *The Elements of Style* (4th ed.). New York, NY: Longman.

Appendix B: Checklist of Items to Include When Reporting a Randomized Trial

This section consists of a slightly shortened version of the checklist from Moher et al. (2001).

TITLE AND ABSTRACT   How participants were allocated to interventions (e.g., "random allocation," "randomized," or "randomly assigned").

INTRODUCTION	Scientific background and explanation of rationale.

METHODS

Participants	Eligibility criteria for participants and the settings and locations where the data were collected.
Interventions	Precise details of the interventions intended for each group and how and when they were actually administered.
Objectives	Subjective objectives and hypotheses.
Outcomes	Clearly defined primary and secondary outcome measures and, when applicable, any methods used to enhance the quality of measurements (e.g., multiple observations, training of assessors).
Sample size	How sample size was determined and, when applicable, explanation of any interim analyses and stopping rules.
Randomization	Method used to generate the random allocation sequence, including details of any restriction (e.g., blocking, stratification).
Allocation concealment	Method used to implement the random allocation sequence (e.g., numbered containers or central telephone), clarifying whether the sequence was concealed until interventions were assigned.
Implementation	Who generated the allocation sequence, who enrolled participants, and who assigned participants to their groups.
Blinding	Whether participants, those administering the interventions, and those assessing the outcomes were aware of group assignment.
Statistical analysis	Statistical methods used to compare groups for primary outcome; methods for additional analyses, such as subgroup analyses and adjusted analyses.

RESULTS

Participant flow:	Flow of participants through each stage (a diagram is strongly recommended). Specifically, for each group, report the numbers of participants who were randomly assigned, received intended treatment, completed the study protocol, and were analyzed for the primary outcome. Describe protocol deviations from study as planned, together with reasons.
Recruitment:	Dates defining the periods of recruitment and follow-up.

Baseline data:	Baseline demographic and clinical characteristics of each group.
Numbers analyzed:	Number of participants (denominator) in each group included in each analysis and whether the analysis was by "intention to treat." State the results in absolute numbers when feasible (e.g., 10/20, not 50%).
Outcomes and estimation:	For each primary and secondary outcome, a summary of results for each group, and the estimated effect size and its precision (e.g., 95%confidence interval).
Ancillary analyses:	Address multiplicity by reporting any other analyses performed, including subgroup analyses and adjusted analyses, indicating those prespecified and those exploratory.
Adverse events:	All important adverse events or side effects in each intervention group.

DISCUSSION

Interpretation	Interpretation of the results, taking into account study hypotheses, sources of potential bias or imprecision, and the dangers associated with multiplicity of analyses and outcomes.
Generalizability	External validity of the trial findings.
Overall evidence	General interpretation of the results in the context of current evidence.

Source: Moher, D., Schulz, K. F., & Altman, D.G., for the CONSORT Group. (2001). The CONSORT statement: Revised recommendations for improving the quality of reports of parallel-group randomised trials. *The Lancet, 357,* 1191–1194.

CHAPTER 13

Reviewing Manuscripts for Scientific Journals

Robert L. Balster

Introduction

One of the main moral principles of virtually all major religions and cultures is the ethic of reciprocity, sometimes known as the Golden Rule: Treat others as you would like to be treated. Show mutual respect. This Golden Rule is also a fundamental principle of the process of peer review, including the review of manuscripts submitted to professional journals. If you have been asked to review someone's article, it is very likely the case that you are an author yourself and will have been subjected to the same peer-review process. Keeping in mind how you expect your own journal submissions should be reviewed, you could readily derive from that experience nearly all of the advice I will be offering you in your role as a reviewer.

Goal

The goal of this chapter is to provide rather specific principles and suggestions on how to be a competent reviewer. Most editors of peer-reviewed journals see the peer-review process similarly, even if some of their specific journal policies differ. It is those commonalities that I will address here, approaching the topic from my various roles as former editor-in-chief of *Drug and Alcohol Dependence*, member of several editorial boards for other journals, and reviewer for

How to cite this book chapter:
Balster, R L. 2017. Reviewing Manuscripts for Scientific Journals. In: Babor, T F, Stenius, K, Pates, R, Miovský, M, O'Reilly, J and Candon, P. (eds.) *Publishing Addiction Science: A Guide for the Perplexed*, Pp. 245–263. London: Ubiquity Press. DOI: https://doi.org/10.5334/bbd.m. License: CC-BY 4.0.

many others. Interested readers may also want to consult prior publications on peer review in general (Moghissi, Love & Straja, 2013; Godlee & Jefferson, 2003) and on peer review of journal submissions specifically (Girden & Kabacoff, 2010; Smart, Maisonneuve & Polderman, 2013; Hames, 2007). Chapters 7, 8, 9, and 12 in this book also provide information that is relevant to journal reviewing.

I should mention that reviewers of journal submissions are sometimes referred to as *referees*. I am not aware of any distinction between being a reviewer and a referee. I have always preferred the term *reviewer* because *referee* conjures images of a sporting contest with winners and losers. When you are a good reviewer, everyone wins—authors, editors, and the scientific community—because reviewing is fundamentally a constructive process. Yes, reviewing involves making judgments and recommendations, but reviewers are not the decision makers in the process. This responsibility falls to the editor. Thus, I will use the words *reviewing* and *reviewer* throughout but just as a matter of personal preference. It seems likely that the words used for the reviewer function in non-English languages have nuances of their own to consider, but that is a matter for another discourse by someone with greater language skills than I.

Brief Overview of the Journal Review Process

If a journal declares itself to be a "peer-reviewed" journal, this normally means that all articles that are eventually published in that journal have been reviewed. Peer review implies review by outside reviewers as well as the editorial staff. Of course, journals differ in their application of the peer-review process, and it is common for journals to publish editorials, commentaries, book reviews, and similar content without peer review, reserving that form of assessment for research reports and critical reviews.

Before moving on to a more detailed discussion of journal reviewing, I want to give a brief overview of the process so readers can appreciate the steps involved. Box 13.1 outlines the basic steps of the review process used by most journals, keeping in mind that editorial structure differs among journals. For most journals today, submissions are made using an interactive Internet-based system, whereby all articles are processed in a centralized editorial office, or at least as email attachments. The editor then decides whether to edit the article or assign it to some kind of associate or assistant editor. Some journals may have more than one submission office or site, depending on where the author comes from or the general topic of the article. Some journals may have more than one editor look at the submission before assigning it to reviewers, whereas others may have an editorial team that considers the recommendations of the reviewers. To simplify this discussion, I will assume there is one "decision editor" who assigns reviewers and makes decisions. I will refer to this individual as the editor, regardless of the specific editorial title assigned by the journal.

1. Editor develop a reviewer database.
2. Authors submit manuscripts to the journal.
3. Editor(s) make an initial assessment to decide if the article is suitable for the journal and if peer review is warranted.
4. Editor selects reviewers and invites them to review.
5. Editor monitors the timeliness of peer review and sends reminders or invites new reviewers if necessary.
6. Reviewers complete the review and provide recommendations and comments to the editor and comments to the authors.
7. Editor makes decision to accept the submission, asks authors for a revision, or rejects the submission.
8. Most journals notify reviewers of the editor's decision.
9. If required, authors revise submissions and return to the editor.
10. Editor decides if further review is needed; if so, the process recommences at Step 4.

Box 13.1: Steps in the Journal Review Process.

The first step in the peer-review process is the assignment of reviewers. In the next section, I will discuss reviewer databases and how editors select and invite reviewers (Steps 1 and 4). All journals have a procedure for monitoring the review process after reviewers have been assigned (Step 5). This has become easier with computer-based systems that notify the editorial team when assigned reviewers decline the invitation to review, when reviews are completed, and when they are late.

At some point, the editor stops the review process and decides whether to accept the manuscript for publication (Steps 6 and 7). Editors carefully consider the recommendations of the reviewers and their comments on the article, but ultimately editors themselves must make the decision based on both the reviews and their own assessment of the submission. Editors usually inform reviewers of the decision (Step 8), and often share the comments of all the reviewers with each other, which I believe is a good practice to strengthen the review process.

There are three basic options open to the editor. First, the editor may accept the submission for publication as submitted. For many journals, it is rare to accept a first submission of an article without asking for any revisions at all, but it does happen.

Second, if the submission seems to the editor to be potentially publishable, he or she will ask the authors to make revisions (Step 9). Most journals divide revisions into minor and major categories. Minor revisions do not change the article very much, and the resubmission usually does not require additional outside reviews. Major revisions typically require significant changes to the article, such as the collection of additional data, a change in the way the existing data are analyzed or presented, or even changes to some of the conclusions of the report. Manuscripts with major revisions are often returned to the same peer reviewers for their comments and recommendations, although at times an editor may send the manuscript to different reviewers (Step 10).

The third option open to editors is rejection of the submission. Some submissions may be rejected without peer review (Step 3), such as when the topic is not appropriate for the journal readership or the form of article (e.g., case report, book review) is not one published by the journal. Editors may also decide that the article's methods or other characteristics give it little or no chance of receiving a positive recommendation in the peer-review process. Giving immediate feedback on such articles may be in the authors' best interest, because it allows them to submit elsewhere without delay. Rejection without review also saves the time of busy reviewers, who are more useful to the journal when reviewing manuscripts with a greater chance of success. Most often, editors base their rejections on negative recommendations in the peer-review process. In such cases, authors usually receive comments from the editor or reviewers that describe some of the weaknesses of their submission. Ideally, authors consider these comments and revise the article before they send it to another journal.

One of the newer developments in journal publishing is the cascading of journal submissions within publishing consortia of individual journals (Barroga, 2013). With the author's permission, editors can forward a rejected submission and the completed reviews of an article to another journal within the consortium where the article may have a better chance of final acceptance. This saves reviewer time and can expedite final publication because the review process does not need to start all over again. A typical consortium includes journals within a publishing company.

How Do Editors Select Reviewers?

As we turn to the principle of reciprocity, who do we, as authors, want assigned as reviewers of our article? To be honest, we probably prefer reviewers who are known to be favorably impressed with our work. But minimally, we want reviewers who are knowledgeable about our field of study and who will be fair in the review process. This is what editors want too; they want competent reviewers with specialized knowledge of the potential advantages and pitfalls of various research approaches. They also want reviewers who are fair and unbiased, without conflicts of interest. Finally, editors want reviewers who complete their work on time and who write constructive comments about the submission.

The process of selecting appropriate reviewers usually involves the use of a database. At a minimum, such a database includes email addresses by which to contact reviewersIn the case of a large, multidisciplinary journal, editors also need some means of matching reviewers with submissions. Smaller, more specialized journals often rely mainly on their normal editorial advisory boards for reviews, but *ad hoc* reviewers carry most of the load for larger journals.

In addition to up-to-date contact information, a typical reviewer database includes some means of identifying their reviewers' areas of expertise. For example, the database may assign to each reviewer one or more keywords or classifications (e.g., molecular genetics, pharmacotherapy, prevention, policy). Commercial editorial software systems are particularly adept at matching the keywords assigned to both submissions and reviewers to provide editors with a list of knowledgeable reviewers. These programs also tell editors if reviewers are currently assigned other manuscripts, the date of their last review, and the number of reviews they have done lately. It is also possible to see if reviewers have defaulted on prior assignments or how long they have taken to complete their prior reviews. Some systems even allow editors to rate reviewers to help them remember who provided useful suggestions and constructive and timely comments in previous reviews.

There are several ways to get one's name added to a journal's database. Perhaps the most common way is by publishing an article in that journal. Editors typically prefer reviewers who themselves have published several articles as senior or corresponding author. If a suitable reviewer is not already in the database, many editors do a quick author search using PubMed or other search tool to see if potential reviewers have other publications, and the editorial software often facilitates this search. Sometimes authors suggest reviewers for their manuscripts. If the editor agrees with the suggestion and the reviewer is not in the database, the reviewer will be added. Editors may ask their editorial advisory boards to suggest new reviewers for the database, and most editors are also pleased to receive self-nominations. Finally, an editor sometimes receives submissions for which no suitable reviewer is available. In such cases, editors use their knowledge of the field or use authors cited in the submission who are clearly doing related work to add new reviewers to the database.

Selecting the best possible reviewer for a submission is one of the most important responsibilities of an editor. In my experience, taking time at this step to ensure a good match between reviewer and submission will often save time—and requests for more reviews—later on. Reviewers, too, prefer to assess articles in areas in which they feel qualified. Each editor goes about the matching process in a different way. In addition to seeking reviewers with expertise in the area, editors also may seek to balance their selection of reviewers for a single submission along several dimensions. For example, editors may seek a methodological balance, in which a reviewer with specific knowledge of a data-analytic approach complements another with knowledge of the content area of the work. It is often good to pair a senior scientist with a less-experienced reviewer because this can serve to train the junior person who subsequently sees the comments of the senior reviewer. Sometimes the editor needs a reviewer known to be unbiased in a particularly controversial area. Editors usually like to have reviewers who provide a broad perspective on the work to ensure geographical, cultural, or gender balance.

In general, only two reviewers are needed for each submission, but there are times when more than two are invited. If the review process becomes delayed because of problems obtaining timely reviews, editors may add a reviewer they know to be particularly reliable. These and many other subtle factors make the reviewer-selection process a challenging one.

In addition to selecting knowledgeable and reliable reviewers, editors do their best to avoid actual or perceived conflicts of interest (COIs) and bias. As I will discuss later, editors cannot know about every possible COI or bias that reviewers may have; therefore, they rely on reviewers to tell them. There are some relatively straightforward methods editors use to try to avoid apparent COIs. They generally exclude as potential reviewers scientific colleagues close to the author, such as reviewers who have co-authored in recent years with any of the authors of the submission or persons known to be part of the same research team. It is usually wise not to assign reviewers from the same institution as the authors, but, in the case of authors based at large, multi-campus institutions, such precautions may not always be necessary. Because editors are usually experts themselves, they may be aware of longstanding disagreements or controversies and take care to select unbiased reviewers. It is impossible to eliminate potential bias completely, but editors do the best they can.

The actual process of inviting reviewers differs among journals. Some editors send a copy of the submission and reviewing instructions directly to the reviewer. Other editors first invite possible reviewers, usually by sending the abstract by email and asking them first to agree to do the review by a certain deadline. If the reviewers agree, they are then assigned the review and provided with access to the full article, review forms, and instructions. One advantage to this invitation process is that potential reviewers can identify COIs they may have or indicate that they lack expertise in the area of study. Some editors will invite several reviewers and then assign only the first two who agree. Computer technology and the use of email have automated some of these steps, making the process faster. It is now possible to receive a submission, invite reviewers, assign them, and get the review process started all in one day.

Why Have Peer Review?

Before further discussion of how to be a good reviewer, I will explain why we use peer review and why someone would want to be a peer reviewer. Peer review has four primary objectives: (a) advise the editorial decision-making process, (b) justify rejections, (c) improve the quality of acceptable manuscripts, and (d) identify instances of ethical or scientific misconduct.

Editorial Decision Making

The most obvious reason to seek reviews of journal submissions is to help the editor make a decision about the article's acceptability for publication in that journal. One should always remember that reviewers only make recommendations; it is the editor who chooses whether to follow those recommendations. This fact often frustrates reviewers, who may feel that the editor ignored their advice. A consideration of some of the factors that affect the editor's decisions can relieve some of that frustration. Editors must make decisions based on all of their reviewers' input, and sometimes different reviewers give conflicting recommendations. The editor may feel that a problem identified by a reviewer could be addressed in a revision or that the author may have good arguments for why the problem is not an important consideration. Indeed, the editor may read the article and disagree with the reviewer's interpretation. An editor may feel that the importance of the article mitigates some problems identified by the reviewers. On the other hand, reviewers may not find any fault with an article that the editor decides carries little impact or belongs in a different journal. Most journals receive more submissions than they can publish and thus editors must choose among potentially publishable articles with no major flaws. All of these factors and others go into the editorial decision-making process. Editors who choose not to follow a reviewer's recommendations may still consider the review excellent. Reviewers who are provided access to the editor's decision letter, or to the comments of the other reviewer, can usually glean the basis for the editor's decision.

Explaining the Basis for Rejection

Nearly all journals ask reviewers to provide comments and suggestions on the submission, which are then sent to the author with the decision letter. When reviewers recommend rejection of a manuscript, they should provide the basis for that recommendation in their comments to the authors. Not only does this give authors the feedback they deserve on their submissions, but it also helps the authors to revise their articles for submission elsewhere and to improve their work in general.

Improving the Quality of Published Articles

Certainly, one of the most important reasons for obtaining constructive comments during the review process is to make the articles that are ultimately published the best they can possibly be. I have seen modest articles transformed during the review process into far better and more important contributions to the field. I received many statements of appreciation by authors for the improvements in their manuscripts brought about by peer review. Of course, not all authors appreciate constructive criticism or relish doing the extra work required to revise their submissions to satisfy reviewers, but I believe that most authors value expert criticism of their work. I suppose that some contributions could be weakened when authors follow reviewer advice, but I cannot recall a specific example where I know this occurred. It is the responsibility of the editor to make judgments about major changes reviewers ask authors to make and to tell authors in a cover letter if the editor does not agree with a reviewer's suggestions. Editors should also carefully consider counter arguments made by authors who prefer not to make some recommended changes.

What is the scientific evidence that peer review improves the quality of scientific publication? Although this chapter is not the place to have a thorough discussion of this topic, it may interest the reader to know that there has been relatively little research on this topic, and the work that has been done does not provide a clear answer to the question (Fletcher & Fletcher, 2003; Overbeake & Wagner, 2003; Jefferson et al., 2007). Nor is there much empirical research on various approaches to peer review (e.g., open vs. blinded review). Criticisms have been published of the peer-review process for both journal publications and research grant applications (e.g., Smith, 2006). Nonetheless, editors in general see articles improve through the peer-review process, but perhaps they have a bias, and peer review is the only system we have for quality control in journal publication.

Identify Areas of Ethical or Scientific Misconduct

Most journals request that reviewers comment on any research subject-protection issues or other ethical concerns they detect in a submission. Reviewers' careful examination of the data may reveal inconsistencies between reported methods and the ways in which data are presented or analyzed or may uncover highly unlikely data sets (e.g., with no variability) that may lead reviewers to suspect errors in data reporting or outright data fabrication.

Reviewers may know of concurrent submissions by the same authors of the same manuscript with the same data to two or more journals or prior publications. It is important for reviewers to communicate their concerns about possible ethical or scientific-misconduct problems to the editor. Typically, this is done in confidential comments to the editor, who then must investigate these concerns. Reviewers do not need proof of these types of problems, just a reasonable basis for concern.

Why be a Journal Reviewer?

In the following, I expand on seven of the main reasons researchers review manuscripts for scientific journals. But keep in mind that the most important reason many people do it is this: They enjoy it. Nonetheless, thinking critically about science, staying informed of the latest advances, and making a contribution to health are what attracted many of us to science in the first place. Reviewing is a scholarly, creative, focused, important activity that is capable of being completed with a few hours of work. What could be better?

Fulfill Your Professional Responsibility

The peer-review process is a simple application of the Golden Rule. You need people to review your articles, and therefore you should review those of others. A system of all authors and no reviewers is doomed. Thus, it is your professional responsibility to be part of the process, both as an author and reviewer. I do not subscribe to the view that senior scientists can be excused from the peer-review process because they "did their duty" earlier in their careers. If you write, you should review. Indeed, the perspective of experienced scholars can be particularly important.

Improve Your Understanding of the Peer-Review Process

Younger scholars particularly need to learn the peer-review process, because much of their career success will depend on it. Many scientific mentors include in their training a gradual exposure to doing peer reviews. A good strategy is for a mentor to ask a junior colleague to prepare a review of a manuscript that has been assigned to the mentor. The mentor should explain fully the confidentiality issues surrounding the review process when asking a junior colleague for help. The novice reviewer then returns the review to the mentor who modifies it as needed and gives feedback to the trainee. The mentor then submits the review to the journal. In such instances of guided peer review, mentors should tell the editor the name of the junior colleague who helped with the review. Mentors might even recommend that the junior colleague be added to the reviewer database once they feel the colleague is ready to do independent work.

Improve Your Critical Thinking

The review of other people's work improves your critical thinking about your own work. Good reviewers attempt to articulate both the strengths and weaknesses of a particular scientific approach. You may be using a similar approach in your own work without thinking about it critically as often as you should. Perhaps you are considering the use of methods similar to those in the article you are reviewing. Thinking about some of the article's weaknesses can lead you to improve the approach.

One of the most useful ways to evaluate your own critical thinking is to see what the other reviewer(s) and editors say about the work. I strongly believe that journals should make all comments to authors and decision letters available to reviewers on an anonymous basis, although not all journals do. When another reviewer or the editor identifies a serious flaw in experimental design or data analysis that you missed, it can be both instructive and a little embarrassing.

Improve Your Own Writing and Data Presentation

Reviewing a manuscript (and comments from other reviewers and the editor) gives you new insights into how to improve your own writing, presentation, and data analyses. By seeing how authors make revisions and respond to reviewer comments, you can improve your own approach to revising articles. Reviewers often advise authors on how articles can be shortened and more sharply focused. As reviewers help authors focus their writing, they likewise learn to do this with their own articles. In addition, proofreading other manuscripts for errors improves the proofreading of your own writing.

Learn More about Research in the Field

Reviewing a scientific article can give a reviewer a much better understanding of the work than does just reading it because of the critical thinking involved in doing a review. In addition, you will probably be asked to review articles that you would not normally read because they are a little outside your specific area of work. The comments and suggestions of the editor and other reviewer(s) are sometimes more interesting and important to you than the article you reviewed. Some of the best scientific writing I have encountered has taken the form of reviews. After all, reviewers are experts in the field who are asked to summarize the salient strengths and weaknesses of a scientific study or new hypothesis in a few paragraphs.

Build Relationships with Journals

The databases that include you as a reviewer generally will also include you as an author. Persons who are regular authors and reviewers for a journal and who are successful in both these roles build a relationship with a journal, becoming good "journal citizens." The editorial team comes to know who you are, appreciate better your areas of expertise, and develop confidence in your work. Your good relationship with a journal may result in your being asked to join its editorial advisory board or take on editing roles yourself.

Fulfill Service Obligations for Promotion

As scientists, our work is constantly being evaluated. This includes assessment for promotion and/or tenure. For most scholars, service to the profession is one of the areas where we are evaluated and judged. As mentioned above, journal reviewing is an important service that should be acknowledged and rewarded. I know that reviewers typically report their journal reviewing activities on their curricula vitae and include them in regular activity reports. Simply being invited to review a manuscript is evidence that you are known in the field and that an editor has some confidence in your expertise. Many journals regularly publish lists of recent reviewers and some use additional incentives (such as small gifts or reduced costs for the publisher's books and journals) to reward good reviewers.

How to be a Good Reviewer

Much of the advice in this section stems from the goals of the review process, as described above. Simply stated, a good reviewer helps the editor achieve the goals of peer review. Another guiding principle is our friend the Golden Rule. You should be the kind of reviewer that you would want to review your work. It is always important for reviewers to try to take an author's perspective and to remember that publishing is vitally important to authors. Sometimes an author needs one more publication to ensure a promotion or to receive a favorable review on a grant application. Authors may be performing research in a highly competitive area where having an article accepted for publication is crucial evidence of their precedence. Reviewers should easily be able to imagine an author's response to a careless review or one that is delayed by months. Reviewers often offer excuses to the editor for late or cursory reviews, but the author feels the delay—and the curt treatment—even more keenly than the editor. I suggest placing the following aphorism on your desk as you participate in the review process: Review others as you would like to be reviewed.

Below, I present what could be viewed as one former editor's advice on being a good reviewer. I have ordered this section essentially in the order of steps in the review process as presented in Box 13.1.

Respond Promptly to Invitations to Review

As I mentioned above, many journals now use email or fax to invite reviewers for a submission. The invitation typically includes only basic information about the article and an abstract. The worst thing you can do with an invitation is ignore it. It takes only a minute or two to decide if you are able to accept the invitation and then notify the editor. If you want to do the review but cannot complete it by the deadline, contact the editor and see if a later deadline is acceptable. If you decide you cannot accept the invitation, you can help the editor with some suggestions for other reviewers, but the main thing is to tell the editor promptly. Failure to reply puts the editor in the difficult situation of deciding how long to wait until contacting other potential reviewers and puts the review process behind schedule. To avoid delays with this invitation step, some editors invite several qualified reviewers and hope that the required number will agree promptly.

Reviewers decline invitations all the time. Editors are used to this. If you are too busy at the moment, have other review assignments to complete, do not feel competent to review the article, or have a COI, editors will understand, especially if you regularly agree to write reviews for that journal. If you will never agree to do a review for that journal, it is best to tell the editor so you can be removed from the reviewer database.

Notify Editor of Any Potential COIs or Previous Reviews You Did of the Article

If you have an obvious COI, you should decline the invitation to review. Often reviewers are uncertain if they should declare what appears to be a slight COI, such as a small collaboration with an author or a collaboration that occurred many years ago. In these cases, it is best to tell the editor about your concerns, who can then decide if it represents a real conflict. Putting such ambiguous conflicts on the record can often lessen an editor's concern. On the other hand, if you feel there might be a conflict that might affect your ability to give a fair and unbiased assessment, then you probably are in conflict and should not accept the invitation to be a reviewer.

More often than one might expect, reviewers are invited to review articles they have already reviewed and rejected for other journals. Many reviewers are in more than one reviewer database, and these journals match reviewers to submissions in much the same way. Some reviewers prefer to decline the second invitation, often stating that they do not want to place the author in double jeopardy. In such instances, I recommend asking the editor what to do. I personally had no problem with reviewers assessing the same manuscript for two different journals. If the new article is identical to the one reviewed earlier and the reviewer feels that the same recommendation and comments are in order, then he or she should submit them again. But if the author has made improvements before resubmitting to the new journal, the recommendation should address this improved manuscript. Authors should be advised that they take a risk if they submit a rejected article to another journal without addressing the concerns raised with the initial submission, because their manuscript may be assigned to the same reviewer.

Maintain Confidentiality

Submissions to journals are confidential information. Reviewers should scrupulously respect the secrecy of the information, including even the existence of the submission. Reviewers who solicit input on a review from a colleague or trainee must first inform the colleague or trainee about the confidentiality of the information. The reviewer is responsible for any disclosures by his or her consultants.

The most unethical use of information in manuscripts under review occurs when the reviewer uses the information to facilitate directly the reviewer's own work, for example by using a new methodology before it has been published or by citing the work in his or her own publication or grant application. Failure to maintain confidentiality and misuse of information obtained in the review process is scientific misconduct.

I strongly advise reviewers not to contact author(s) with questions or offers to negotiate some changes in the manuscript or for any other reasons related to the submission. This applies even if the journal uses an unblinded review process, whereby the authors know the identities of their reviewers. Such communications usually go badly for both reviewer and author, who can end up arguing about the article, and they improperly exclude the editor from the decision-making process, perhaps concealing from him or her important aspects of the review process. It also can damage important scientific relationships among the parties involved.

Complete the Review on Time

Editors give reviewers a fixed period within which to return a review recommendation. Reviewers who are late ultimately disrespect the author. When a reviewer accepts an invitation to review with a specified deadline, there is little excuse for tardiness. If you know you are going to be late with a review, notify the editor, who can decide whether to wait for your review or invite someone else. One of the most disagreeable aspects of being a journal editor is the need to remind reviewers, often several times, of late reviews. Computerized reviewer databases keep track of how long it takes reviewers to complete reviews so that chronically late reviewers can be removed from the database.

Make a Publication Recommendation

Completing a review requires at least two steps: making a publication recommendation and writing comments to the authors (and editor, when needed). Both are important, but I begin with some advice about making recommendations. Read the reviewer instructions carefully to learn on what basis the journal wants you to make a recommendation. How does the editor want you to balance technical merit versus importance to the field, etc.? Most journals ask for your overall recommendation based on the review criteria specified. Although journals differ on this, they usually want one of four possible decisions: accept, minor revision, major revision, or rejection. Because many journals receive more technically acceptable articles than they can publish, editors want recommendations that also consider the importance of the information and whether it covers new ground or applies a novel perspective. In my experience, reviewers are least comfortable with judging the importance of submissions, but as experts in the area, they may be in the best position to make this judgment.

Provide Confidential Comments to the Editor

Nearly all journals give reviewers the option of providing comments to the editor. These comments are confidential and not shared with the author or other reviewer(s). There is no need to reproduce your comments to the author, but do include things you believe the editor should know besides your comments to the author. The following are some of the matters you would want to bring to the confidential attention of the editor.

Identify COIs not previously reported. If you have some relationship to the authors or have some financial or personal interest in the work you are reviewing, you should tell the editor. The editor will take this information into account when making decisions based on your recommendations. If the editor believes the COI precludes you from being a reviewer, your review may not be considered in making a decision and your comments to authors may not be sent on. Situations like this are rare, but it is better to tell the editor too much than too little.

Identify your areas of expertise. There may be some aspects of the article about which you do not feel competent to provide a review. As I mentioned before, you may have been invited to complement another reviewer who lacks expertise in your area. Tell the editor if there are parts of the article for which another expert is needed. It is far better to place statements about your areas of expertise in the comments to the editor than in your comments to the author.

List concerns about ethics or scientific misconduct. As I have already mentioned, you need not be certain of ethical problems to report your suspicion. If you have sufficient cause for question, tell the editor.

Provide other comments. Reviewers may have additional information or some options the editor should consider that are not appropriate to send to the authors. Reviewers should feel free to help editors in any way they can to make the right decision and seek changes in a submission.

Examples include when a reviewer is familiar with a controversy surrounding the reported research or when there is excessive overlap with other reports coming from the same research group.

Complete Questionnaires

Most journals ask reviewers to complete questionnaires, which may include items about COIs, ethics, technical merit, significance, language usage, or other matters. Some of these items (e.g., COI and research subject protection) may be crucial to the review process. It should not take you long to complete the questionnaire, and the information you provide will help the editor.

Provide Comments for Authors

Here again, the Golden Rule tells us how to act. As an author, you would undoubtedly want to know the basis for an editor's decision; so, too, do the authors of the articles you review. Comments to authors are a crucially important part of the review process and should be written for the edification of both the author and the editor. There is no standard length for these comments. In my experience, one to two pages is usually sufficient, but if the key basis for your recommendation can be stated in a paragraph or two, that is fine. Some reviewers do fairly detailed page-by-page suggestions for improvements; these are much appreciated by editors if they are constructive. Below are guidelines for writing a good review.

State the article's main strengths and weaknesses. Start the comments to authors with your views of the main strengths and weaknesses of the manuscript. In general, after reading this section the editor should understand the basis for your final recommendation (which you should *not* include in the comments to authors). Make your statement of weaknesses as constructive as possible and suggest possible avenues by which the author might address the problems in a revision. Weaknesses that are inherent in the design and execution of the study cannot be fixed in a revision; therefore, you can expend less effort in telling authors how they should have done their study. Even if you recommend rejection, you should provide constructive suggestions for improving the article in the event that the editor gives the author an opportunity to address the weaknesses. Many reviewers organize the comments to authors by describing the strengths and weaknesses of each of the sections of the article (e.g., introduction, methods, results, and discussion).

Try to balance technical merit with scientific significance. This balance has been the subject of debate in the grant-review process for many years, with many believing that technically competent but scientifically unimportant or uncreative proposals have an advantage, although they may lack significance. It is important for reviewers of journal publications to identify particularly important or creative articles and, conversely, to indicate if in their judgment an article represents only a minor advance in the field.

Provide specific suggestions for improvements. Detailed suggestions for improving the writing in the article, the figures, or the tables are very helpful to both the editor and the author. Identify paragraphs or sentences that are unclear, point out areas where information is missing, and explain how the writing can be clearer. If you have trouble understanding the article, there is a good possibility that other readers will as well. Reviewers familiar with the journal's formatting requirements can point out departures for correction. Reviewers also can identify relevant publications that should have been cited by the author.

It is important for reviewers to appreciate that journal space is often limited and that articles should be as short as they can be while still covering the necessary material. Reviewers can be especially helpful in pointing out how to shorten articles, perhaps by eliminating tables or figures or summarizing data in text. Introductions are often longer than they need be: Point out nonessential background information that can be removed. Similarly, discussions can be too speculative or focused on minor aspects of the study results. Chapter 12 provides advice on writing articles for addiction journals; reviewers can offer similar advice in their comments to authors. Reviewers are often exceptionally generous with their help, especially to inexperienced authors. I have been truly impressed and am grateful to the many reviewers who see the peer-review process as a mentoring opportunity.

Having advised reviewers to provide both general and specific suggestions for improving manuscripts, I would not want authors to see this as a rationale for submitting rough drafts or articles that have not been proofread so they can get reviewer feedback. Submission of unpolished drafts shows a lack of respect for the reviewers and editors and is not a good application of the Golden Rule. Authors should submit articles that they would be happy to review. Reviewers often become exasperated when reviewing sloppy submissions—it colours their recommendations and discourages them from providing detailed suggestions. Authors should submit their very best work.

Comment on language issues. Chapter 4 discusses the problems faced by authors having to submit their articles to English-language journals when English is not their first language. Reviewers cannot be expected to correct language problems. When assigned an article obviously written by a non-native English speaker, reviewers should do their best to focus on the science being presented and simply point out areas where language usage needs to be improved. There are various means by which authors and editors can handle this problem; therefore, reviewers should not be biased in their recommendations based on poor English language usage. Instead of addressing language issues in peer review, some journal publishers provide language assistance, or the editor can request that the author seek help from a native English speaker or a language editing service if the article is likely to be accepted.

Avoid unconstructive comments. Some things do not belong in comments to authors. Foremost are pejorative comments about authors or the work. Reviewers should strive to be constructive at all times. Editors may remove overly personal criticisms or other material that insults the authors, their institution, or their geographical location.

Jokes and witty remarks also do not belong in comments to authors. Editors may appreciate them in comments to the editor, but for authors this is serious business. They may perceive such remarks as a lack of serious intent by the reviewer.

Unless the journal uses unblinded reviews, reviewers should not reveal their identity in the comments to authors. Personal pronouns can provide clues to the reviewer's identity ("We did a study that showed. . ."). Often, reviewers provide an author with several of the reviewers' own publications that they feel should have been cited, leading the author to suspect who the reviewers are.

Reviewing Revisions

Most of what I have said about reviewing applies to reviewing revisions of manuscripts. I feel that reviewers have a special obligation to agree to review revised articles that they reviewed in an earlier version. Reviewers should evaluate how the authors addressed the weaknesses they identified in the earlier version. If authors have failed to address the concerns successfully, it should be stated in this subsequent review. On the other hand, if the authors were successful in their revisions or, alternately, convinced you that your concerns were unwarranted, this should also be stated in your comments. In any case, you should check to make sure the authors actually made the changes in the manuscript. Addressing problems only in a cover letter is not constructive.

It is also helpful for reviewers to look at the concerns of the other reviewer(s) and see how the authors addressed them. If you disagree with the other reviewer(s) and agree with the authors' explanation and defense of their original article, help out the editor by discussing these issues. If you feel that another reviewer was way off base, it is probably best to discuss your concerns in the comments to editor rather than sharing your views with the authors.

Sometimes when authors clarify their methods or data analyses, you see that the submission is even weaker than you initially thought. If that is the case, state your new observations clearly, and make your recommendation accordingly. Sometimes in reading a revision of an article, reviewers will identify weaknesses they missed in the first review. It is fine to voice these new concerns in your written review, but ideally you should have identified these problems in your earlier review. Depending on the scope of additional changes you and the editor request in a second revision, the editor may return the same article to you a third time for your recommendation.

What Do Editors Do with Peer Reviews?

After reviews of a manuscript are completed, the editor evaluates the article and the reviews and makes a decision. As I mentioned, with some journals this decision-making process may involve multiple editors and/or members of the editorial advisory board; others may involve one editor's decision. Many factors go into making a decision on a submission. The primary factor is the recommendations of the reviewers, but other factors include the editor's assessment of the importance of the article and how well it fits with the journal's area of coverage, additional concerns the editor may have that were not identified by the reviewers, the likelihood that the submission can be successfully revised to address its weaknesses, and the overall rejection rate for the journal.

Addiction journals differ greatly in their rejection rates with some able to publish much less than half of the submitted articles. Journal editors far prefer the opportunity to select only the best articles for their journal over not having enough acceptable submissions to fulfill their page budget. Journals with high rejection rates usually have higher impact factors because they choose only top-quality articles, and better and more important articles tend to be cited heavily. It is important for reviewers to understand this dynamic to appreciate why their reviews are so important to both the authors and the journals.

Summary and Conclusions

I have tried to provide background and helpful advice on reviewing articles for scientific journals. I provided an overview of the peer-review process, reasons for having peer review, why scholars would want to be peer reviewers, and tips on how to be a good reviewer. I have done this primarily from the point of view of a former journal editor but have emphasized authors' points of view as well. I encouraged you to apply the Golden Rule and be the type of reviewer you would want to review your submissions to journals. If you take an author's point of view, it will greatly help you be a good reviewer. Peer review is a very important part of the scientific process. Your work as reviewers is greatly appreciated and necessary. I hope this chapter helps you improve your reviewing skills.

Please visit the website of the International Society of Addiction Journal Editors (ISAJE) at www.isaje.net to access supplementary materials related to this chapter. Materials include additional reading, exercises, examples, PowerPoint presentations, videos, and e-learning lessons.

References and Further Reading

Barroga, E. F. (2013). Cascading peer review for open access publishing. *European Science Editing, 39*(4), 90–91.

Fletcher, R. H., & Fletcher, S. W. (2003). The effectiveness of journal peer review. In T. Jefferson & F. Godlee (Eds.), *Peer review in health sciences*, 2nd edition (pp. 62–75). London, England: BMJ Books.

Girden, E. R., & Kabacoff, R. I. (2010). *Evaluating Research Articles*, 3rd edition. Thousand Oaks, CA: Sage Publications.

Godlee, F., & Jefferson, T. (Eds). (2003). *Peer review in health sciences*, 2cd edition. London: BMJ Books.

Hames, I. (2007). *Peer review and manuscript management in scientific journals: Guidelines for good practice.* Hoboken NJ: Wiley-Blackwell.

Jefferson, T., Rudin, M., Folse, S. B., & Davidoff, F. (2007). *Editorial peer review for improving the quality of reports in biomedical studies.* Cochrane Database of Systematic Reviews. DOI: https://doi.org/10.1002/14651858.MR000016.pub3

Jefferson, T., & Godlee, F. (Eds.). (2003). *Peer review in health scheinces*, 2cd edition. London: BMJ Books.

Moghissi, A. A., Love, B. R., & Straja, S. R. (2013). *Peer Review and Scientific Assessment.* Alexandria, VA: Institute for Regulatory Science.

Overbeke, J., & Wagner, E. (2003). The state of evidence: What we know and what we don't know about journal peer review. In F. Godlee and T. Jefferson (Eds.), *Peer review in health sciences* (2nd ed.). London, England: BMJ Books.

Smart, P., Maisonneuve, H., & Polderman, A. (Eds.). (2013). *Science Editor's Handbook*, 2cd edition. Cornwall, UK: European Associatoin of Science Editors (especially Section 4).

Smith, R. (2006). Peer review: A flawed process at the heart of science and journals. *Journal of the Royal Society of Medicine, 99,* 178–82.

SECTION 4

Ethics Matter

CHAPTER 14

Dante's *Inferno*: Seven Deadly Sins in Scientific Publishing and How to Avoid Them

Thomas F. Babor, Thomas McGovern and Katherine Robaina

"Relinquish all hope, ye who enter here."
Dante Alighieri, *The Divine Comedy, Inferno*, Canto III, 9

Some 700 years ago, Dante Alighieri (1265–1321) wrote an epic poem about a man's journey through the afterworlds of hell, purgatory, and heaven. In his *Divine Comedy*, he catalogued the vices and virtues of people who had passed into those spiritual domains, in part to provide a valuable insight to us, the living. Dante described hell as a very unhappy and inhospitable place that had nine different levels ranging from the blazing inferno of the eternally damned to a rather benign area, called the First Circle, which was reserved for worthy individuals who were born before the world was redeemed and therefore could not enter the gates of heaven (Alighieri, 1947).

Within this general metaphor, this chapter will take the reader on an educational journey through the various levels of scientific misconduct, from unintentional but questionable research practices, such as citation bias, to serious scientific fraud, such as the fabrication of data. Our purpose is not to scare the fear of God into the gentle hearts of our readers. Rather, like Dante on his journey through the netherworld, we too should see the mortal consequences of scientific misconduct so that we can learn how to avoid them. Table 14.1 shows the seven types of misconduct this chapter explores. In addition to describing

How to cite this book chapter:
Babor, T F, McGovern, T and Robaina, K. 2017. Dante's Inferno: Seven Deadly Sins in Scientific Publishing and How to Avoid Them. In: Babor, T F, Stenius, K, Pates, R, Miovský, M, O'Reilly, J and Candon, P. (eds.) *Publishing Addiction Science: A Guide for the Perplexed*, Pp. 267–298. London: Ubiquity Press. DOI: https://doi.org/10.5334/bbd.n. License: CC-BY 4.0.

these various "sins" and the people who commit them, we also discuss their relative seriousness, the punishments that can result, and how to prevent these kinds of problems before they arise. In Chapter 15, we discuss the same issues within a framework of ethical decision making, using case studies to illustrate each topic.

The first issue is carelessness, exemplified by unconscious or conscious citation bias, misrepresenting the accomplishments or findings of others, and neglecting to reference findings that an informed reader would need to know to interpret the author's conclusions. In its most benign form, this problem consists of a failure to read and understand the articles one cites. A more serious offence is the distortion of others' work so that their ideas or findings support a preconceived point of view that the author is trying to advance. Carelessness can also be manifested in poor management or inaccurate presentation of data.

The second ethical issue is dual and redundant publication, which occurs when two or more articles share any of the same data without full cross-referencing.

The third issue we consider is unfair or irresponsible authorship. According to standard Ethical Practice Guidelines published by the International Society of Addiction Journal Editors (ISAJE) and similar guidelines of other organizations (e.g., Committee on Publication Ethics (COPE)), all persons named as

	Sin	Exampes	Punishments
1	Carelessness	Citation bias, understatement, negligence	Request for correction, letter to editor
2	Redundant and duplicate publication	Same tables or literature reported without noting prior source, same article published in different journals	Rejection of manuscript, copyright infringement, retraction
3	Unfair authorship credit	Failure to include eligible authors, inclusion of honorary authors, use of ghost writer	Angry colleagues, complaints to editor or employer
4	Undeclared conflict of interest	Failure to cite funding source	Letter to editor, public apology
5	Human/animal subject violations	No ethical approval	Rejection of manuscript, notification of employer
6	Plagiarism	Reproducing others' work or ideas as one's own	Retraction of manuscript, notification of employer
7	Scientific fraud	Fabrication or falsification of data, misappropriation of others' ideas or plans given in confidence	Retraction of manuscript, notification of employer, publication ban

Table 14.1: The seven deadly sins and punishments of those who engage in publishing misconduct.

authors should have made a major contribution to the work, not just a token contribution.

Failure to declare a conflict of interest (COI) is the fourth ethical issue considered in this chapter. A COI is a situation or relationship in which professional, personal, or financial considerations could be seen by a fair-minded person as potentially in conflict with the researcher's or author's independence of judgment.

The fifth ethical violation is the failure to conform to minimum standards of protection for animal subjects or human research participants. The latter includes confidentiality of patient records and other data, informed consent, and proper explanation of the risks of research participation. Abiding by standards set by national and institutional boards for the protection of animal or human subjects is an important aspect of research under this rubric.

Plagiarism is the sixth issue. Plagiarism literally means the act of "literary theft" by using or closely imitating the language and thoughts of another author as if they were one's own.

The final level is scientific fraud. This form of misconduct consists of the deliberate fabrication of data or the alteration of findings to make a study more credible and acceptable for publication.

A meta-analysis of survey studies conducted by research scientists and their student trainees representing a range of disciplines indicates that up to 33.7% admit to engaging in questionable research practices, and 0.3%–4.9% have fabricated or falsified data. Misconduct is reported more frequently by medical and pharmacological researchers than those from other disciplines (Fanelli, 2009). How prevalent are these various ethical problems among addiction scientists? ISAJE conducted an informal survey of its members to learn about the kinds of ethical misbehavior of most concern to journal editors (Stenius & Babor, 2003). Duplicate publication in various forms and inappropriate citations were the most common problems encountered by journal editors in their routine processing of manuscripts. A substantial number of journals had experienced at least some of the more serious forms of scientific misconduct, such as plagiarism and failure to declare COI. Authorship problems were also noted quite often. Although most problems were considered infrequent occurrences by the editors, it is likely that these issues are often hidden from the eyes of busy editors and reviewers. For example, editors and reviewers are unlikely to detect scientific fraud in the normal editorial process because data fabrication can be easily hidden in lab records and computer files that are inaccessible during the review process. Skilled reviewers are more likely to detect plagiarism and citation bias, but there is a general suspicion that the cases of identified and provable misconduct are the tip of an iceberg.

In the following sections of this chapter, each of these ethical improprieties is discussed in terms of its relative importance, possible consequences, and strategies for avoidance. Table 14.2 provides definitions of the various types of ethical problems discussed in the chapter.

Citation bias	A form of carelessness that ranges from a rather benign failure to read the articles one is citing to distorting the meaning of others' work.
Copyright	The legal right granted to an author, publisher, or distributor to exclusive publication, production, sale, or distribution of a scientific work.
Divided publication	Information from a single research study is divided for publication in two or more articles. Also called "salami science."
Duplicate publication	Re-publication of the same article in two places without clear reference to the other publication.
Fabrication	Presenting data in a research report that have not been obtained in the manner or by the methods described in the report.
Fractionally divided publication	Reporting in a single article only a fraction of the data that have been or will be reported in their entirety in another article.
Ghost authorship	A published article fails to acknowledge the original writers' contributions.
Guest authorship	A researcher is invited to add his or her name to a study or publication without fulfilling authorship criteria.
Misappropriation	Illicitly presenting or using in one's own name an original research idea, plan, or finding disclosed in confidence.
Misrepresentation (falsification) of findings	Altering or presenting original findings in a way that distorts the result in a scientifically unjustified way or by omitting results or data pertinent to conclusions.
Partial repetitive publication	Repeatedly publishing parts of the same information in modified form.
Plagiarism	Presenting someone else's manuscript, article, text, or idea as one's own.
Redundant/repetitive publication	Publishing the same information two or more times (e.g., in journal articles and book chapters).
Self-plagiarism	Copying and presenting one's own text or article without properly attributing its original source.
Unethical authorship	Authorship which violates the principle that all persons named as authors should have made a major contribution to the work reported and be prepared to take public responsibility for it. Similar to guest authorship.

Table 14.2: Definitions of terms referring to various forms of scientific misconduct.

Negligent Carelessness and Citation Bias

The first Deadly Sin described in Table 14.1 refers to minor forms of negligent carelessness and citation bias that are likely to mislead readers and distort the value of scientific research. Perhaps the most benign and most prevalent form of ethical impropriety, negligent carelessness, is characterized by such deficiencies as a failure to adequately review the literature on a topic, lack of candor or completeness in describing one's research methods, or presentation of data that are based on faulty statistical analyses. A related problem occurs when an author cites articles taken from other reports or from published abstracts without having read the primary sources.

A more serious form of carelessness in scientific writing is citation bias. One form of this bias is the selective citation of only those articles that support a particular point of view, ignoring or understating the importance of articles that contradict that viewpoint. For example, a study of all therapeutic intervention studies included in meta-analyses published between January 2008 and March 2010 in the Cochrane database found that studies with statistically significant findings were cited twice as often as nonsignificant studies (Jannot et al., 2013). A citation bias favoring significant results is also evidenced in the psychiatric literature (Nieminen et al., 2007). Within the addiction field, Etter and Stapleton (2009) found that randomized controlled trials for nicotine replacement therapy that included positive and statistically significant results were more often cited than articles that did not ($N = 41$ vs. 17, $p < .001$). In addition, a meta-analysis of 42 studies reporting smoking among people with schizophrenia found that the actual average prevalence of smoking among this population is 62%, as opposed to the 80%–90% rate frequently reported. The analysis also found that, for every 10% increase in prevalence reported in a study, there was a 28% increase in the likelihood of that study being cited. These higher rates were also inaccurately reported in publically available information and by the media (Chapman et al., 2009). The intention to deceive others may not be operative in all or even most cases, but this does not make this practice any less unacceptable.

Another form of citation bias is selective citation to enhance one's reputation, epitomized by self-citation. We discuss these issues in Chapter 10 in terms of various deviations from ideal citation practice. A case analysis of these practices in Chapter 15 further illustrates the ethical dimensions of such transgressions.

Consequences

If the effect of these practices is to mislead or misinform the reader, then they are considered a form of scientific misconduct, even if they only occur at the drafting stage when they are often detected by observant colleagues or reviewers who are likely to request a more balanced literature review or the correction of obvious mistakes. In some cases, an editor may reject an otherwise acceptable manuscript if reviewers raise questions about the author's objectivity or intellectual sloppiness. The consequences could be more serious if carelessness or citation bias is detected only after the article is published. If readers of a published article detect a statistical mistake, a clear bias in the formulation of a research question, or the selective reporting of the literature, they may write letters to the editor pointing out the problem. Editors in turn may ask for corrections to the text or the data analyses, which are subsequently published as a special note to readers. Beyond these embarrassing consequences, failure to cite relevant studies and bias in the interpretation of previous research is likely to create a negative impression of the author among his or her colleagues. The institution with which the author is affiliated may experience criticism and damage of reputation. Furthermore, if articles showing favorable results with large effect sizes are cited more often, readers can be misled into thinking treatments or interventions are more effective than they really are. This may affect the health of individuals and the way services are organized for the public, or it could have other policy implications. Figure 14.1 provides an illustration of how citation bias could have adverse policy and clinical implications.

Figure 14.1: Citation bias and its potential consequences.
Source: Chapman et al. (2009).

Prevention

As discussed in Chapters 7, 8, and 10, the best way to avoid these problems is to follow appropriate citation practices, conduct a thorough review of the literature (by searching for positive as well as negative outcomes), read all of the articles you cite, present research findings accurately, and interpret them objectively. Locating unpublished studies and/or outcomes may also help to reduce bias. Authors who collaborate on multi-authored articles have a special responsibility to read all drafts of a manuscript with extreme care to make sure these problems are detected during the early stages of the publication process. Even when several authors divide responsibility for writing different sections of a research report, authors should always check each other's work.

Duplicate and Redundant Publication

Authors wishing to reach the widest possible audience, or a variety of specific audiences, may seek to report a single definable body of research in more than one article, in repeated reports of the same work, in fractional reports, or in reports in more than one language (Huth, 1986). But there are also less noble motives for duplicate and redundant publication, including the desire for multiple publications to enhance one's reputation.

Redundant publication occurs when two or more articles share the same data without full cross-reference (COPE, 1999). Duplicate (or dual) publishing, according to the International Committee of Medical Journal Editors (ICMJE, 2013, p. 8), refers to the "publication of a paper that overlaps substantially with one already published without, clear, visible reference to the previous publication." In general, journal editors expect authors to ensure that no significant part of the submitted material has been published previously and that the article is not concurrently being considered by another journal. Meta-analyses (Choi et al., 2014; Gotzsche, 1989) indicate that repetitive publishing practices have become a serious problem, and the evidence suggests that this is true across nations, accounting for 18.1% of all retractions of articles published on PubMed between 2008 and 2012 (Amos, 2014). In Finland and China, the rate of retractions for duplicate publication is much higher (37.5% and 29.4%, respectively) (Amos, 2014). Therefore, many journals now require authors to state in writing whether the data have been previously reported in part or in whole (ICMJE, 2013).

As indicated in Table 14.2, a number of different terms have been used to describe this phenomenon. Although there are some important differences among prior, duplicate, repetitive, fragmented, and redundant publication, they are all part of a common problem. Duplicate and redundant publication and their variants consume valuable resources that otherwise might be devoted to other authors who are publishing original data or ideas. Because of limited journal space, the publication of one person's article means that another's article will be rejected. If there are questions about the extent of the overlap between two articles, editors and reviewers need to take extra time to review several publications to determine the extent of redundancy and whether it violates any copyright agreements.

Regardless of whether the repetition occurs with data or ideas (e.g., repetitive review articles), the information from duplicated sources is sometimes inadvertently cited in a way that implies that the findings or conclusions are independent of each other, when in fact they are based on the same source. Without full disclosure in the original sources, authors of subsequent meta-analyses and review articles based on these source articles may come to biased conclusions because the effect of a given finding is multiplied or distorted.

Instances of Acceptable Secondary Publication

As Huth (1986) has noted, some types of repetitive publication are legitimate and should not be considered scientific misconduct. This is particularly the case in the publications associated with large data sets that involve multiple investigators across many sites. Often, the collaborating investigators have included measures related to a particular hypothesis or methodology, which could and should be reported in separate articles even though the article presents the same subjects, methods, procedures, and even some of the same data as other articles. Such publications may be intended to highlight the relevance of particular clinical findings for a particular audience, especially if they have been first published in a technical journal that did not permit the reporting of particular findings or the discussion of clinical implications. Articles presented at scientific conferences or meetings but not published in full may also be submitted to journals for publishing. In such cases provide an explanatory letter along with copies of related materials (ICMJE, 2013).

It is also acceptable to re-publish ideas, data, or review findings when journal editors or book editors request that a popular author write a topical review or commentary for their publication, as long as the author tells the editor about previously published material and cites all relevant reports in the commissioned article.

Another possibly acceptable variant is publication of the same article, often in its entirety, in two languages when the editors of both journals agree to it and when the translated version cites the original version as the primary publication (which cannot be submitted simultaneously). Submitting the same article to two journals may also be justifiable if the two journals are in very different disciplines and the publication is intended for different groups of readers, the authors have received permission from the editors of both journals, the title indicates it is a secondary publication of a primary publication, *and* the reviewers' comments bring about considerable changes to the manuscript (ICMJE, 2013).

Self-Plagiarism

A special case of redundant publication is "self-plagiarism," a topic on which relatively little has been written. According to Griffin (1991), this occurs when an author re-uses text from his or her own previously published article in a way that fails to give proper acknowledgement to its source and its owner. By *owner*, we mean the person who or organization that owns the copyright (see Table14.2 for definition), which is often the publisher of the previous version of the borrowed text, not the original author. This problem typically occurs when authors re-use text from a literature review or the methods section of an article either without changing the wording or by quoting the original text. Unlike the re-use or re-publication of original data, self-plagiarism is something that is more the result of laziness than dishonesty. It can also be a form of self-aggrandizement.

Consequences

If a duplicate publication constitutes a copyright infringement, it may result in a reprimand for the author, a retraction of the article, or an apology to the journal editors and the publishers involved. Editors, likely embarrassed by the need to publish a retraction, have adopted policies and regulations to prevent this questionable research practice. *Recommendations for the Conduct, Reporting, Editing, and Publication of Scholarly Work in Medical Journals* (previously known as *Requirements for Manuscripts Submitted to Biomedical Journals)*, which is endorsed by over 500 medical journals, cautions the following: "Authors who attempt duplicate publication without such notification should expect at least prompt rejection of the submitted manuscript. If the editor was not aware of the violations and the article has already been published, then the article might warrant retraction with or without the author's explanation or approval" (ICMJE, 2013, p. 9).

Furthermore, redundant articles may mislead researchers because of the duplicate counting of subjects in a meta-analysis, as illustrated in the case described in Box 14.1 (Tramèr et al., 1997). And when instances of scientific misconduct like this are reported to the public, they diminish the reputation of scientists and their work. In general, an author is not allowed to re-use previously published material when the rights have been assigned to the publisher, which occurs in most instances of scientific journal publications. Reprinting more than one or two sentences verbatim without proper attribution may constitute a violation of copyright and could result in legal sanctions, although this rarely occurs in cases of minor copyright violations.

The negative consequences of self-plagiarism may be less obvious and editors are unlikely to consider small amounts (the *BMJ* uses a baseline of 10%) of "borrowing" to be a major problem, but if an observant reviewer detects

Investigators conducting a meta-analysis of randomized controlled trials for the medication ondansetron found that 17% were duplicate publications that had not been cross-referenced, resulting in a 28% duplication of patient data. Furthermore, the duplicated randomized controlled trials reported greater efficacy than nonduplicated studies. If duplicated data were included in the meta-analysis, the efficacy of ondansetron would be overestimated by 23% (Tramèr et al., 1997).

Box 14.1: Case study: Impact of duplicate publication on meta-analysis.

widespread self-plagiarism, the editor may reject the article. Nevertheless, the more that authors re-use text without proper quotation or attribution, the more they risk adverse consequences from editors and publishers, ranging from a reprimand to legal action for copyright violation.

Prevention

Authors of overlapping articles would be seriously remiss in failing to cite their previously published work (see Jerrells, 2001, for a discussion of this problem) or submitting the same article to two different journals while intending the piece to be recognized as two original articles. The fault in this sin does not insomuch lie with the duplicate publication itself but with the author's intent to deceive. When there is any possibility of repetitive publication, authors must notify editors to explain the connection between the current article and its predecessors. Ideally, the author should submit all related publications to the editor along with an explanation of the potential overlap and the reasons for the new report. Second, all versions of related articles must contain appropriate citations and complete references to the related articles so that readers and editors can evaluate the implications of the repetition and overlap. This includes citing illustrations or tables reprinted or adapted from other journals. When publishing an article in two different journals, each publication should clearly state, "This paper is also published as 'Title of paper' in the *Title of Journal,* Vol x(x), pp. x." (Bretag & Mahmud, 2009, p. 194), or secondary publication should refer to the primary one in the title (ICMJE, 2013).

A survey on redundant publishing (Yank & Barnes, 2003) found that both editors and authors believe that journals do not do enough to expose, condemn, and penalize this publishing sin. Authors also felt that that redundant publications occur because the practice is not condemned by academic leaders and because authors do not understand how redundant reporting distorts the aggregation of data (i.e., meta-analyses). Therefore, editors, authors, and academic leaders should clarify and enforce mutually acceptable standards on redundant publication (e.g., *Recommendations for the Conduct, Reporting, Editing, and Publication of Scholarly Work in Medical Journals,* Section III.D.2 and IV.B).

Regarding self-plagiarism, set off short quotations from a previously published article in quotation marks, and cite the original version. Permission must be requested from the publisher or other copyright holder when large sections are reproduced. When there is a need to repeat the information contained in a previously published literature review or a methods section, the best solution is to change some of the wording in each sentence and to refer the reader to relevant sources for previously published material (e.g., "As discussed in our previous report [give author names and year of publication]," etc.). It has also been suggested (Bretag & Mahmud, 2009) that authors could use text-matching software to ensure that they have appropriately described all previously published work; however, it should be noted that there has been some concern over potential ethical and legal issues surrounding the software (McKeever, 2006).

Unfair Authorship

Authorship of a scientific report refers not only to the writing of a manuscript but also to the origin of a writing project, any experimentation or other research connected with it, and the substantive kinds of work that led up to it. According to the ISAJE Ethical Practice Guidelines (www.isaje.net) and other codes (COPE, 1999; ICMJE, 2013), all persons named as authors should have made a major contribution to the work reported and be prepared to take public responsibility for its contents (in proportion to the credit they claim on the author list). An editorial (Huth, 1982, p. 613) in the *Annals of Internal Medicine* defines relevant terms as follows:

> *Responsibility* means the ability and willingness to defend the content of the paper if it is challenged by readers. *Public* means that authors are willing to carry out this responsibility in a published defense, such as a signed letter to the editor; private defense in private correspondence would not reach the scientific public.
>
> *Content* means not simply packages of data but also the conceptual framework on which they are hung; the justification for a study or clinical observations; the basis for the study design; methods for collection of valid data; the analysis and interpretation of the data; and the logic that led to the conclusions.

The ICMJE's (2013) four criteria for authorship are also relevant in this context: (a) substantial contributions, (b) drafting the work or revising it critically, (c) final approval of the version to be published, and (d) agreement to be accountable.

There are a number of ways in which authorship decisions can result in ethical improprieties. First, some persons who have made significant contributions to an article may not receive sufficient credit or may receive no credit at all. This occurs when an article is drafted without the knowledge or consent of someone who made a substantive contribution earlier in the process. It also occurs when a decision to list the order of contributions is not made fairly with the full agreement of the co-authors, as when a major contributor is listed after a minor contributor to enhance the ego or career of the minor contributor. Another instance of inappropriate credit occurs when a co-author, such as a science writer, is not listed because the research group might be embarrassed to admit that someone else wrote the article, such as a science writer hired by a drug company to expedite the publication of favorable findings. This is called *ghost authorship* because the real author's identity is unknown to those who read the article. Ghost writers are used by drug companies (Moffatt & Elliott, 2007) and were used by the tobacco industry (see Box 14.2) (Davis, 2008). By contrast, *guest authorship* occurs when articles are prepared by hired writers but published under the names of academics or scientists who allow themselves to be listed (sometimes for a payment or other incentives) without satisfying authorship criteria (Stern & Lemmens, 2011). The concern with this unethical practice lies with COIs and the potential for bias, as evidenced by ghostwritten articles on hormone replacement therapy, Vioxx (an anti-inflammatory drug that was withdrawn amid safety fears) and Fen (a popular diet drug withdrawn for safety reasons). A less serious form of ghost writing can occur when researchers, who are either too busy or poor writers, employ professional science writers to draft manuscripts of original research. For the purposes of this discussion, we concentrate on the former definition.

An analysis of tobacco industry documents and transcripts of tobacco litigation testimony showed that British American Tobacco ghost-wrote the International Advertising Association (IAA) report titled "Tobacco Advertising Bans and Consumption in 16 Countries," originally published in 1983 and again as a revision in 1986. J.J. Boddewyn, a marketing professor, served as "guest" editor of the reports. The reports concluded that tobacco advertising bans did not result in a reduction of tobacco use. These reports were then publicized in print materials, media campaigns, and legislative hearings during the 1980s and later. The Tobacco Institute, the major trade association representing the major U.S. cigarette manufacturers at the time, helped arrange for Boddewyn to present the findings to the U.S. Congress and the media.

Box 14.2: Case study: Ghost writing by the tobacco industry.

A second type of authorship problem arises when some persons are listed as co-authors even though they made no substantive contribution to the article or the research. A common example is the practice of listing the head of a department or a research center director, often at the end of the author list, a custom known as gratuitous, honorary, or gift authorship. Again, in the light of this practice, one must question the ethical climate in research settings that allows such behavior to exist. Ethical guidelines, appropriately crafted and implemented, might deter such transgressions.

Between these two extremes, there are a number of related infractions, such as the failure to give proper recognition to a person's contribution by listing him or her inappropriately low in the author list, or the tendency to award co-authorship for minor contributions based on personal or political considerations. A more complete discussion of authorship issues is provided in Chapters 5 and 11, which also describe procedures to minimize ethical and interpersonal problems related to authorship credits. Our purpose here is to discuss the seriousness and consequences of this type of misconduct and to summarize the steps that can be taken to prevent its occurrence.

Consequences

Authorship credits may be one of the most contentious issues in scientific publishing. At the level of collaborating research groups, the consequences range from hurt feelings to formal complaints made to a scientist's unit director or institutional authority. In between these extremes, there are likely to be recriminations, perceptions of unfairness, and poisoned working relationships, which could damage the reputations of some of the parties involved. In the case of ghost writing, the funder (for example a drug manufacturer) obtains the credibility and prestige attached to the guest author, which may translate into distorted perceptions of the evidence base and affect public health. When instances of unfair authorship credit are detected, the editor's response could range from the rejection of a pending manuscript to the call for a correction to a published article. Some journals (e.g., *PLoS Medicine*), call for a formal retraction if unacknowledged ghostwriting is discovered after publication and reporting of authors' misconduct to institutions, in addition to banning the guest author from future submission (*PLoS Medicine* Editors, 2009). But these questionable research practices rarely come to the attention of editors unless there is a case of scientific fraud, where co-authors might claim that they were not sufficiently involved in the writing of the article to detect the fabrication in the first place. Some have called for academic sanctioning (Moffatt & Elliott, 2007), and because ghostwritten articles have been used in litigation to support drug companies' claims, others (Stern & Lemmens, 2011) argue that a guest author's claim for credit of an article written by someone else constitutes legal fraud.

Prevention

How can authors best deal with ethical issues related to authorship? As noted in Chapter 11, we advise early agreement on the precise roles of the contributors and collaborators and on matters of authorship and publication. The ICMJE (2013) has attempted to control unfair authorship practices by requiring that journals ask detailed questions about each author's contributions. The lead author should periodically review the status of authorship credits within a designated working group by having open discussions of substantive contributions with all prospective collaborators. To avoid disputes, lead authors should distribute and discuss authorship guidelines with all potential collaborators on a manuscript. Those who may have been listed as an "honorary author" should instead be mentioned in the acknowledgments and have their contributions specified. An open discussion of authorship should be on the academic agenda of research centers. Involving an institutional ethics committee in drawing up institutional guidelines might also be helpful. Open and ongoing conversation about these issues, combined with institutional policies, is the best way to avoid problems.

Undeclared Conflict of Interest

When a gift or gesture of any size is bestowed, it imposes on the recipient a sense of indebtedness. The obligation to directly reciprocate, whether or not the recipient is conscious of it, tends to influence behavior (Katz et al., 2003).

A COI is a situation or relationship in which professional, personal, or financial considerations compromise, or could be seen by a fair-minded person as potentially compromising independence of judgment (ISAJE, 1997). This problem has become exacerbated by closer relationships between government and industry (e.g., Bonner & Gilmore, 2012), industry-civil society partnerships, and cuts to government funding which encourage the procurement of industry sponsors.

Real, Apparent and Potential COIs

Real (or actual) COIs should first be distinguished from "apparent" and "potential" conflict situations (See Table 14.3), as a COI only indicates the potential for bias, not the likelihood. A *real COI* means that the author, or the administrative unit with which the author has an employment relationship, has a financial or other interest that could unduly influence the author's position with respect to the subject matter being considered. An *apparent COI* exists when an interest would not necessarily influence the author but could prompt others to question the author's objectivity. Sometimes a conflict may exist, but the link is not so clear, as was the case with a young investigator who failed to declare funding

Real or actual COI	A direct conflict exists between professional judgment/objectivity and private interests
Apparent COI	It appears or could be perceived that competing interests are improperly influencing the professional's judgment, whether or not that is actually the case
Unapparent COI	A conflict may exist, but the link is unclear
Potential COI	Private interests are not but could come into direct conflict with professional judgment

Table 14.3: Conflict of interest (COI) situations.

from the Institute for Research on Pathological Gambling. When contacted by the journal about her failed declaration, the researcher reported that she had no idea that the Institute's funding came from the gambling industry. *Unapparent COIs* such as these occur when sponsorship is provided through an industry-funded social aspects organization or another third party, or when the recipient of the funding is unaware of the funding source. A *potential COI* involves a situation that may develop into a real COI.

One's perception of COI is just as important as COI itself, as even paid travel, honoraria, or other relationships can subconsciously "create strong dispositions or obligations to reciprocate" (Mauss, 1967). As explained by Katz et al. (2003) "When a gift or gesture of any size is bestowed, it imposes on the recipient a sense of indebtedness. The obligation to directly reciprocate, whether or not the recipient is conscious of it, tends to influence behavior." This means that one does not necessarily need to have a financial interest in the outcome of one's research to constitute having a COI.

COIs can be financial, personal, political, or academic. *Financial interests* can include employment, research funding, stock or share ownership, payment for lectures or travel, consultancies, or company support for staff (COPE, 1999). These kinds of conflict are most often discussed in ethics codes and reports on research integrity because they are easier to document and quantify. *Personal conflicts* might include a vendetta against another researcher whom the author dislikes. *Political conflicts* exist when researchers distort their findings or interpretation to conform to a specific political idea or ideology. *Academic conflicts* include the attempt to validate "pet" theories that support one's own ideas. These kinds of conflict are difficult to detect, but authors should nevertheless consider them when evaluating their own work. Authors in the past received little guidance in evaluating and responding appropriately to issues of regarding COIs. The existence of compliance offices in research settings is helpful, but these institutions themselves will not solve the problem. Researchers and research groups need appropriate training about the ethical dimensions involved as well as about opportunities for ongoing dialogue and conversation (Institute of Medicine, 2002).

One way to determine whether a COI exists is to ask the following question: If the situation or relationship were revealed to the editor or the reader only after the article was published, would it make a reasonable person feel misled or deceived? COI is not in itself wrongdoing. However, scientific misconduct does occur when there is a failure to declare real or potential conflicts to an editor, one's co-authors, and the readers of an article, to the extent that potential conflicts are very important in the evaluation of any piece of scientific work. As discussed in more detail in Chapter 16, the potential for COI in the addiction field is enhanced by any relationship or funding connected to the tobacco industry, the alcohol beverage industry, for-profit health care systems, private hospitals, the pharmaceutical industry, or "social aspect organizations" that receive their primary support from industry sources. For example, in the search for medications that may be used to treat tobacco, alcohol, or illicit drug dependence, scientists involved in research on a particular product may have financial ties with companies that have a business interest in that product.

The alcohol and tobacco industries have also funded researchers to conduct policy studies or policy-related program evaluations.

Sometimes the industry funds studies directly; other times, it funds studies indirectly through social aspect organizations, think tanks, or other third parties that receive support from industry sources (see Box 14.3 for a list of these organizations). In addition to research funding, industry ties can include paid consultancies, conference presentations, stockholding, advisory board membership, or patent holding.

Two major questions regarding the need for COI policies and precautions are whether industry funding affects the quality and eventual publication of research and whether the effect is deleterious. Bias toward "positive" results may exist even among articles that disclose financial ties to industry (Cho, 1998). For example, pharmaceutical industry–supported medication studies are significantly more likely to report "positive" findings (i.e., that the manufacturer-associated medication is better than the placebo) than non–industry-funded

- Foundation for Alcohol Research (formerly ABMRF)
- Institut de Recherches Scintifiques sur les Boissons (IREB)
- National Center for Responsible Gaming (NCRG)
- Center for Consumer Freedom (CCF)
- European Foundation for Alcohol Research (ERAB)
- International Alliance for Responsible Drinking (IARD, formally ICAP)
- Alcohol Information Partnership

Box 14.3: Organizations receiving industry support.*
This list is not exhaustive.

studies (Stelfox et al., 1998). Several examples of such biases have been observed in the addiction field. One analysis found that industry-supported studies were more likely than non–industry-funded studies to conclude that secondhand smoke has no health effects (Lambe et al., 2002). In reviewing all randomized controlled trials on nicotine replacement therapy included in the Cochrane database, Etter et al. (2007) found that industry-supported trials were more likely to produce statistically significant results when compared with independent trials. Researchers (Cataldo et al., 2010) conducting a meta-analysis on the link between smoking and Alzheimer's disease found that, in tobacco-industry-affiliated studies, smoking was associated with a significantly decreased risk for Alzheimer's disease, whereas those with no industry affiliation demonstrated a significant increased risk. Such instances of COI could be made worse by publication bias, in which industry-favorable studies are more likely to get published than are unfavorable ones.

There are several possible mechanisms to explain how conflicts, especially those connected with industry ties, may lead to publication bias (see Cho, 1998). One is suppression of publication, whereby negative findings are not published because either the author fears loss of funding from industry sponsors or the industry itself imposes restrictions on publication. Another mechanism is self-selection or industry selection of researchers who are more likely to get positive results. Even when grants are awarded by industry-funded organizations that convene expert review panels, the panel members themselves may be influenced by receipt of honoraria, travel funds and invitations to speak at industry-supported conferences. A third possibility is industry control of the research agenda, so that funding is only provided for topics that are not likely to threaten an industry's financial interests. A final possibility is that even when the funding source has no influence on the findings, researchers compromise their own credibility by being associated with industries that have a vested interest in the outcomes of the research.

From the literature reviewed in this section, we conclude that industry funding can affect the nature, quality, and credibility of research, and the effect is likely to be deleterious.

Consequences

The existence of a COI does not mean that the conflict will result in adverse consequences. However, people with a conflict often fail to realize the extent to which the conflict has affected their judgment, because this can occur subconsciously. Another consequence of having competing financial interests is the possible limitation of publication options. Although most journals do not ban publication of articles because of their authors' financial interests, some journals have now begun to prohibit authors of editorials and review articles from publishing if the author has a substantial financial interest in the product discussed in the editorial or review (Relman, 1990). This policy does not apply to authors of scientific reports that present original data.

Undeclared COIs, when detected, may have serious consequences, such as the rejection of a pending article, the retraction of a published article, or the author's need to publish an apology. A more subtle effect of real or apparent COI is the perception by one's scientific colleagues that one's scientific work is biased because of a personal or financial interest. Industry relationships can also threaten the integrity of the author's host institution itself.

Prevention

Researchers must first be made aware of the ethical issues that arise when exploiting COIs. Many schools are requiring ethics classes that include education on COIs, and many academic institutions and medical centers have adopted rules governing financial support for faculty activities. These rules describe when faculty must disclose particular interests and when they must divest themselves of particular financial interests. In 2013, an expert Task Force convened by the Pew Charitable Trusts published COI best practice recommendations for academic medical centers, which can be read in Table 14.4.

COI committees, when they operate as part of ethical review committees, are a part of institutional compliance oversight and hold promise in this respect.

COI area	Best practice recommendation
Disclosing COIs	Required to disclose all industry relationships that relate to academic activities in teaching, research, patient care, and institutional service.
Acceptance of gifts and meals	Prohibited
Industry-funded speaking	Prohibited
Industry-sponsored fellowships	Clinical training: prohibited Research training: permitted
COI curriculum	Required
Consulting and advising relationships	Marketing: prohibited Scientific activities: permitted
Industry support of accredited continuing medical education	Should not be supported
Ghostwriting and honorary authorships	Prohibited

Table 14.4: Recommended best practices in medical conflict of interest (COI) policies.
Note: Adapted from Pew Charitable Trusts (2013).

Authors should pay close attention to the guidelines issued by these commit-tees. As noted in Chapter 16, the scientific community has issued warnings about the advisability of accepting any funding from the tobacco and alcohol industries and has suggested rigorous adherence to voluntary ethical codes when such funding is accepted.

According to Loue (2000), the best way to avoid problems associated with potential COI is self-elimination from participation in potentially conflicting activities. Short-term consulting arrangements with the tobacco, alcohol, and pharmaceutical industries are often not worth the questions the researcher must face about his or her objectivity. Arrangements with industry can be par-ticularly problematic when the researcher is asked to sign a restrictive con-tract regarding the ownership of data, the sponsor's control of the data, and the investigator's right to publish them.

Even when these guidelines have been followed appropriately, however, authors should declare to the editor any real, potential, or apparent COI with respect to their involvement in a particular publication. Authors should declare conflicts between (a) commercial entities and authors personally and (b) com-mercial entities and the administrative unit with which the authors have an employment relationship.

Authors should also declare sources of funding for a study, review, or other publication in a way that can be clearly understood by the reader, even if the journal does not require authors to do so. A footnote or an acknowledgment is the most appropriate mechanism. Describe funding sources in sufficient detail so that an average reader can recognize potential COIs. If a funding source is a social aspect organization with an ambiguous name such as The Alcohol and Health Fund, the reader should be informed that, for example, the organization is supported by a group of beer companies.

Disclosure alone will not necessarily eliminate publication bias. Research-ers who are serious about avoiding even the appearance of COI are advised to dilute the conflicting relationship by getting funding from both industry and nonindustry sources and by refusing to sign industry agreements that do not guarantee the researcher's right to publish the results regardless of the study's outcome. Other management strategies include avoiding additional financial ties that are not absolutely necessary to the pursuit of the research, such as the acceptance of advisory board memberships, stock options, or consulting fees from companies sponsoring research (Cho et al., 2002).

Human/Animal Subjects Violations

Addiction research involving human and animal subjects has been conducted for over a century. During this period, regulations governing human and ani-mal experimentation have developed into a very complex set of procedures that are typically governed by appointed committees located at institutions involved in biomedical research. These procedures include ethical review of research protocols, safety monitoring of animals and human research participants, and informed consent requirements for human participants. These procedures were developed out of concern for the rights of research participants following a series of well-publicized medical experiments in which human subjects were exposed to harmful agents or had effective treatments withheld without their knowledge or consent (Loue, 2000). It has now become customary, if not mandatory, to submit proposed research for independent review by an ethical research committee to determine its ethical acceptability from the perspective of the local community and the researcher's institution (Federman et al., 2003).

Such boards focus primarily on the protection of research participants by assuring that the study's procedures minimize risks of unwarranted harm to participants. Although regulations regarding types of study requiring ethical approval vary across the world, formal international standards developed to guide experimentation involving human participants have been put forth in the 1964 Declaration of Helsinki, which states that in medical research involving human participants, the well-being of the individual research subject takes precedence over all other interests. In particular, the 1975 and 1983 revisions emphasized the importance of voluntary informed consent to participate in research (Loue, 2000).

Additional ethical issues may also have to be considered for certain types of research involving individuals who are substance dependent. Does drug or alcohol dependence in combination with other factors limit capacity to give informed consent? What other factors—intoxication, withdrawal, chronic recidivism? Do the criteria for dependence imply impaired decision making? If someone is using drugs despite reoccurring problems and does not seek treatment, should he or she be categorized as not exhibiting concern for his or her welfare and therefore incapable of providing informed consent?

Genetic research raises similar if not even more challenging ethical issues. Genetic research in relation to addiction exposes subjects, their families, and the broader social community to additional risks (Chapman et al., 2012). Risks to subjects include the loss of privacy and the loss of control over sensitive personal information. Financial remuneration for research participation may increase the use of drugs or alcohol if adequate precautions are not taken. Incentive payments to parents to encourage them to enroll their children in genetic studies are unacceptable because of the risk of coercing children to participate. Editors and authors have a duty to make sure that published research is subject to rigorous ethical review.

Nevertheless, in some cases, particularly the social sciences, there is the perception that ethical review has gone too far in its attempts to minimize risks that may not be present. As explained by Mäkelä (2006): (a) social research is generally much less invasive than medical research; (b) its impact on research participants involves different casual chains; (c) social research design tends to be more open ended; and (d) in social research, the context of the relationship between researcher and participant is closer to that of a journalist and a minister rather than that of a doctor and patient.

Consequences

Failure to follow recommended or required journal procedures regarding protection of human and animal research subjects could have several important consequences. Although most journals do not ban publication of articles because they have not been submitted for ethical review, some journals now require authors to state whether their research conforms to the minimum standards outlined in the Declaration of Helsinki, a set of ethical principles regarding human experimentation developed by the World Medical Association. In particular, social and behavioral research such as survey studies and research involving archival records may not require stringent informed consent procedures. However, it would be an error to rely on this perception. Surveys, on occasion, have resulted in significant harm to individuals and to institutions. It is safer to submit all research for institutional review and to let the committee decide whether the researcher is exempt or not. Failure to obtain ethical approvals or informed consent from research participants may lead an editor to question the purpose and value of the research and could result in a decision not to send the manuscript out for review or, when the failure is detected during peer review, to decline the manuscript. Another consequence could be the notification of an official from the author's institution.

Prevention

It is always wise to mention both in the cover letter to the editor and in the text of a submitted manuscript that the researchers have followed appropriate ethical review procedures. If there are any questions regarding the applicability of human subjects requirements, these should be raised with the editor in the cover letter or in a telephone call or email message before submission of a manuscript. Often these questions can be resolved by consulting the journal's website or instructions to authors. The ICMJE (1991, p. 339) has provided the following guidance regarding ethical issues:

> When reporting experiments on human subjects, indicate whether the procedures followed were in accordance with the ethical standards of the responsible committee on human experimentation (institutional or regional) and with the Helsinki Declaration of 1975, as revised in 1983. Do not use patients' names, initials, or hospital numbers, especially in illustrative material. When reporting experiments on animals, indicate whether the institution's or the National Research Council's

guide for, or any national law on, the care and use of laboratory animals, was followed.

Scientific journals also have an important role to play in the protection of human and animal research subjects. Journals are responsible for the dissemination of research findings. They "are obligated to publish research that meets high ethical standards . . . for which the authors have attested to their compliance with regulatory and ethical standards" (Federman et al., 2003, p. 205). A number of journals have implemented policies requiring authors to certify compliance with informed consent procedures, and ISAJE (1997) subscribes to these policies.

Plagiarism

Plagiarism refers to both the theft of intellectual property, such as ideas and images, and the copying of unattributed textual material. Plagiarism ranges from the unreferenced use of others' published and unpublished ideas, including research grant applications, to submission under "new" authorship of a complete article, sometimes in a different language. It can also include copying of another's work verbatim or nearly verbatim in a way that misleads the ordinary reader about the author's own contribution. Table 14.5 provides examples of instances that can be constituted as "clear plagiarism," such as copying an entire article, as well as less serious forms like the "minor copying" of a string of words (COPE, 2011).

It may occur at any stage of planning, research, writing, or publication. It applies to both print and electronic versions of a publication. The Office of Research Integrity, an office within the U.S. Department of Health and Human Services that monitors investigation of research misconduct, considers plagiarism to include both the theft or misappropriation of intellectual property and the substantial unattributed textual copying of another's work, such as sentences, paragraphs or even entire manuscripts, in a way that misleads the ordinary reader regarding the contribution of the author.

	Least serious ◄─────────────────────────► Most serious		
Extent	Few words	Whole paragraph	Whole article
Originality	Commonly used	Used by small number of authors	Original idea
Referencing	Full and accurate referencing		Not referenced
Intent	Unintentional deception		Intentional deception

Table 14.5: Features of plagiarism identified by the Committee on Publication Ethics.
Note: Adapted from COPE (2011).

Consequences

Developments in text-matching software (e.g., CrossCheck, eTBLAST) have made detecting instances of plagiarism much easier. The consequences of plagiarism can be serious, ranging from an editor's reprimand to a formal hearing and loss of employment after an allegation is reported to the author's institutional officials. The US Office of Research Integrity generally does not pursue the limited use of identical or nearly identical phrases that, for example, describe a commonly used methodology or previous research, because these are not considered to be substantially misleading to the reader or of great significance. Journal editors can be unrelenting and at times unforgiving if they detect instances of plagiarism. The typical approach is first to request a written explanation from the author soon after the plagiarism has been discovered. Most often, these instances are discovered by knowledgeable and vigilant reviewers or by readers who sometimes report that their own words, sentences, paragraphs, or articles have been misappropriated. If the author's explanation is credible and the amount of copying is small, the consequences may be nothing more than a letter of reprimand and possibly the rejection of the manuscript. More extensive types of plagiarism may result not only in the rejection of the manuscript, but also in the publication of a correction or retraction if the

1. Cite idea sources and identify the contributions of others without exception, even when paraphrasing or summarizing.
2. Use quotation marks for any verbatim text taken from another author.
3. Clarify for readers which ideas are the author's own and which are derived from another source.
4. Be familiar with copyright law.
5. Paraphrasing and summarizing requires authors to produce the same meaning using their own words.
6. Paraphrasing and summarizing requires authors to possess a comprehensive understanding of the material.
7. Refer to the primary literature, as opposed to a secondary source.
8. Always double check citations and reference section.
9. If uncertain as to whether an idea or fact is common knowledge, cite the original source.
10. Do not partake in ghostwriting.

Box 14.4: Guidelines for avoiding plagiarism.
Source: Roig (2013).

material has already been published, and authors may be banned from submitting to the journal in the future (COPE, 2011).

Studies indicate that retractions for this deadly sin are increasing in recent years, accounting for 9.8%–17.0% of retractions (Fang et al., 2012). More importantly, such matters may then be referred to the author's institutional employer, who typically will have responsibility for dealing with allegations of scientific misconduct. This is discussed in more detail in the next section. Although failure to attribute the original source of a sentence or paragraph may constitute a copyright infringement and could result in civil proceedings, such cases are rarely prosecuted.

Prevention

The US Office for Research Integrity offers *26 Guidelines on Avoiding Plagiarism* (Roig, 2013), which focus on disclosing all sources through appropriate citation or quotation conventions (see Box 14.4 for relevant guidelines). If the author plans to use a large amount of other people's written or illustrative material, he or she must seek permission to reprint the material (COPE, 1999). Legal definitions may vary from country to country regarding plagiarism, copyright, and intellectual property rights. The author should review these with the editor when there is any question (Roig, 2013).

A more common problem that may result in an embarrassing revelation is the unintentional copying of small amounts of textual material or the borrowing of others' ideas or concepts without appropriate attribution. These cases are usually the result of negligence, sloppiness, or laziness, as when an author fails to use quotation marks or paraphrases someone else's ideas without stating the source. In these instances, the best prevention method is the careful documentation of all source documents in the course of note taking and the development of writing habits that allow ample time to prepare a manuscript. Authors can ensure they have appropriately cited their work using text-matching software recommended by the Office of Research Integrity.

Other Types of Scientific Fraud

According to various ethical authorities (e.g., Committee on Publication Ethics, 2011), scientific fraud is manifested in the following forms:

- fabrication or falsification of data, that is, presenting data in a research report that have not been obtained in the manner or by the methods described in the report or altering or presenting original findings in a way that distorts the result in a scientifically unjustified way, or by omitting results or data pertinent to conclusions;

- plagiarism, that is, presenting someone else's manuscript, article, or text as one's own;
- misappropriation, that is, illicitly presenting or using in one's own name an original research idea, plan, or finding disclosed in confidence; and
- noncompliance with legislative and/or regulatory requirements.

Although the terms fraud and misconduct are often used interchangeably, it is important to note that fraud implies intentional deception. Fraud can occur in the course of proposing, conducting, or reporting research. It is most often detected at the time of publication, primarily because reviewers, editors, and readers of scientific articles are very critical and skeptical by nature and profession. In the course of this chapter, and in other parts of this book (see Chapters 5, 10, and 11), we have described several of the less serious instances of scientific

In December 2003, the Court of Justice of the Canton of Geneva gave its sentence in an (in)famous case of scientific fraud. A Swedish professor at The University of Geneva and formerly of Gothenburg University had charged two tobacco activists with libel after they accused him of 'unprecedented scientific fraud' concerning the risks of passive smoking. The court dismissed the case, stating that "Geneva has indeed been the platform of a scientific fraud without precedent in the sense that. Professor Ragnar Rylander has acted in his capacity of associate professor at the University, taking advantage of its influence and reputation and not hesitating to put science at the service of money, in disregard of the mission entrusted to this public institution." According to the court, for thirty years the professor had had a close but secret relationship with Philip Morris, which included substantial financial rewards. Thus he lied when he stated to The *European Journal of Public Health* that he had never had contact with Philip Morris. In his research on passive smoking and in several conferences on the topic he questioned the risks connected with passive smoking. According to the Court, the professor "did not hesitate to deceive the general public in order to show himself favorable to the tobacco company." In particular, the Court reported as apparently fraudulent a study on respiratory diseases in children in which he altered the database so that no link could be made between passive smoking and the frequency of respiratory infections.

Box 14.5: An example of scientific fraud from the tobacco field.
Sources: Domstol i Geneve slår fast svenskt vetenskapsfusk (Court in Geneva gives sentence on Swedish scientific fraud). *Svenska Dagbladet*, 16.12.2003 www.prevention.ch/rypr151203.htm, accessed 11 June 2004.

misconduct, such as the selective interpretation of others' findings, inappropriate citation practices, unfair authorship practices, selective reporting of data, or use of inappropriate statistics. The problem with these questionable research practices and the more serious forms of fraud (e.g., data fabrication) is the damage it does to the scientific enterprise, to the extent that it misleads other scientists and establishes a false record that may be misinterpreted by the public, policymakers, or clinicians. Box 14.5 provides an example of scientific fraud from the field of addiction research.

Consequences

Journal editors, funding agencies, and academic institutions take allegations of scientific misconduct seriously, especially those institutions that depend on public support for their research. Typically, an editor who receives information about possible fraud or who suspects it during the course of a manuscript review has a limited number of options, starting with the notification of the author. Many scientific and academic institutions have procedures to deal with allegations of fraud and misconduct; therefore, an editor can begin by passing the allegation and the author's response to an appropriate institutional official or review committee for further action if the allegation seems credible. In general, the process begins with a preliminary investigation, followed by a more formal inquiry if the allegation has sufficient substance or importance. In such cases, the withdrawal or rejection of the manuscript, or the publication of a correction in the case of an already published article, is the least of the author's worries. Fraud can lead to disciplinary action, banishment from advisory committee or review boards, and the re-review and possible retraction of previously published articles. As is the case with the previous publishing sins, fraud also distorts research findings and can erode the public's trust in research (Gupta, 2013).

Prevention

There can be no substitute for careful mentoring and training of scientists in the prevention of scientific misconduct. Most scientists have such high respect for the values of science that they would never deliberately fabricate data or mislead their colleagues about the data they have collected or its interpretation. Milder forms of scientific misconduct may result from ignorance, so that deliberate exposure to ethical training may help individual scientists avoid these kinds of problems. Researchers are encouraged to review the resources listed in Table 14.6. Because scientists typically work in groups along with research support staff, the best way to prevent fraud is to check the data as well as colleagues' work carefully at every stage in the process of conducting a research project and preparing a scientific report. *BMJ* goes so far as to require investigators to submit full data sets to accompany trials that are published in that journal.

1. *Code of conduct for social science research*
UNESCO [undated] http://www.unesco.org/new/fileadmin/MULTIMEDIA/HQ/SHS/pdf/Soc_Sci_Code.pdf.
2. *Guidelines for research ethics in the social sciences, law and the humanities*
The National Committee for Research Ethics in the Social Sciences and the Humanities (NESH), 2006 https://graduateschool.nd.edu/assets/21765/guidelinesresearchethicsinthesocialscienceslawhumanities.pdf
3. *The concordat to support research integrity*
Universities U.K., 2012 http://www.universitiesuk.ac.uk/highereducation/Documents/2012/TheConcordatToSupportResearchIntegrity.pdf
4. *European code of conduct for research integrity*
European Science Foundation (ESF) and ALLEA (All European Academies), 2011 http://www.esf.org/fileadmin/Public_documents/Publications/Code_Conduct_ResearchIntegrity.pdf
5. *Singapore statement on research integrity*
2nd World Conference on Research Integrity, 2010 http://www.singaporestatement.org/statement.html
6. *Teaching the responsible conduct of research in humans (RCRH)*
Koreman, S. G., Office of Research Integrity, 2006 http://ori.hhs.gov/education/products/ucla/default.htm
7. *Declaration of Helsinki—Ethical principles for medical research involving human subjects*
World Medical Association, 1964 http://www.wma.net/en/30publications/10policies/b3/index.html

Table 14.6: Resources.

Finally, we encourage readers to come forward with good-faith allegations of scientific misconduct and remind readers about protections for whistleblowers, for example, those endorsed by the US Office of Research Integrity, Department of Health and Human Services (2014) in the *Whistleblower's Bill of Rights*.

Conclusion

At various times in its short history, addiction research has had its credibility damaged because of ethical breaches in its research and publication practices. Today the field is experiencing an even greater crisis in values, caused by increasing pressure to publish, COIs, and ethical committee restrictions on research

(Babor, 2009). Furthermore, questionable research practices may be implicitly encouraged by publication practices that focus on significant findings.

This situation has been exacerbated by researchers and organizational entities such as journals and professional societies not having a consistent framework of ethical standards and ethical decision making that can protect authors, the scientific community, and the public from the ethical problems that arise in research and scientific writing. A practical, case-based approach with appropriate ethical analysis, designed to address the realities of research and publishing, follows in Chapters 15 and 16.

In most countries, the general public rates biomedical and social scientists highly in terms of their occupational prestige and credibility. When scientific misconduct is detected and publicized, scientists violate this trust and science loses public support. By following the preventive measures described in this chapter, researchers can avoid most of the major and minor ethical dilemmas associated with scientific misconduct. But the obligation of ethical conduct in reporting research in journal publications does not rest with the authors alone. The Institute of Medicine (2002) report affirms what this chapter espouses in terms of the integrity of individual authors (researchers) by advocating "above all a commitment to intellectual honesty and personal responsibility for one's actions and to a range of practices that characterize the responsible conduct of research" (p. 5). This report also notes that individuals can only flourish in institutions that "establish and continuously monitor structures, processes, policies and procedures [that support] integrity in the conduct of research and use this quality improvement" (Institute of Medicine, 2002, p. 5). There is no one strategy that can be relied on to fully overcome questionable research practices or instances of serious research misconduct. Therefore, a multipronged approach is required by researchers, academics, journal editors, peer reviewers, funders, ethics committees, and regulatory authorities. Such an approach would not only go a long way in preventing the Seven Deadly Sins, it would also remove the need for punishments meted out in the Circles of Hell.

Please visit the website of the International Society of Addiction Journal Editors (ISAJE) at www.isaje.net to access supplementary materials related to this chapter. Materials include additional reading, exercises, examples, PowerPoint presentations, videos, and e-learning lessons.

References

Alighieri, D. (1947). *The Divine Comedy* (translated by Laurence Binyon in *The Portable Dante*, The Viking Portable Library #32). New York, NY: Viking Press.

Amos, K. A. (2014). The ethics of scholarly publishing: Exploring differences in plagiarism and duplicate publication across nations. *Journal of the Medical Library Association, 102,* 87–91.

Babor, T. F. (2009). Alcohol research and the alcoholic beverage industry: Issues, concerns, and conflicts of interest. *Addiction, 104,* 34–47.

Bonner, A., & Gilmore, I. (2012). The UK Responsibility Deal and its implications for effective alcohol policy in the UK and internationally (Editorial). *Addiction, 107*(12), 2063–2065.

Bretag, T., & Mahmud, S. (2009). Self-plagiarism or appropriate textual re-use? *Journal of Academic Ethics, 7,* 193–205.

Cataldo, J. K., Prochaska, J. J., & Glantz, S. A. (2010). Cigarette smoking is a risk factor for Alzheimer's disease: An analysis controlling for tobacco industry affiliation. *Journal of Alzheimer's Disease, 19,* 465–480.

Chapman, A. R., Kaplan, J. M., & Carter, A. (2012). Summary and recommendations: Ethical guidance for genetic research on addiction and its translation into public policy. In A. R. Chapman (Ed.), *Genetic research on addiction: Ethics, the law, and public health* (pp. 232–245). New York, NY: Cambridge University Press.

Chapman, S., Ragg, M., & McGeechan, K. (2009). Citation bias in reported smoking prevalence in people with schizophrenia. *Australian and New Zealand Journal of Psychiatry, 43,* 277–282.

Cho, M. K. (1998). FUNDamental conflicts of interest. The shift toward industry funding biomedical research. *HMS Beagle: The BioMedNet Magazine, 24,* 1–8.

Cho, M. K., Shohara, R., and Rennie, D. (2002). What is driving policies on faculty conflict of interest? Considerations for policy development. In N. H. Steneck & M. D. Scheetz (Eds.), *Investigating Research Integrity: Proceedings of the First ORI Research Conference on Research Integrity* (pp. 127–132). Retrieved from http://ori.hhs.gov/documents/proceedings_rri.pdf.

Choi, W. S., Song, S. W., Ock, S. M., Kim, C. M., Lee, J., Chang, W. J., & Kim, S. H. (2014). Duplicate publication of articles used in meta-analysis in Korea. *SpringerPlus, 3,* 182.

Committee on Publication Ethics (COPE). (1999). *The COPE Report 1999: Guidelines on good publication practice.* London: BMJ Publishing Group.

Committee on Publication Ethics (COPE). (2011). *How should editors respond to plagiarism? COPE Discussion Document.* Retrieved from http://publicationethics.org/files/Discussion%20document.pdf.

Davis, R. M. (2008). Special communication: British American Tobacco ghostwrote reports on tobacco advertising bans by the International Advertising Association and J J Boddewyn. *Tobacco Control, 17,* 211–214.

Etter, J. F., Burri, M., & Stapleton, J. (2007). The impact of pharmaceutical company funding on results of randomized trials of nicotine replacement therapy for smoking cessation: A meta-analysis. *Addiction, 102,* 815–822.

Etter, J., & Stapleton, J. (2009). Citations to trials of nicotine replacement therapy were biased toward positive results and high-impact-factor journals. *Journal of Clinical Epidemiology, 62,* 831–837.

European Science Foundation (ESF) and ALLEA (All European Academies). (2011). The European Code of Conduct for Research Integrity. Retrieved from http://www.esf.org/fileadmin/Public_documents/Publications/Code_Conduct_ResearchIntegrity.pdf.

Fanelli, D. (2009). How many scientists fabricate and falsify research? A systematic review and meta-analysis of survey data. *PLoS ONE, 4*(5), e5738.

Fang, F. C., Steen, R. G., & Casadevall, A. (2012). Misconduct accounts for the majority of retracted scientific publications. *Proceedings of the National Academy of Sciences of the United States of America, 109,* 17028–17033.

Federman, D. D., Hanna, K. E., & Rodriguez, L. L. (Eds.). (2003). *Responsible research: A systems approach to protecting research participants.* Washington, DC: National Academies Press.

Griffin G. C. (1991). Don't plagiarize—even yourself! *Postgraduate Medicine, 89,* 15–16.

Gøtzsche, P. C. (1989). Multiple publication of reports of drug trials. *European journal of clinical pharmacology, 36*(5), 429-432.

Gupta, A. (2013). Fraud and misconduct in clinical research: A concern. *Perspectives in Clinical Research, 4,* 144–147.

Huth, E. J. (1982). Authorship from the reader's side [Editorial]. *Annals of Internal Medicine, 97,* 613–614.

Huth, E. J. (1986). Irresponsible authorship and wasteful publication [Editorial]. *Annals of Internal Medicine, 104,* 257–259.

Institute of Medicine. (2002). Integrity in scientific research: Creating an environment that promotes responsible conduct. Washington, DC: National Academy of Sciences Press.

International Committee of Medical Journal Editors (ICMJE). (1991). Uniform requirements for manuscripts submitted to biomedical journals. *British Medical Journal, 302,* 338–341.

International Committee of Medical Journal Editors (ICMJE). (2013). *Recommendations for the conduct, reporting, editing, and publication of scholarly work in medical journals.* Retrieved from http://www.icmje.org/recommendations/.

International Society of Addiction Journal Editors (ISAJE). (1997). The Farmington Consensus. *Addiction, 92,* 1617–1618.

Jannot, A., Agoritsas, T., Gayet-Ageron, A., & Perneger, T. V. (2013). Citation bias favoring statistically significant studies was present in medical research. *Journal of Clinical Epidemiology, 66,* 296–301.

Jerrells, T. R. (2001). Duplicative publication of data and other ethical issues in publishing of scientific findings. *Journal of Substance Abuse Treatment, 20,* 111–113.

Katz, D., Caplan, A. L., & Merz, J. F. (2003). All gifts large and small: Toward an understanding of the ethics of pharmaceutical industry gift-giving. *American Journal of Bioethics, 3,* 39–46.

Koreman, S. G. (2006). Teaching the responsible conduct of research in humans (RCRH). Office of Research Integrity. Retrieved from http://ori.hhs.gov/education/products/ucla/default.htm.

Lambe, M., Hallhagen, E., & Boethius, G. (2002). Cyniskt spel inom tobaksindustrin. Tvångspublicerade interna dokument avslöjar mångåriga ansträngningar att förneka eller tona ner tobakens negativa hälsoeffekter. [The cynical policies of the tobacco companies—internal documents reveal deliberate cover-up of the health hazards of smoking]. *Läkartidningen, 99,* 2756–2762.

Loue, S. (2000). *Textbook of research ethics: Research and practice.* New York, NY: Kluwer.

Mäkelä, K. (2006). Ethical control of social research. Nordisk alcohol och *Narkotikatidsskrift, 23*(6; English supplement), s5–s19.

Mauss, M. (1967). *The gift: Forms and functions of exchange in archaic societies* (translated by Ian Cunnison). New York, NY: Norton.

McKeever, L. (2006). Online plagiarism detection services—saviour or scourge? *Assessment & Evaluation in Higher Education, 31,* 155–165.

Moffatt, B., & Elliott, C. (2007). Ghost marketing: Pharmaceutical companies and ghostwritten journal articles. *Perspectives in Biology and Medicine, 50,* 18–31.

National Committee for Research Ethics in the Social Sciences and the Humanities (NESH). (2006). Guidelines for research ethics in the social sciences, law and the humanities. Retrieved from https://graduateschool.nd.edu/assets/21765/guidelinesresearchethicsinthesocialscienceslawhumanities.pdf.

Nieminen, P., Rucker, G., Miettunen, J., Carpenter, J., & Schumacher, M. (2007). Statistically significant papers in psychiatry were cited more often than others. *Journal of Clinical Epidemiology, 60,* 939–946.

Office of Research Integrity, Department of Health and Human Services. (2014). *Whistleblower's Bill of Rights.* Retrieved from http://ori.hhs.gov/Whistleblower-Rights.

Pew Charitable Trusts. (2013). *Task force recommendations on relationships with industry.* Retrieved from http://www.pewhealth.org/reports-analysis/data-visualizations/task-force-recommendations-on-relationships-with-industry-85899525795.

PLoS Medicine Editors. (2009). Ghostwriting: The Dirty Little Secret of Medical Publishing That Just Got Bigger. *PLoS Medicine, 6*(9), e1000156.

Relman, A. S. (1990). New 'Information for Authors' — and readers [Editorial]. *New England Journal of Medicine, 323,* 56.

Roig, M. (2013). *26 Guidelines at a glance on avoiding plagiarism.* Office of Research Integrity. Retrieved from http://ori.hhs.gov/plagiarism-0.

Stelfox, H. T., Chua, G., O'Rourke, K., & Detsky, A. S. (1998). Conflict of interest in the debate over calcium-channel antagonists. *New England Journal of Medicine, 338,* 101–106.

Stenius, K., & Babor, T. F. (2003). *Ethical problems and ethical guidelines of addiction journals: A report to the ISAJE membership.* Unpublished manuscript, International Society of Addiction Journal Editors.

Stern, S., & Lemmens, T. (2011). Legal remedies for medical ghostwriting: Imposing fraud liability on guest authors of ghostwritten articles. *PLoS Med, 8*(8), e1001070.

Tramèr, M. R., Reynolds, D. J., Moore, R. A., & McQuay, H. J. (1997). Impact of covert duplicate publication on meta-analysis: A case study. *BMJ, 315,* 635–640.

Yank, V., & Barnes, D. (2003). Consensus and contention regarding redundant publications in clinical research: Cross-sectional survey of editors and authors. *Journal of Medical Ethics, 29,* 109–114.

World Medical Association. (1964). Declaration of Helsinki – Ethical Principles for Medical Research Involving Human Subjects. Retrieved from http://www.wma.net/en/30publications/10policies/b3/.

CHAPTER 15

The Road to Paradise: Moral Reasoning in Addiction Publishing

Thomas McGovern, Thomas F. Babor and
Kerstin Stenius

With the gesture of a guide, whose goal's in sight,
She spoke: "We from the greatest body move,
Emerging in the heaven that is pure light;
Light of the understanding, full of love,
Love of the true good, full of joy within,
Joy that transcends all the heart conceiveth of."

Dante Alighieri (1947), *Paradiso*, Canto XXIX, 37–42

Introduction

The descent into the various levels of the Inferno, described in Chapter 14, resulting from the capital sins (vices) associated with varying degrees of scientific misconduct, is replaced in this chapter by a description of an ascent into Paradise, a realm of enlightened ethical conduct and decision making. This transformation is achieved by the practice of virtue and by adherence to ethical principles and moral reasoning. The ascent into the heavenly spheres is achieved by those virtuous ones who exemplify fortitude, prudence, justice, and temperance in the overarching context of faith, hope, and love. "Being good" (character ethics) exemplifies this state, and this is a quality found in society,

How to cite this book chapter:
McGovern, T, Babor, T F and Stenius, K. 2017. The Road to Paradise: Moral Reasoning in Addiction Publishing. In: Babor, T F, Stenius, K, Pates, R, Miovský, M, O'Reilly, J and Candon, P. (eds.) *Publishing Addiction Science: A Guide for the Perplexed*, Pp. 299–321. London: Ubiquity Press. DOI: https://doi.org/10.5334/bbd.o. License: CC-BY 4.0.

institutions, and individuals. Moral reasoning and ethical reflection inculcate attitudes that promote good behavior, the moral stance of "being good."

Reflection and conversation are at the heart of ethical dialogue in any setting. The road to the unethical publication pitfalls described in the previous chapter is paved with good intentions. Perhaps "good conversation" is equally a culprit, as discussion about ethical issues in research does not seem to have influenced actual behavior. Kass (2002), a seasoned veteran in the field of bioethics, complained that "in bioethics at the present, the action is mostly talk" (p. 57). In our search for the best solution to ethical problems, we have lost sight of the original goal of ethics: to improve the quality of our behavior. How then can we provide meaningful direction for researchers wishing to avoid the seven deadly sins pertaining to scientific writing and fraudulent research and at the same time devise a virtuous path leading to responsible research?

Let us abandon ethics as good conversation, an approach advocated by many respected authorities (Brody, 1990; Glaser, 1994), and concentrate exclusively on practical recommendations to guide the behavior of authors and researchers. The Greeks, in their wisdom, saw virtue—the quality of being good in any human endeavor—as the condition of being poised between the two extremes (vices) of any given situation. In proposing an approach to ethical issues in addiction research and publishing, we embrace the advice of the Greeks, which accounts for both talk and action in fashioning practical responses to moral questions.

In pursuing a path that leads to ethical behavior guided by moral reasoning, we find guidance in the initiatives promoted by the International Society of Addiction Journal Editors (ISAJE) and by the Committee on Publication Ethics (COPE). ISAJE was organized in 1997 and has authored the Farmington Consensus (1997) and *Ethical Practice Guidelines in Addiction Publishing: A Model for Authors, Journal Editors and Other Partners* (ISAJE Ethics Group, 2002). Since its own inception in 1997, COPE has published guidelines, policy statements, and more than 500 case reports that provide guidance in promoting ethical standards in all aspects of publication while at the same time addressing the pitfalls of unethical behavior and the associated vices described in Chapter 14. Table 15.1 describes the types of the cases covered by COPE in its ethical analysis of moral problems associated with overall research, including addiction research. The data show that a wide variety of ethical problems have precipitated inquiries for ethical analysis, especially with regard to questionable and unethical research. Most of these issues have already been discussed in Chapter 14. COPE also provides a variety of policy statements and guidelines that address major issues in publication from the perspective of publishers and editors and issues resulting from the case reports described above.

We begin with some reflection about ethics as the human endeavor in addiction research. This approach to ethics is characterized by twin goals: "to be good" and "to do good." Being good is at the heart of "virtue" or "character"

Type of case	Frequency	Percentage
Questionable and unethical research	167	21.1%
Redundant and duplicate publication	113	14.3
Data sloppiness, fabrication, etc.	105	13.3
Misconduct/questionable behavior	99	12.5
Correction of literature	82	10.3
Conflict of interest	61	7.7
Plagiarism	55	7.0
Miscellaneous	54	6.8
Peer review	54	6.8
TOTAL	790	100%

Table 15.1: Frequencies, percentages, and types of ethical cases covered by COPE.
Source: Frequency data obtained from COPE website: http://publicationethics. org/cases, accessed June 21, 2015.

ethics and espouses qualities such as integrity, honesty, and compassion for others. Doing good is the basis for principle-based ethics, such as *autonomy, beneficence,* and *justice* (see Table 15.2 for definitions of italicized ethical terms).

We combine both goals in the ethical discussion in this chapter and bring them to bear on two basic questions that are interwoven in every research enterprise: (a) Can we do it (the technical or research question), and (b) should we do it (the ethical or moral question)? Both questions, individually and collectively, are challenging: They hold us to equally rigorous scientific and ethical standards in all of our research and publishing undertakings. The materials produced by COPE are also an invaluable resource in this undertaking.

Autonomy	Respect people's choices, and do not obstruct their actions unless those actions are harmful to others.
Beneficence	Do good: Provide competent and compassionate care and maximize benefits to individuals, institutions, society.
Nonmaleficence	Do no harm: Minimize risks to individuals, institutions, society.
Justice	Give each person his or her due.
Fairness	Avoid discrimination and exploitation.
Stewardship	Use resources efficiently and justly.

Table 15.2: Key ethical principles used in moral reasoning and decision making.

The Ethical Challenge

Scientific issues in the addiction field, as elsewhere, embrace three ethical realms: the individual, the institution, and society (Glaser, 1994). In affording respect to each of these realms, researchers honor what Kass (2002) describes as "the rich broth of our social, civil, cultural, and spiritual life together and of the ways in which it seasons us without our knowledge" (p. 65). The well-being of the individual, of institutions, and of society as a whole is at stake in assessing the ethical issues that arise in addiction research, as illustrated by recent studies of corporate social responsibility programs, research on chronic drug users, and the use of animals involved in addiction research (Casswell, 2013; Fisher, 2011; Lynch et al., 2010; Miller et al., 2011). Consider the following scenario as an invitation to apply our discussion up to this point to the realities of a possible research publication situation:

> A university research team wishes to examine drug use in a poor, disadvantaged minority neighborhood with an identifiable ethnic population. The intent of the study is to test a new treatment for addiction that holds great promise for society as a whole. The political climate in which the research is conducted is one that is willing to provide research support for biological and social research but is not prepared to address the deeper societal issues underlying drug problems. In addition, the community in which the research is to be conducted sees drug use as both a matter of choice and best controlled through stringent and oppressive legal measures. Furthermore, the larger community views the minority drug-using group with suspicion and distrust. The individuals who will constitute the research population are disadvantaged, have little education, and are a vulnerable population that can be easily exploited in a research endeavor.

Can such research be conducted in a manner that meets appropriate scientific standards? The answer is yes: Many measures can be taken to assure its appropriateness. For instance, researchers can offer guarantees that ensure respect for the dignity of the research participants. Researchers can also safeguard the vulnerability of the individuals involved, together with the community as a whole, by meeting the standards of ethical review committees and other governmental and institutional regulations on research. At first sight, the ethical and scientific standards for responsible research seem to be met at the individual level.

But what of the larger community and societal implications of this research? How will the individuals involved be treated by the larger community if the study shows a high prevalence of drug dependence in the population? Conceivably, an increase in discrimination and oppression might occur (McGovern, 1998), a result researchers would want to avoid. Another consideration centers on who shall benefit from the favorable outcomes of the research: the individuals in the poor neighborhood or the more privileged members of society? Balancing individual, institutional, and societal concerns can lead to a better understanding of the risks and benefits of research in such situations. Whether or not such research should be undertaken determines whether or not it is published.

A helpful perspective on research and publishing as a whole, as well as the case under consideration, can be found by applying the theories and principles associated with "doing good." In analyzing the proposed scenario, a utilitarian approach might seek to maximize the good and minimize the harm. In its most simplistic application, utilitarianism is based on the maxim that the end justifies the means. One could argue from this perspective that the benefit accruing to the majority of the population outweighs the harm to the individual research participants. A duty-driven or deontological approach would counter by arguing that humans can never be used as a means to an end, that their basic dignity must be valued as an end in itself. Two very different responses to the legitimacy of the research and of its subsequent publication thus result from invoking the utilitarian and the deontological positions. One must always be skeptical of research and publications that are justified on the basis of utility or expediency. Grave harm can be inflicted on minority populations and on persons unable to adequately protect their basic dignity (Elwood, 1994). Such a caveat needs to be heeded by authors and editors alike.

In following the road to ethical paradise in research and in publication, then, it is helpful to remember a number of principles derived from ethical theory. *Autonomy,* or respect for persons, obliges the researcher in our scenario and those who oversee research to respect the dignity of those involved in the research project. This is guaranteed by safeguarding privacy and confidentiality and by receiving informed consent—with special attention given to assure that research participants fully understand the risks and benefits involved in the study. Likewise, the principle of *nonmaleficence*—that is, doing no harm to individuals, communities, and society as whole—is of the utmost importance. Conducting research in a competent and compassionate fashion is embodied in the principle of *beneficence.* Although often criticized as the basis for a paternalistic approach, this principle is indispensable in addressing the needs of vulnerable individuals and vulnerable communities, as in the scenario under consideration.

The ethical principle of *justice* guarantees persons their due and guards against discrimination. We would invoke this principle to ensure that we do not expose the research population to undue risks for the benefit of another population. *Fairness,* as a guiding principle, is difficult to invoke in a society overzealous in its defense of individualism and autonomy, without equal attention to the common good (see Ross et al., 1993, pp. 17–28, for discussion of these principles). Finally, *stewardship* demands that investigators use resources responsibly and efficiently.

Toward a Problem-Solving Approach

The first step in the development of an effective problem-solving approach to ethical dilemmas in addiction publishing is to create a code of professional practice for use by research organizations and scientific journals. Such a code now exists in the form of the ethical practice guidelines developed by ISAJE (ISAJE Ethics Group, 2002). The ISAJE guidelines articulate values and define the boundaries of appropriate and inappropriate conduct in addiction research. As such, they provide a moral compass authors can use to guide ethical decision making. One should also note the very significant contributions of COPE in providing moral direction in our research undertakings.

However, the most enlightened and practical direction might be found in the comprehensive analysis of actual situations, especially if they can be considered paradigm cases. This approach finds expression in casuistry, with its ancient roots in moral philosophy and in theology, which provides a consistent focus on individual moral behavior (Jonsen & Toulmin, 1988). It values broad consensus, the development of maxims based on practical wisdom, and the acceptance of probable certitude as the ultimate outcome. Casuistry is attractive because it most closely resembles how we approach moral issues in day-to-day living. Brody (1990) argues that if we examine any ethical situation in research or publishing from every possible angle, we will be able to arrive at a consensus and, in doing so, cover all the various ethical approaches, including theory and principles. The case reports and ethical analysis provided by COPE, which have been previously referenced, are also an invaluable resource in this respect.

Another necessary step toward ethical decision making is to learn how to apply these codes in a practical way. To this end, the main part of this chapter is devoted to the analysis of a set of case studies. These cases are presented in the form of short vignettes that describe a situation or problem, followed by an analysis of the ethical principles involved and the appropriate course of action to be taken by the author. The vignettes have a touch of humor in their presentation, intended as a relief from the doom and gloom of traditional moral analysis. We have also organized the incidents depicted in the vignettes according to the following topics:

1. citation bias: a selective reporting of the literature;
2. redundant publication: when two or more articles share any of the same data or text without full cross-referencing;
3. unethical authorship: all persons named as authors should have made a major contribution to a publication and be prepared to take public responsibility for its contents;
4. undeclared conflict of interest;
5. failure to conform to minimal standards of protection for animal or human subjects;
6. plagiarism: unreferenced use of others' published and unpublished ideas; and
7. scientific fraud.

The analyses provide guides to action, rather than definitive decisions, by deriving conclusions about the most appropriate course of action from sound (and, for the most part, universal) ethical principles such as autonomy, beneficence, justice, honesty, conscientious refusal, stewardship, and nonmaleficence.

A Synthetic Model for the Analysis of Ethical Dilemmas

In their book, *Critical Incidents: Ethical Issues in the Prevention and Treatment of Addiction*, White and Popovits (2001) describe a synthetic model for ethical decision making that borrows from the major traditions and ethical principles described above. The goal is not to provide definitive answers to difficult ethical choices but rather to stimulate thinking about ethical complexity and to suggest options for an ethical course of action. The model involves the application of three questions:

1. *Whose interests are involved, and who can be harmed?* Stating this question in another way, who are the potential winners and losers? In the situations described in this chapter, the main parties likely to be involved are the authors of a particular journal article, the editor of the journal, the author's co-workers, the institution with which the author is affiliated, the professional community of addiction researchers, and society at large. By reviewing the interests and vulnerabilities of these different stakeholders, it becomes possible to identify areas of conflicting interest, where the benefits to one party must be balanced against the harm that could be done to another party or institution.

2. *What universal or culturally specific values apply to this situation, and what course of action is suggested by these values?* According to White and Popovits (2001), this question requires one to explore how widely held ethical values (defined in Table 15.2) can be applied to guide the best course of action in a particular situation. The identification of values that may be in conflict (e.g., honesty vs. loyalty) is an important part of this process, leading to a resolution of the conflict by choosing the higher value. White and Popovits indicate that "the higher value is often determined by the degree of good to be achieved or the degree of harm to be avoided [as] identified through the first question" (p. 27).

3. *What standards of law, professional propriety, organizational policy, or historical practice apply to this situation?* The third step in this process involves the review of established standards of professional conduct, which prescribe or proscribe certain actions for the situation in question. These standards include legal mandates (e.g., copyright laws), professional practice standards, human-subjects requirements, and institutional policies.

Incident/situation_____

1. Whose interests are involved; who can be harmed, how serious is the potential harm? Which interests, if any, are in conflict?

	significant	moderate	minimal/none
Your own interests			
Co-workers			
Research participants			
Your institution			
Professional field or science			
Society			

2. Application of universal values. Check all that apply to your case.
   ____ Autonomy (freedom over one's own destiny)
   ____ Beneficence (do good, help others)
   ____ Nonmaleficence (do not hurt anyone)
   ____ Justice (be fair, distribute by merit)
   ____ Obedience (obey legal and ethically permissible directives)
   ____ Conscientious refusal (disobey illegal or unethical directives)
   ____ Gratitude (pass good along to others)
   ____ Competence (be knowledgeable and skilled)
   ____ Stewardship (use resources wisely)
   ____ Honesty and candor (tell the truth)
   ____ Fidelity (keep your promises)
   ____ Loyalty (do not abandon)
   ____ Diligence (work hard)
   ____ Discretion (respect confidence and privacy)
   ____ Self-improvement (be the best that you can be)
   ____ Restitution (make amends to persons injured)
   ____ Self-interest (protect yourself)
   ____ Other culture-specific values

3. What laws, standards, policies, practice guidelines, and historical practices should guide us in this situation?

Box 15.1: Checklist for analysis of critical incidents.
Adapted from White and Popovits (2001).

Case Studies

In this section, we present seven case studies, each dealing with an important ethical dilemma. Following each case are a series of discussion questions that draw attention to the moral reasoning issues covered. After considering these questions, the reader should follow the outline shown in Box 15.1, which provides further guidance about how to resolve a particular dilemma. Then compare your responses with the ethical analysis that follows each case, which is conducted according to the moral reasoning procedures proposed by White and Popovits (2001). A further source of case discussion and ethical analysis is found in the materials published by COPE, which provides ready access to case materials, including ethical analysis, under the following headings: authorship, conflict of interest, consent for publication, contributorship, data, editorial independence, funding/sponsorship, miscellaneous (books, social media, legal issues), misconduct/questionable behavior, mistakes, peer review, and plagiarism. In the discussion of the cases that occur in this chapter, ethical opinions from COPE are included in the ethical analysis of the cases we have chosen.

Case 1. Selective Reporting of the Literature

Mr. C. Lective is a graduate student in clinical psychology at Orgone University who has just finished his doctoral dissertation under the direction of his mentor, the prominent clinical psychologist Prof. Ann Dorphin. The dissertation topic was based on Prof. Dorphin's Theory of Addiction Reflection, which proposes that drug users' brainwaves give off an aura of escaping endogenous opiates that can be captured by perceptive therapists and recycled to form a therapeutic alliance. After several promising quasi-experimental studies and case reports of Addiction Reflection therapy, all published by Prof. Dorphin or her students, two independent randomized trials produced negative results. A review article was then published questioning the validity of the theory as well as the unorthodox research methods used at Orgone University. Consistent with previous studies at Orgone University, Mr. Lective's dissertation has produced positive but unimpressive results in support of the theory. Prof. Dorphin strongly suggests that the results be published and collaborates in the drafting of an article that recommends that Addiction Reflection therapy be adopted widely in routine clinical practice. The article is submitted to a small psychotherapy journal. After receiving the reviews, the editor of the journal writes the following letter to Mr. Lective:

"I have now received two reviews of your manuscript. The first reviewer liked the article and has few recommendations for revision. The second reviewer, however, notes that your literature review fails to describe recent studies of Addiction Reflection therapy, including a highly critical review article, and thereby presents an inaccurate and misleading characterization of the current status of the theory. Although your study does not seem to contain any fatal flaws, I have decided

not to accept the article because of the reviewer's criticism that the background, rationale, hypotheses, and discussion are all in need of major revision, and the level of scholarship reflected in the article's introduction suggests that the authors are either unfamiliar with recent research on the topic or are being unusually biased in their reporting of the background to their study."

Discussion Questions

1. What could Mr. Lective and Prof. Dorphin have done to avoid this situation?
2. Who is responsible for the selective reporting of the literature, the first author (Mr. Lective), the second author (Prof. Dorphin), or both?
3. Whose interests are involved, and what ethical principles apply to this case?

Ethical Analysis

The responsibility for providing a complete account of the literature and research pertaining to Addiction Reflection therapy rests with both authors, with Prof. Dorphin shouldering most of the responsibility because of her supervisory position. Selective reporting of the literature to support a particular point of view is a significant ethical infraction. It clearly deviates from accepted standards of citation, as described in Chapter 10. Using the White–Popovits grid (see Box 15.3) for the analysis of critical incidents as a guide, this ethical violation has significant moral implications for the authors, their institution, the addiction field, and society as a whole. The reprimand that the authors received from the editor, together with the rejection of the manuscript and the accompanying professional embarrassment, is minor inconvenience compared with the greater harm that might have resulted from the publication of their work. Consider how their faulty research might have harmed the well-being of clients being treated by service providers who, in good faith, followed the researchers' clinical recommendations.

The authors' actions, probably motivated by self-interest, violated the ethical principles of nonmaleficence and justice. There is a clear mandate to "do no harm" enshrined in the principle of nonmaleficence. Mr. Lective and Prof. Dorphin's lack of honesty in espousal of self-interest has the potential to endanger the well-being of all clients and institutions involved with the new therapy. In addition, the principle of justice (fairness) becomes relevant when one considers the fruitless expenditure of scarce resources on a futile mode of treatment. In addition, Prof. Dorphin is clearly in a position to violate the student's autonomy (self-determination) by bringing undue pressure on him to publish his research in a manner supportive of her original theory. This form of coercion, which is clearly unethical, is often ignored in research situations, with consequences for everyone involved when this is uncovered. Much of the harm, real and potential, involved in this situation could have been avoided by following the established standards of citation practice—that is, to present all sides of the related literature, as described in Chapter 10. COPE provides further insight into the ethical issues raised by this case in their discussion of the potential fabrication of data in primary studies included in articles for publication (http://publicationethics.org/cases; Case number 14-01 2014).

Case 2. Redundant Publication

A junior faculty member, Dr. Salame Science, is approaching tenure review at a large university that places great emphasis on the number of first-authored publications as the main criterion for promotion. Dr. Science, who has been working with three other investigators on a large collaborative survey study, suggests that the investigators report their findings separately for each of 16 drugs, thereby giving each of the investigators four first-authored publications. Dr. Science develops a template in which the literature review, methods, and statistical analyses are virtually the same for each article, with only the name of the drug being changed for the 16 articles. When one of the articles dealing with a new rave drug is submitted to a journal for review, the authors fail to advise the editor of the other 15 articles under review at different journals, and do not cite any of these articles in their report. Moreover, the co-authors all sign an ethical statement required by the journal indicating that the article has not been published in whole or in part by another journal and is not under consideration by another journal.

Discussion Questions

1. What should Dr. Science and her co-investigators have done with the reporting of the survey findings?
2. What, if anything, should they have told the editor at the time they submitted the manuscript?
3. Whose interests are involved, and what ethical principles apply to this case?

Ethical Analysis

As noted in Chapter 14, a place in Hell is reserved for those guilty of promoting their own self-interest in the practice of redundant publication, in violation of accepted ethical norms. Dr. Science and her three collaborators find themselves in this unholy situation by submitting material that is (partially) under consideration by another journal and by using verbatim material without quotation marks or attribution. By signing the journal's ethical statement, they have blatantly lied about the existence of the other articles and their relationship to the rave drug study.

Thus, however inadvertent it initially appears, the deception involved in failing to disclose the relationship between the articles has serious ethical implications. Referencing again the White–Popovits analysis grid (see Box 15.1), several types of harm can result at professional, clinical, and societal levels. First, if all 16 articles were in fact published (as opposed to one or two comprehensive articles), the authors would deny as many as 15 competing and perhaps equally worthy authors of the opportunity to publish in the same journals, because many journals have limited space and must reject a high proportion of submitted articles. Second, the task of reviewing and processing these redundant articles creates unnecessary work for reviewers and editors, most of whom volunteer their time as a service to the peer-review system. Whether the possible harm rises to the level of significant in the White–Popovits grid is debatable; it is certainly moderate, in terms of harm inflicted by any standard of ethical analysis. Clearly, the authors' actions have violated the standards of honesty, candor, fidelity, and diligence. The decision of the authors to lie in their ethical declaration attacks the basic trust that undergirds the scientific enterprise and has the capacity to inflict the type of "irreparable damage to scientific investigators, editors, and the community" described in Chapter 14.

By following established standards for citing the interrelationships involved in their collaborative studies, and by responding honestly to the statement required by journal editors and publishers, the authors could have avoided both the ethical and legal censure resulting from their deception and dishonesty.

A case report from COPE (number 06-22 2006) provides further insights into the ethical issues created by redundant publications (http://publicationonethics.org/cases/).

Case 3. Authorship Credits

Dr. Mary Doogood is a postdoctoral fellow at the prestigious National Addiction Research Collaborative (NARC). She is conducting research on prescription-drug addiction under the direction of her mentor, Dr. Arthur Stringalong. After a preliminary analysis of the findings, Dr. Stringalong (who helped design the study, secure grant funding, and analyze the data) suggests that they prepare an article for submission to the Journal of Irreproducible Results.

When Dr. Doogood finishes the first draft, Dr. Stringalong insists on two additions to the list of authors: (a) the scientific director of NARC, who had nothing to do with the study or the writing of the manuscript, and (b) the research assistant who conducted the interviews, entered the data, and did a literature search but who otherwise had little involvement in the study design, data analyses, interpretation of findings, or drafting of the manuscript. Dr. Stringalong tells Dr. Doogood that with the NARC director as last author, the article would have a better chance of being accepted by the Journal of Irreproducible Results. *He also suggests that the research assistant, Ms. Day Tamanager, deserves to be listed as a reward for her hard work; a publication credit will help her application for admission to graduate school.*

Discussion Questions

1. Whose interests are involved, and what ethical principles apply to this case?
2. What should Dr. Doogood do about the suggestion to add the name of the scientific director of NARC?
3. What should Dr. Doogood do about the suggestion to add the name of the research assistant?

Ethical Analysis

One could argue that this situation has significant ethical implications for Drs. Doogood and Stringalong on an individual basis and moderate implications for the scientific director and the research assistant. Dr. Stringalong violates Dr. Doogood's autonomy as first author by insisting on the addition of the extra names, although he would not violate her autonomy if he merely suggested it. Dr. Stringalong's insistence is all the more egregious because of the implications of the duress deriving from his position of authority. There are also issues of doing no harm and of fairness, understood as distribution of credit according to merit. Ms. Tamanger, the research assistant, may have some claim to co-authorship from a fairness perspective but does not really meet the criteria for authorship described in Chapter 11 of this book. Of course, Dr. Doogood could include both in the acknowledgment section without violation of the rule of appropriate attribution-of-authorship credit. Should the names be included as co-authors, an argument could be made that the profession, the field, and society could be moderately damaged.

Dr. Stringalong might counter, from a utilitarian viewpoint, that using the scientific director's name to assure the publication of the data would work toward the betterment of individuals and society and, thereby, outweigh the harm involved by including the additional author. He might likewise remind us that names are regularly added to lists of authors without being seen as a major ethical violation.

The counter-argument points to the damage, certainly moderate and possibly significant, inflicted on the field by the violations of honesty, equity, fidelity, and loyalty involved in this practice of gift authorship. It is clearly contrary to the practice guidelines endorsed by journal editors over the past several decades. In summary, the issues raised in this case involve ethical violations at the individual, institutional, and societal levels and therefore cannot be justified.

Case 4. Undeclared Conflict of Interest

Dr. Boyam I. Greedy was asked by the editor of the Journal of Neuropsychopharmacoepidemiology (NPPE), *Dr. Tom Naïve, to submit a review article on the subject of anti-dipsotropic medications. Dr. Naïve based his invitation on Dr. Greedy's expertise in the pharmacological treatment of craving and his widely cited articles on a new anti-craving drug called Payola. Dr. Greedy prepared the review and submitted it to the journal editor. In the article, Dr. Greedy cited both published and unpublished reports to support his contentions that:*

- *anti-craving drugs like Payola reduce drug craving and substance abuse;*
- *a large multi-center clinical trial of Payola is currently underway by the manufacturer, Chemical Therapeutics, Inc.; and*
- *methods to deliver Payola via patch technology have been developed.*

Because the Journal of NPPE *has no formal policy, Dr. Greedy was not asked to declare any real or apparent conflicts of interest. In addition, in the acknowledgements section of the article, Dr. Greedy included pertinent information about the people who helped him prepare the article. But neither his communications with the editor nor the acknowledgements section revealed the following information:*

- *Dr. Greedy holds U.S. Patent 6,375,999 on "Methods and Devices for Transdermal Delivery of Payola."*
- *Dr. Greedy is a member of the scientific advisory board of Chemical Therapeutics, Inc., and as such received an option to purchase 7,000 shares of stock at 5 cents per share. When the projected initial public offering of shares by Chemical Therapeutics, Inc., occurs in the near future at the corporation's estimated share price of $25.00 per share, Dr. Greedy's equity will be valued at $175,000.*
- *Dr. Greedy received substantial consulting payments from Chemical Therapeutics, including first-class airfare to numerous international meetings, where he spoke about his research on Payola.*

Discussion Questions

1. What ethical issues could arise in this convergence between Dr. Greedy's role as a scientist writing a review article and his connections with the drug company, Chemical Therapeutics, Inc.?

2. To what extent does Dr. Greedy stand to gain financially by gratuitously promoting his patented Payola patch?
3. To what extent does Dr. Greedy stand to gain financially from the interest that his positive assessment of Payola might generate for Chemical Therapeutics, Inc., in advance of a public stock offering?
4. What are the real or apparent conflicts of interest in this case?
5. What are Dr. Greedy's ethical obligations in this case?

Ethical Analysis

Dr. Greedy has many personal, professional, and financial interests embedded in the promotion of Payola. His ability to influence a wider public and to advance the acceptance of the new drug is closely tied to the publication of his review article. A real conflict of interest exists and a host of ethical concerns arise at the individual, institutional, and societal levels.

At the outset, it is important to establish the stakeholders—that is, those who are likely to benefit or lose from the publication of a review article that fails to acknowledge the author's financial stake in Payola's development. First, the author stands to profit in many ways from the publication of the review, although the extent of this benefit depends partly on the prestige of the journal and its influence on readers. Second, patients experiencing addiction stand to gain if knowledge of the efficacy of the new medication becomes widespread following the article's publication.

In his defense, Dr. Greedy might say that the promotion of the new product was the province of the advertising arm of Chemical Therapeutics, Inc., and that neither he nor the company would benefit unduly from the publication of the review article itself. He might even add that his ownership of the patent and his financial ties to the company were matters of public record and these activities are perfectly legal and ethical (even in academic circles) in his role an entrepreneur-scientist. He made his decision to publish his findings solely out of respect for the editor, Dr. Naïve. If the journal had a disclosure policy about conflict of interest, he would have had the option of either complying with it or declining the invitation to publish his data.

Another important set of stakeholders in this case includes the journal itself, its editor, and the publisher. An objective bystander might question the professional and ethical judgment of the editor, Dr. Naïve, in inviting Dr. Greedy to submit an article without first consulting the editorial board. Here Dr. Naïve has failed in his fiduciary responsibilities to the author, the publisher, the journal, and its readers. Even if Dr. Greedy's review were fair, balanced, and critical, deserving of the broadest possible dissemination, the integrity of both the journal and the field are nonetheless called into question by Dr. Naïve's lack of responsibility. The absence of a conflict-of-interest disclosure policy excuses neither the editor nor the author. In a like vein, neither Dr. Greedy nor

Dr. Naïve should claim that the possible good resulting from the publication of the review article outweighs the harm done. One could further argue that if this practice of nondisclosure became widely accepted, irreparable harm could result for patients, the publishing field, and society as a whole.

This case gives us pause when we acknowledge a certain reluctance on the part of the entire scientific community—in its individual, academic, and research components—to provide full disclosure. The relationship among research, industry, and publishing outlets is a necessary one, but ethical standards are needed to manage conflicts between self-interest and concern for the common good.

COPE, in many of its case reviews and related publications, emphasizes the importance of addressing conflict of interest as an ongoing issue of ethical concern in the publication of research. In a case titled "Multiple failure to declare a relevant conflict of interest" (case number 07-33 2007), it provides excellent guidance on how to deal with situations like this.

Case 5. Human Subjects Requirements

Dr. X. Ploit, a clinical psychologist working at the Department of Parole, hears about a dataset consisting of clinical records, demographic information, and rearrest data for parolees (i.e., convicted criminals who are released to the community under close supervision) who were exposed to a new substance use disorder treatment program. Because the program could not accommodate all parolees, only people being released from prison on alternate weeks were assigned to the program. The others received no treatment. When Dr. Ploit learns of this "natural experiment," he concludes that the data could comprise a very valuable contribution to the literature, because the parolees were, in effect, randomly assigned to treatment and control conditions and were not pre-selected for participation in a research project. Because of his lack of ethical training, Dr. Ploit is unaware of the need to obtain ethical review board approval to access these kinds of records for research purposes, even though he has legitimate access to the same records because of his clinical responsibilities. Thus, he obtains the names of the selected paroled prisoners, looks up their remand records, and conducts a statistical analysis. The analysis reveals that the parolees who were exposed to treatment were significantly less likely to return to prison for parole violations associated with alcohol and other drug use. Dr. Ploit writes up the results and submits them to the Journal of Drug Criminalization.

When he submits the article, Dr. Ploit is asked to sign a form stating that the study had received all necessary human subjects approvals by an ethical review board. Although Dr. Ploit feels conflicted about signing the statement, he decides to lie about his failure to seek ethical approval, reasoning that (a) the results do not identify individual prisoners and (b) the ethical review board would probably have given him permission to access the data anyway. Dr. Ploit also hesitates to seek post hoc permission from the ethical review board at this point, because they might now deny permission. He reasons that the value of the findings for society and the prisoners far outweighs his minor ethical transgression.

Discussion Questions

1. Why did the editor require Dr. Ploit to submit documentation that he had met ethical review requirements for the study?
2. What is the function of institutional and editorial requirements regarding the treatment of human participants?
3. Do compliance standards in themselves assure ethical behavior in research?

Ethical Analysis

In this case, it is appropriate to emphasize the vulnerability of persons with addictions in all aspects of their well-being, including treatment and research, and the intensification of such vulnerability in particular environments, such as correctional facilities. Such concerns are central to Dr. Ploit's research, which describes the response of parolees to an innovative treatment program. Even though the way in which participants were originally assigned to the new treatment arose out of limited resources, ethical review is very important to make sure that coercion was not a factor. These questions arise in the presence or absence of a research protocol.

The question of ethical approval, requested by the editor as a condition for accepting this piece for review, is an important one. Ethical review gives some assurance that the research itself meets basic ethical standards and also includes the expectation to provide oversight of the ongoing research in terms of participant well-being in a research environment. The ethical review board, if it had been involved in the discussion of this research, could have decided that the research enjoyed exempt status under the rubric of quality assurance and chart review. On the other hand, it may have required full compliance with all the requirements of a regular research protocol. In addressing a journal's ethical concerns about compliance with ethical review committees or other supervisory bodies, the nature of Dr. Ploit's work changes when it becomes research. The editorial board could reasonably restrict Dr. Ploit's research to data gathered subsequent to approval.

Compliance with regulatory bodies generally satisfies legal requirements in research undertakings and guarantees that basic ethical standards are in place. The regulatory research bodies share with journal editors a concern for the promotion of good and the avoidance of harm at the individual, institutional, and societal levels. The author has a fiduciary relationship with the ethical review board and with the editor, and all parties are mutually dependent on each other acting in good faith and in compliance with a commonly accepted ethical framework that promotes the common good. Compliance standards in and of themselves guarantee minimum protection for stakeholders in research undertakings; ethical standards often espouse a higher degree of care.

The ethical dimensions involved with the protection of human subjects have societal, institutional, and individual implications. This has been discussed in the ethical analysis of this case, and further insight into this analysis is provided by the COPE publication on inadequate assurance of human research ethics for questionnaires, case number 12-33 2012 (http://publicationonethics.org/cases/, retrieved June 3, 2015).

Case 6. Plagiarism

Wilhelm Reicht and Ena G. Orgone are new doctoral students working on a project at the University of Freudberg that explores the impact of the therapist–patient relationship in psychoanalytic treatment for female abusers of prescribed psychotropics. Reading the background literature, they find a very good article by Professor Eve N. Id in one of the big U.S.-based psychoanalytic journals. In the article, Dr. Id explores how the angle of the analyst's sofa can influence the level of subconsciousness that the patient is able to reach in therapy. The article establishes the so-called Divanaltitude theory.

The two ambitious students decide to submit an article to the Bayerische Zeitschrift für Psychoanalytische Alkoholstudien *to demonstrate that they are on the cutting edge of current research. Their article, written in German, presents the Divanaltitude theory along with some findings from a small, local survey that the students conducted to learn what alcohol and other drug therapists think about the design of sofas in therapeutic settings. Reicht and Orgone inform the editor that they consider their text to be an overview and not a piece of original research.*

The editor, who is not familiar with the Divanaltitude theory, sends the text to a referee. The referee's critique comes back after two weeks. She has discovered that the introduction is a direct translation of Professor Id's abstract. Several subtitles and the structure of the first part of the article are identical to Dr. Id's. That the authors have one reference to Dr. Id's article in the second paragraph of the text is obviously not enough; the referee considers this to be a case of plagiarism.

The editor subsequently sends a letter to the young authors stating that he cannot accept the article for publication because large sections of the text are identical to an already published article. He states that their submission breaches internationally accepted ethical rules of publishing and demands an explanation. The editor also informs the authors that he will send a copy of his letter to the head of their department at Freudberg.

Discussion Questions

1. How could the students have avoided the reprimand of the journal editor and the possible censure of their chair and university?
2. What harm, real or potential, could result from the students' action?
3. Could the students claim that they were unfamiliar with the ethical rules of publishing? If they were unfamiliar, whose obligation was it to inform them?

Ethical Analysis

The students' plagiarism has important implications, with the possibility of harm for the students themselves, the original author, the research institution, the addiction field, and for society as a whole. The students, according to the White–Popovits grid, exposed themselves to the risk of possible dismissal from their doctoral program as punishment for their violation of accepted ethical norms. It is conceivable, however, that they acted out of ignorance and that they had not received appropriate ethics training from their professors or their institution. Had the individual professors and the institution been remiss in providing appropriate direction for the students, then the institution and its representatives would be as culpable as the students.

The actions of the students obviously involved a form of theft where Dr. Id's work is concerned, but any damage to her reputation will be moderate or minimal according to the White–Popovits scale. Their transgressions also present the possibility of injuring the professional field and society as a whole, especially if such actions were to become commonplace in the publishing field. According to the White–Popovits scheme of universal values, the students violated the values of justice, honesty, and diligence in their failure to acknowledge the work of the original author. They acted out of self-interest, with lack of regard for established ethical and professional guidelines. They might be accused of violating the original researcher's autonomy by denying her the opportunity to control her own work through appropriate citations. If the students failed to receive appropriate ethical formation and direction from their institute, then the administrators and professors at the institute would be in violation of the principles of beneficence and nonmaleficence. Institutions have a moral responsibility to provide an environment in which integrity and honesty are an essential part of their research undertaking (Institute of Medicine, 2002). Stewardship also enters into the equation because, from a societal perspective, institutions have a social responsibility to use resources wisely.

The need to address plagiarism in its many forms, including self-plagiarism, is central in maintaining the integrity of research publication, with ongoing attention to the ethical dimensions addressed in the analysis of this case. In examining a report of possible self-plagiarism, COPE case number 14-10 2014 provides further insight into this important issue (http://publicationonethics.org/cases/, retrieved on June 3, 2015).

Case 7: Scientific Fraud—"Data Trimming"

Dr. Frank N. Stein is a junior faculty member in the Department of Anatomical Protuberances at a large Transylvanian medical school. His latest research project deals with the effects of brain transplants on addiction careers. Preliminary analysis of the data on the first 10 transplants shows an interesting trend, but the p value is just shy of statistical significance. Dr. Stein's statistician, Mr. Igor Numbers, suggests they conduct a few more transplants to increase statistical power and then add an equal number of cases to the control group (without the benefit of random assignment). Igor also suggests they conduct a one-tailed test to get a more favorable alpha level and drop some of the covariates to increase the degrees of freedom. After Dr. Stein and Mr. Numbers have made all these protocol changes, they submit their article for publication as a true random assignment study with significant differences between groups. One of the reviewers questions the use of a one-tailed test, suggesting that the authors include more covariates in their analyses and asks the editor to obtain more detailed information from the authors (Dr. Stein and Mr. Numbers) about the way they assembled their samples. Dr. Stein's institution has granted appropriate approval for the research. In addition, the research enjoys societal approval through funding that provides appropriate resources for good scientific work.

Discussion Questions

1. Was it ethical for Dr. Stein to use the one-tailed test?
2. How should Stein respond to the editor?

Ethical Analysis

The stakeholders are the recipients, the scientists, the medical school, and society as a whole. The good espoused by Dr. Frank N. Stein's research is the enhancement of the addiction field through the advancement of knowledge about the effects of brain transplants. Whether to continue this research depends on outcome studies, largely dependent on the findings of Dr. Stein and Mr. Numbers, who are convinced that the changes in their statistical analysis are minor and ethical. They feel that the continuation of their work will confer immense benefits on all involved and especially people with addictive disorders. Their decision to use the new statistical analyses, together with their justification of this approach in their response to the review process, shows they believe the end justifies the means. After all, this is a new cutting-edge enquiry where data trimming on a minor scale may be considered no more than a minor peccadillo.

The researchers, despite their idealism and good intentions, are blind to the implications of honesty, stewardship, and fairness in their decisions. Their dishonesty impinges on the well-being and safety of the recipients of brain transplants. In addition, they are not good stewards of the funds that supported this research. Furthermore, their unethical use of funds constitutes disservice to the other, unfunded scientists whose requests for funding are based on honest and responsible findings.

Our tongue-in-cheek response to this fanciful scenario uncovers many ethical pitfalls resulting from what might appear *prima facie* as minor adjustments in one's statistical approach. Rigorous honesty must inform the research itself, and authors must be candid with editors about methods and outcomes. The relationship between the two parties is a fiduciary one, and the engendered trust touches the basic integrity of scientific publishing. Using the White–Popovits grid, one could award this case a perfect score of "significant" on all the interests and vulnerability items.

Fraud, as we have identified in the ethical analysis of this scenario, is the most egregious violation of professional integrity in research undertakings. COPE, in its analysis of case number 14-05 2014, again provides excellent insights into the implications of fraud in research situations (http://publicationethics. org/case/fraud-or-sloppiness-submitted-manuscript). Distinguishing between fraud and sloppiness is difficult to determine, and this case analysis is helpful in this respect.

Conclusion

The intent of this chapter was to illustrate an ethical framework that provides practical guidance for investigators in publishing responsible and trustworthy research. Central to this understanding is a high degree of trust, as demonstrated in the case analyses. A fiduciary relationship is at the heart of the assurance whereby researchers address the well-being of individuals, institutions, and the overall common good.

In a climate of self-interest, often nurtured by a high regard for an exaggerated form of individualism (which is inimical to the common good), it is difficult to develop a consistent appreciation of the place of trust in research undertakings, as is the case elsewhere in society (Institute of Medicine, 2002). Societal safeguards need to be in place, as envisaged by ethical review committees and other regulatory agencies, to ensure that the trust that individuals, institutions, and society afford to research is well placed and respected. Research communities and regulatory agencies need to establish the highest level of collaboration as a first step in creating and maintaining a climate of trust.

Regulatory agencies in and of themselves cannot ensure ethical behavior in research or publishing, both of which have trust as their foundation. Other forces are in play, such as virtue or character considerations. Individuals, institutions, and publishing enterprises should ideally encompass qualities such as integrity, fairness, and trust in their undertakings evaluating the presence or absence of virtue in larger bodies is not easy. It is difficult to determine if an institution is virtuous based on an analysis of the goodness of the institution where the research occurs. Other forces are equally important, such as virtue or character considerations involving individuals and institutions in the research and publishing enterprises. Inserting virtue ethics by encompassing qualities such as integrity, fairness, and trust is not an easy task. Equally difficult is the infusion of like qualities into the culture of institutions where research occurs. Many centuries ago, in his dialogue with Socrates, Plato wrestled with this problem as recounted in his work, *Meno:* "Can you tell me Socrates, is virtue something that can be taught? Or does it come by practice? Or is it neither teaching nor practice that gives it to a man, but natural aptitude or something else?" (translation by Thompson, 1980). In fashioning a character-based ethic to guide the behavior of researchers and authors, traditional wisdom might prompt one to respond "all of the above" in answer to Plato's questions.

The "something else" to which Plato alludes is intriguing and invites comment as a concluding thought for this chapter. Perhaps Plato was hinting, for our present-day edification, that the fullest ethical analysis of persisting contemporary issues in research and publication, along the lines of the case studies in this chapter, is that "something else." Ongoing conversation about actual issues is the best assurance that an ethical climate will inform research ethics and promote responsible publishing behavior.

Please visit the website of the International Society of Addiction Journal Editors (ISAJE) at www.isaje.net to access supplementary materials related to this chapter. Materials include additional reading, exercises, examples, PowerPoint presentations, videos, and e-learning lessons.

References

Alighieri, D. (1947). *The Divine Comedy* (translated by Laurence Binyon in *The Portable Dante*, The Viking Portable Library #32). New York, NY: Viking Press.

Brody, H. (1990). Applied ethics: Don't change the subject. In B. Hoffmaster, B. Freedman, and F. Raser (Eds.), *Clinical ethics: Theory and practice* (pp. 183–200). Clifton, NJ: Human Press.

Casswell, S. (2013). Why do we not see the corporate interests of the alcohol industry as clearly as we see those of the tobacco industry? *Addiction, 108,* 680–685. DOI: https://doi.org/10.1111/add.12011

Elwood, W. (1994). *Rhetoric in the War on Drugs: The triumph and tragedy of public relations.* Westport, CT: Praeger.

Fisher, C. B. (2011). Addiction research ethics and the Belmont Principles: Do drug users have a different moral voice? *Substance Use & Misuse, 46,* 728–741. DOI: https://doi.org/10.3109/10826084.2010.528125

Farmington Consensus. (1997). *Addiction, 92,* 1617–1618. DOI: https://doi.org/10.1080/09652149736332

Glaser, J. W. (1994). *Three realms of ethics: Individual, institutional, societal.* Kansas City, MO: Sheed and Ward.

Institute of Medicine. (2002). *Integrity in scientific research: Creating an environment that promotes responsible conduct.* Washington, DC: National Academy of Sciences Press.

ISAJE Ethics Group. (2002). *Ethical practice guidelines in addiction publishing: A model for authors, journal editors and other partners.* London, England: International Society of Addiction Journal Editors. Retrieved from: www.isaje.net.

Jonsen, A. R., & Toulmin, S. (1988). *The abuse of casuistry: A history of moral reasoning.* Berkeley, CA: University of California Press.

Kass, L. R. (2002). *Life, liberty and the defense of dignity: The challenge of bioethics.* San Francisco, CA: Encounter Books.

Lynch, W. J., Nicholson, K. L., Dance, M. E., Morgan, R. W., & Foley, P. L. (2010). Animal models of substance abuse and addiction: Implications for science, animal welfare, and society. *Comparative Medicine, 60,* 177–188.

McGovern, T. F. (1998). Vulnerability: Reflection on its ethical implications for the protection of participants in SAMSHA programs. *Ethics and Behaviour, 8,* 293–304.

Miller, P. G., de Groot, F., McKenzie, S., & Droste, N. (2011). Vested interests in addiction research and policy. Alcohol industry use of social aspect public relations organizations against preventative health measures. *Addiction, 106,* 1560–1567.

Ross, J. W., Glaser, J. W., Rasinski-Gregory, D., McIver Gibson, J., & Bayley, C. (1993). *Health care ethics committees: The next generation.* Chicago, IL: America Hospital Publishing.

Thompson, E. S. (Ed.). (1980). *The Meno of Plato.* New York, NY: Garland Publishing.

White, W. L., & Popovits, R. M. (2001). *Critical incidents: Ethical issues in the prevention and treatment of addiction* (2nd ed.). Bloomington, IL: Lighthouse Institute.

CHAPTER 16

Relationships with the Alcoholic-Beverage Industry, Pharmaceutical Companies, and Other Funding Agencies: Holy Grail or Poisoned Chalice?

Peter Miller, Thomas F. Babor, Thomas McGovern,
Isidore Obot and Gerhard Bühringer

Introduction

The ethical dimensions of the relationships among researchers, research organizations, journal editors, and the various industries that profit from addictive substances and behaviors are complicated and extensive. They embrace the individual, institutional, and societal dimensions of ethical reflection. In a way, this chapter is a case study on a grand scale that calls for profound ethical analysis. The forces and interests involved are of necessity interwoven, and researchers are dependent on many funding sources as a mainstay for their research. These will be covered in detail as the chapter unfolds. At the heart of the ethical conversation is an issue of trust for individuals and institutions. Ultimately, there are no simple guidelines to help an investigator decide which funding sources to accept or reject. However, it is vital that researchers go through an ethical assessment to consider the issues involved. In this chapter, we will explore the ways in which different interest groups have influenced the research process before demonstrating the use of the PERIL (purpose, extent, relevant harm, identifiers, link) analysis (Adams, 2007), an ethical decision-making framework

How to cite this book chapter:
Miller, P, Babor, T F, McGovern, T, Obot, I and Bühringer, G. 2017. Relationships with the Alcoholic-Beverage Industry, Pharmaceutical Companies, and Other Funding Agencies: Holy Grail or Poisoned Chalice? In: Babor, T F, Stenius, K, Pates, R, Miovský, M, O'Reilly, J and Candon, P. (eds.) *Publishing Addiction Science: A Guide for the Perplexed*, Pp. 323–352. London: Ubiquity Press. DOI: https://doi.org/10.5334/bbd.p. License: CC-BY 4.0.

developed specifically to address ethical decision-making. We will extend this previous work to challenge even this framework by asking whether it is simply enough just to question the intentions of vested interests in their funding of research. We will close by stressing the importance of understanding corporate political activity in the context of how vested interests are capable of undermining evidence-based policy at local, state, national, and international levels.

A high proportion of an active researcher's workload is spent applying for grant income. Successful receipt of grant monies is seen as an independent measure of a scientist's worth to the field. But the successful awarding of research money can occasionally be a "poisoned chalice" because of the problems engendered by an association with a funding agency. Such problems include having commercial or other vested interests set the research agenda, determine the way in which research is conducted, or define when and where research is published. Contracts that might seem reasonable when the cash is being waved under one's nose may prevent entire studies from being published or, even worse, result in selective publication that does not portray the actual findings accurately. These types of experiences can devastate individual researchers, both personally and professionally. From the outset, we want to emphasize that individual researchers cannot deal with these issues alone but need support from senior colleagues, their institutions, professional associations, and academic journals.

A Growing Concern

In a climate of self-interest, often nurtured by a high regard for an exaggerated form of individualism (which is inimical to the common good), it is difficult to develop a consistent appreciation of the place of trust in research undertakings, as is the case elsewhere in society.
(McGovern et al., Chapter 15).

Concerns about the integrity of the evidence base of addiction science have been raised in a number of forums recently (e.g., Adams, 2007; Babor & Robaina, 2013; Hall, 2006a; Miller, 2013; Miller et al., 2006; Stenius & Babor, 2010). Many of the authors expressing these concerns have reminded us that, although safeguards such as ethical review committees and other regulatory agencies are in place, ensuring the integrity of the evidence is an ongoing task that requires an awareness of new players (e.g., energy-drink producers) seeking to influence the evidence base, as well as awareness of new technologies for doing so (Hall, 2006a), such as paid contributions to edited books that look scholarly but often have a hidden political agenda. On the other hand, there have been strong developments in the study of such industries and the way in which they use research to muddy the waters of evidence and influence the political process (Hawkins & Holden, 2014; Savell et al., 2014). This will be discussed later in the chapter in regard to assessing the purpose of industry-funded research.

Miller et al. (2006) highlighted the influence that major funding bodies (e.g., pharmaceutical companies and governmental departments) can have on research findings and the information-dissemination process. This was considered important from two angles: (a) keeping true to the ideal of science and (b) adhering to the ethical principle of *beneficence* (Chapter 15). Maintaining the ideal of science was seen as essential for the field, in terms not only of sustaining public trust (as mentioned above) but also of ensuring that the field moves toward the most-effective interventions available. Adhering to the ideal of beneficence (the obligation to maximize possible benefits and minimize possible harms) was viewed as equally important when considering whether research (which may be censored, be partially reported, or go unpublished) could truly be said to be in the best interests of the research participants.

The debate within academic journals and subsequent commentaries has added substantially to our knowledge of how funding bodies influence research both directly and indirectly (Adams, 2007; Ashcroft, 2006; Babor, 2006; Babor & Miller, 2014; Hall, 2006a, 2006b; Hough & Turnbull, 2006; Khoshnood, 2006; Lenton & Midford, 2006). The observations collected from various authorities and presented in Box 16.1 highlight some of the main issues and point to

---

"Because . . . research may adversely affect the reputations of governments and government departments, 'project management' has become an increasingly central part of contractual arrangements between researchers and funders" (Hall, 2006b, p. 240).

"[I]n the current funding climate, universities and research centres have incentives not to adhere rigorously to these norms" (Ashcroft, 2006, p. 238).

"In recent years almost all [Australian] state and federal funded drug education research has been commissioned according to funder specifications, rather than being investigator driven" (Lenton & Midford, 2006, p. 244).

"Certainly, too, government departments set research agendas—and specify research methodologies to suit their own interests, rather than to contribute in a disinterested way to the body of knowledge that relates to policy issues. Government departments do not intentionally commission research that will embarrass their ministers" (Hough & Turnbull, 2006, p. 242).

"Senior academic researchers should be prepared to 'out' funding bodies for bad behaviour. Researchers with seniority and the protection afforded by tenure should be prepared to protect junior researchers and advocate for an unencumbered right to publish research results" (Hall, 2006b, p. 240).

Box 16.1: Observations about research funding from different commentators.

the fact that influences on the research process go far beyond industry-related funding bodies alone.

Types of Adverse Influence

Miller et al. (2006) identified five major avenues through which funding bodies can regulate research in an adverse way: (a) direct censorship (where material is edited or dissemination is interfered with), (b) limiting access to data (either affecting some point or to be used as coercion for favorable interpretation), (c) ongoing funding insecurity (attaching conditions to subsequent funding if previous findings have been awkward or unwelcome), (d) using under-qualified or easily-influenced researchers (which allows funders to control the quality of investigation being carried out, even before the research has commenced), and (e) setting research agendas or *dilution* (whereby decisions are based on the political, financial, or ideological interests of the funder). For example, pharmaceutical companies overemphasize studies that examine the efficacy of pharmacotherapeutic solutions to drug-related problems, which could make the evidence base appear to be overly favorable for such an intervention (Wagner & Steinzor, 2007). Other authors (e.g., Gruning et al., 2006; Kassirer, 2005) have provided similar, although slightly different, descriptions of the ways in which interest groups have influenced health policy and scientific research (Box 16.2).

The Tobacco Industry

The best known example of the way a funding body can act to undermine research integrity and muddy the waters surrounding a topic of public health interest is the concerted campaign by the tobacco industry first to deny the links between smoking and lung cancer and then more recently to support programs that attribute responsibility to the individual smoker rather than to the tobacco companies.

Investigations into tobacco companies continue to identify new ways in which the industry seeks to encourage smoking and at the same time divest itself of responsibility for the subsequent health costs (Drope et al., 2004; Iida & Proctor, 2004; King, 2006; Muggli et al., 2004; Ong & Glantz, 2000). There are numerous examples of how tobacco companies have acted to undermine or adulterate health initiatives. The tobacco industry has been found to influence research using every one of the techniques discussed earlier (e.g., Hirshhorn et al., 2001; King, 2006). According to one authority, "perhaps research grants coming from tobacco companies should carry their own Surgeon General's warning. Caution: Tobacco industry sponsorship may be hazardous to the public's health" (Parascandola, 2005, p. 549).

Gruning and colleagues (2006) identified five ways in which the tobacco industry in Germany distorted science:

- *Suppression,* through actions such as closing the German Industry Research Institute (which it funded) when its head published results unfavorable to the industry and having subsequent scientists in its employment guarantee that unfavorable results would not be published;
- *Dilution,* through selective funding of research and the recruitment of scientists who had doubts about the adverse health effects of smoking or whose previous work had found no links, as well as funding research projects designed to find no association between smoking and disease (e.g., Wander & Malone, 2006);
- *Distraction,* by selecting and supporting a large number of "confounder studies," which are research projects aimed to distract attention from smoking by investigating other potential causes of smoking-related diseases;
- *Concealment,* using third-party scientists whose connection to the industry was hidden to increase the credibility and impact of the studies published; and
- *Manipulation,* the vetting of articles and presentations by the industry before publication or presentation.

Box 16.2: The tobacco industry in Germany.
Source: Gruning et al. (2006).

One example of this is the tobacco industry's support of scientific research and their use of academics as expert witnesses in court cases. As many senior researchers in the addiction field are occasionally asked to serve as expert witnesses for a defendant or a plaintiff, it is instructive to examine cases where such testimony could have implications for public health, especially when it proves to be wrong. Can direct payment of a scientist bias that person's opinions and even sworn testimony in a court case?

Until 1998, most of the tobacco industry funding for research on nicotine and tobacco came through Council for Tobacco Research (CTR) and the Center for Indoor Air Research (CIAR). These two organizations were established and maintained by funding from the tobacco industry. They played a central role in the lawsuits brought against the tobacco industry in the 1990's, when it was found that industry-funded research contradicted the conclusions of independent scientists (Shick and Glantz, 2007). A US judge presiding over two state cases described CTR as "nothing but a hoax created for public relations purposes with no intention of seeking the truth or publishing it." (Janson, 1988). The Master Settlement Agreement (MSA) in 1998 dissolved the CTR and CIAR, as they were implicated in a conspiracy of massive fraud. Tobacco companies also agreed to pay $206 billion over the first twenty-five years of the agreement to compensate the States for taxpayer money spent for health-care costs connected to tobacco-related illness.

In a series of court cases and depositions, then Professor Emmanuel Rubin testified that the research conducted by the CTR was of high scientific quality and that its scientific review adhered to widely recognized scientific standards. For example, in 2000 testimony for Philip Morris Inc. (p. 29) he stated:

"In my opinion the Council for Tobacco Research was an affective (sic), efficient, generous and thoroughly honest organization that provided funds for excellent biomedical research. It acted in an independent fashion that was no different from other agencies that provided grants. I think that the research that was funded by CTR contributed significantly to understanding the issues of tobacco and health. And, for that reason, I have no objections to funding by the CTR."

Box 16.3 provides excerpts taken from Dr. Rubin's deposition in 2000 during a case brought by a health insurance company against Philip Morris for the costs connected to tobacco smoking. The line of questioning begins with questions of financial payments received by Dr. Rubin. It then continues to explore Dr. Rubin's opinions about the qualifications of members of the Scientific Advisory Board who were senior executives of RJR Tobacco Company, and the practice of having grant applications screened initially by industry lawyers before they were submitted for scientific review. Given the outcome of the trial, Dr. Rubin's testimony provides a good example of how financial COIs may influence the opinions of scientists who serve as expert witnesses.

The Alcohol Industry

Using terms of justification such as "corporate social responsibility" and "partnerships with the public health community," the alcoholic-beverage industry (mainly large producers, trade associations, and "social-aspects" organizations) funds a variety of "scientific" activities that involve or overlap with the work of independent scientists using techniques that range from efforts to influence public perceptions of research to the direct commissioning of research that is consistent with their public-relations priorities (Babor & Robaina, 2013).

There are at least three organizations funded predominantly by alcohol-industry sources for the primary purpose of conducting scientific research on alcohol: the European Research Advisory Board, the ABMRF/The Foundation for Alcohol Research, and the Institut de Recherches Scientifiques sur

Q*. It appears to me that you've given deposition testimony in six smoking and health litigations and have given trial testimony in one. . . . . . Can you give me an estimate on how much money you have been compensated for performing as an expert witness in the various tobacco and health litigations in which you have done so?

A*. I haven't kept records and that is for, you know, all of this time. I'm not in business, but I'd estimate all of those things, $500,000, $600,000.

Q. Over a six year period?

A. Yes.

———————

Q. Dr. Rubin, you just testified that it would not be proper for the president of CTR to send grant applications to CTR's lawyers for legal review solely on the basis of the fact that the research-called for could implicate cigarette smoking as a cause of human disease, correct?

If you were shown evidence that that, in fact, did happen, would that change any of the expert opinions that you've expressed in your expert report?

A. Well, I'd like to know the circumstances...

———————

Q. Did your opinions change if you were shown evidence to indicate that this was a continuing, regular practice, at CTR?

A. You would have to show me the evidence.

*Q. refers to questions asked by attorneys for Blue Cross and Blue Shield of New Jersey (Plaintiffs). A. refers to answers provided by Dr. Rubin, expert witness for Philip Morris, Inc.

Box 16.3: Excerpts from Dr. Emmanuel Rubin's Testimony in Blue Cross and Blue Shield of New Jersey vs. Philip Morris, Inc.
Source: Blue Cross and Blue Shield of New Jersey, et al., Plaintiffs, vs. Philip Morris, Incorporated, et al., Defendants. Case no. 98 CIV 3287 (JBW) Videotaped deposition of Emanuel Rubin, M.D., April 12, 2000, Bates Number: 522994762-522994916. pp 47; 110-111. Available at http://industrydocuments.library.ucsf.edu/tobacco/docs/yqnk008347.

les Boissons. Although some consider the operations of these organizations as a model of the way industry should contribute to alcohol science, questions have been raised about the way they operate and their influence on the scientific process (Babor & Robaina, 2013). For example, the Institut de Recherches Scientifiques sur les Boissons commissions its own studies in addition to funding investigator-initiated projects, thereby increasing the possibility that industry-favorable topics are promoted. It has also been suggested that a scientist's objectivity might be compromised by receipt of the honoraria and travel funds involved, as well as through the opportunities to fraternize with industry executives at international meetings. Each of these organizations also funds research on industry-favorable topics such as the health benefits of moderate drinking, which then are used as a part of the marketing strategies by the wine and beer industries or as reasons why regulation and taxation should not be imposed on the alcohol industry (Stenius & Babor, 2010).

In addition to indirect support of research through third-party organizations, there have been several instances in which individual alcohol producers or industry-supported social-aspects/public-relations organizations provide direct support to university-based scientists engaged in alcohol research. The most-notable examples include the Ernest Gallo Clinic and Research Center established by the Gallo Winery at the University of California to study basic neuroscience and the effects of alcohol on the brain; Anheuser-Busch's support of social norms research at seven U.S. universities; and a research center on youth binge drinking funded by Diageo Ireland, part of Diageo PLC, the world's largest producer and distributor of alcohol (Babor, 2006; Babor et al., 1996).

Little is known about the internal marketing research conducted by the alcohol industry and contract research organizations because the information is not shared with the public, the scientific community, or public health professionals. In the case of tobacco, previously secret internal industry documents have revealed that independent analysis of research on sensory perception was used to inform product design for targeted segments of the cigarette market, including young adults (e.g., Carpenter et al., 2005), and there is evidence that the alcohol industry does similar research (Babor, 2009). Contract research requires the services of social and behavioral scientists; therefore, it may pose ethical problems to the extent that such research could facilitate the marketing of products (e.g., alcopops) that are misused by vulnerable populations.

These kinds of funding initiatives not only have the potential for competing interests, but they may also affect the objectivity of independent scientists and the integrity of science. At best, the scientific activities supported by the alcohol industry provide financial support and small consulting fees for basic and behavioral scientists engaged in alcohol research. At worst, they confuse public discussion of health issues and policy options, raise questions about the objectivity of industry-supported alcohol scientists, and provide industry with a convenient way to demonstrate "corporate responsibility" in its attempts to avoid taxation and regulation (see Box 16.4 for further examples of industry activities).

ICAP is an industry-funded, social-aspects/public-relations organization located in Washington, D.C., USA. It was founded in 1995 by a consortium of alcohol companies, including MillerCoors, which at that time was part of tobacco giant Phillip Morris. According to an article on the early history of ICAP (Jernigan, 2012), MillerCoors's primary interests in the creation of ICAP were purely commercial, that is, to aid their planned international expansion by managing worldwide issues and thereby assisting their sales and marketing group in an increasingly competitive marketplace.

Despite ICAP's original mission to promote understanding of the role of alcohol in society and help reduce the abuse of alcohol worldwide, there is strong evidence that ICAP has evolved primarily into an industry public-relations organization dedicated to the advancement of industry-favorable alcohol policies (Anderson & Rutherford, 2002; Babor & Robaina, 2013; Bakke & Endal, 2010; Foxcroft, 2005; Jernigan, 2012; McCreanor et al., 2000; Room, 2005). For example, ICAP sponsored conferences and governmental consultations in a number of African countries in which industry-invited representatives helped governmental officials draft national policy plans for their countries. In one analysis of this initiative (Bakke & Endal, 2010), the national plans—ostensibly designed to fit the specific needs of four different African countries—were found to be virtually identical, with all documents originating from the MS Word document of a senior executive of SABMiller, one of the ICAP's funders.

There is also evidence that ICAP-supported research is of poor quality and is biased in favor of industry positions supporting alcohol education over more-effective alcohol policies (Babor & Xuan, 2004). ICAP also pays scientists to edit and write chapters for commissioned books that have been criticized for their bias toward industry-favorable positions on alcohol policy (Caetano, 2008; Stimson, et al., 2006).

Any pretense of ICAP's objectivity and independence was abandoned in 2014 with their announced merger with the Global Alcohol Producers Group, a major industry lobby organization. With this merger, ICAP was renamed the International Alliance for Responsible Drinking (IARD). Since its inception in 2005, the Global Alcohol Producers Group has spent more than USD$1.15 million on lobbying the World Health Organization (OpenSecrets.org, 2015), taking positions that seem to be diametrically opposed to those recommended by the international public health community.

Box 16.4: The research pedigree of the International Center for Alcohol Policies (ICAP), now called the International Alliance for Responsible Drinking (IARD).

The Pharmaceutical Industry

The pharmaceutical industry has become more interested in the discovery and evaluation of medications that can be used for the treatment of addiction, including opiate-substitution therapies and nicotine-replacement therapies. As such, pharmaceutical companies represent a different type of research funder from those, such as the tobacco industry, who sell dangerous consumables. The pharmaceutical industry commissions and funds legitimate research that has genuine benefit for the treatment of substance-related disorders. However, this industry also produces psychotropic substances like analgesics, hypnotics and sedatives. They are helpful treatment options when adequately prescribed but there is also increasing concern about prescribed and over-the-counter non-medical use of these substances, caused by aggressive marketing and inadequate prescriptions by primary care doctors. Examples include the dramatic increase of prescribed opioid analgesics in Canada and the United States, leading to severe negative health consequences and premature death (Fischer et al., 2011; Fischer et al., 2013), or the fact that in many western countries the number of substance use disorders for these classes of drugs is as high as the number of alcohol use disorders (e.g. for Germany: Kraus et al., 2013)

Pharmaceutical companies are as profit driven as the tobacco and alcohol industries and have demonstrated a willingness to engage in such activities as suppression, through delayed or nonpublication of null or negative findings, and dilution, through the selective funding of certain types of research (Kassirer, 2005). There is also evidence that some industry-supported research is biased (Brennan et al., 2006; Kassirer, 2005; Singer, 2008). In an interesting case study that combines pharmaceutical companies and tobacco, Etter et al. (2007) assessed whether the source of funding affected the results of trials of nicotine-replacement therapy for smoking cessation. They found that, compared with independent trials, industry-supported trials were more likely to produce statistically significant results and larger odds ratios.

In general, it has been found that researchers who report a financial competing interest are more likely to present positive findings (Friedman & Richter, 2004). Such behavior has not been documented within the addictions field, although medications used by many addicted patients for other complaints such as depression and anxiety have been the subject of controversial research practices.

The Gambling Industry

Problem gambling has been strongly linked to a range of personal and social problems (Gupta & Derevensky, 1998). The opportunities for addiction scientists to receive funding from gambling-industry sources have increased significantly over the last decade, raising a number of ethical and organizational risks similar to those associated with accepting funding from other dangerous consumption industries (Adams, 2007).

As in the case of relationships with the tobacco and alcohol industries, relationships with social-aspects/public-relations organizations have been used to mitigate potential negative associations with gambling problems and to give the impression either that the activity leads to public good or that they have at least attempted to rectify potential harm (Adams & Rossen, 2006). In countries such as Australia and New Zealand, a governmental or quasi-governmental agency has been created to manage voluntary funds in a way that appears independent of the source. Adams and Rossen point out that the major problem with such arrangements "is the perception that donor organizations should still retain a significant say in how the money is used" (p. 11). This culture leads to uncritical acceptance of gambling-industry perspectives and misrepresents the industry's willingness to trade profits for public health. This has meant in the past that industry officials were "consistently instrumental in ensuring that activities that might threaten the consumption of gambling were unlikely to receive significant funding (this particularly applied to research, health advocacy, and public health initiatives)" (Adams & Rossen, 2006, p. 12). This may explain why there have been few studies of the role of the gambling industry in the promotion of gambling behavior and pathological gambling.

It has been proposed that government-mandated contributions provide an alternative option to support research and provide a way to mollify criticism. In this arrangement, governments enact legislation that requires gambling providers to allocate a portion of their net income to projects, including research, with a community purpose. The major difficulty with this arrangement is the risk of increasing financial dependency, leading scientists to avoid criticizing gambling interests (Adams & Rossen, 2006). Likewise, the responsibility of governments to regulate gambling and prevent gambling problems may be compromised by the possibility that governments have themselves become "addicted" to the tax revenues derived from gambling.

Governmental Agencies

Albert Einstein (1934) once said that the "pursuit of scientific truth, detached from the practical interests of everyday life, ought to be treated as sacred by every government, and it is in the highest interests of all that honest servants of truth should be left in peace." Einstein's plea, directed at the fascist government of Mussolini, has been honored by most government funding agencies, but there are many cases in which the interests of government are prioritized over scientific pursuit of truth. In a situation similar to that of the pharmaceutical companies, national and international governmental bodies fund many valuable research studies. However, as seen in earlier examples, research has sometimes been used to achieve political or financial goals, such as supporting current budget allocations, protecting policy makers who have made bad decisions, or undermining more-effective strategies because they are unpopular and politically risky. Miller et al. (2006) identified two examples in which governmental funders acted to distort research findings in Australia and the United Kingdom, particularly regarding more-controversial activities such as needle and syringe programs. Similar observations have been made about the difficulty in obtaining funding for research into the effectiveness of needle and syringe programs and other forms of harm reduction in the United States (Pollak, 2007; Small & Drucker, 2006;).

Other Funding Agencies

Increasingly, charitable organizations such as the Robert Wood Johnson Foundation in the United States, the Joseph Rowntree Foundation in the United Kingdom, and the Millennium Trust in Australia have taken on agenda-setting roles that include research. Although most do not have profit imperatives akin to those seen in the tobacco, alcohol, and pharmaceutical industries, some nonetheless have their own agendas, and only a worthy few use transparent peer review. For example, the Wates Foundation in the United Kingdom has previously funded only research that supports abstinence-only approaches. Nepotism and personal competing interests can also come into play when trustees back projects supported by their friends or projects in which they are personally involved. This lack of peer review and external accountability means that such organizations may end up skewing the evidence base by supporting research into only certain types of intervention. Although some of this might be balanced by different foundations having different interests, the reality is that these funders have the potential to, at times, favor ideologically and politically simple and popular interventions. For example, although a small number of trusts, such as the Soros Foundation, have funded research into harm reduction and drug-policy reform, there are many more foundations that will fund only abstinence-based programs or programs aimed at abstinence, such as education programs. Although there are many reasons for this, most revolve around trustees not being knowledgeable about the available evidence and theory. In addition, many trustees and directors are politically aware individuals who are in the public spotlight. They may be reluctant to become associated with politically sensitive topics. All of this means that researchers should be aware of the possible consequences of applying for funding from such organizations, because even limited research might contribute to the overall publication bias in the field.

Other Interest Groups

Funding bodies are not the only groups to control research findings. For instance, Hall (2006b) identified the possibility of drug-user groups and socially conservative members of ethics committees prioritizing their own interests at the expense of the integrity of the research. Members of ethics committees hold very powerful positions when it comes to rejecting, delaying, or modifying research proposals. Although most declare financial competing interests, ideological positions are different, and indeed many would not identify strongly held beliefs as being competing interests. For example, individual members of ethics committees who are strongly attached to abstinence-only programs may block or delay research into controlled-drinking interventions in the belief that they cannot be morally justifiable.

There is also substantial room for competing interests inherent in the current peer-review framework (Hall, 2006a). With increasing competition over scarce resources, editors or reviewers may thwart the publication of research articles that counter their own theories or may thwart the publication of findings of their major competitors for funding. Although some journals have begun to publish ethical statements for editors, similar statements for reviewers of articles and funding applications may soon be required. Similarly, we should not forget that most researchers have their own pet theories, which can result in skewed research findings, particularly when those theories align with the interests of others such as professional societies, governments, or industry bodies. As noted in Chapter 14, these kinds of competing interests are difficult to detect, but they should nevertheless be considered by authors when evaluating their own work.

Other social groups that might seek to influence research include professional associations, fellowship groups, religious organizations, and even service providers. Professional associations (e.g., medical societies) have traditionally sought to maintain or increase their influence regarding any number of areas of knowledge and practice (Willis, 1989). Each discipline produces its own literature base. The size and complexity of this literature base helps to determine differential power structures within treatment settings. In the alcohol and drug sector, medicine and psychiatry (with the support of the pharmaceutical industry) dominate the literature base, resulting in the medical model (and pharmacotherapies) having the strongest evidence base. In a different type of influence, some fellowship groups may influence research findings through nonparticipation (e.g., Wilton & DeVerteuil, 2006).

Service providers are also not disinterested parties. Almost all (with a few notable exceptions) derive their income (and some of their *raison d'etre*) from treating addiction. This has substantial implications for the politics of treatment and the vested interests many people bring to the research enterprise. The political and economic weight of mantras such as "treatment works" bear little relation to the complex evidence base and far more to the pragmatic needs of governments and service providers. Although many service providers use the discourse of charitable objectives, they are invested both financially and existentially in the perceived success of the treatment they provide. This raises substantial ethical issues when conducting program-evaluation research in treatment settings, especially if the evaluation is funded by the service provider or its funding body. Ethical considerations such as the true reporting of findings (even when negative), full editorial control of research projects, and the assurance of adequate dissemination should be negotiated before research commences. Such issues require that researchers, reviewers, and journal editors within the field apply a strong critical gaze to research and encourage an ethos of independence, even when such independence may not be economically prudent.

Funding Issues in the Developing World

All of the examples discussed thus far describe the situation in the developed world. However, the issues facing researchers in the developing world are likely to be even more complicated and are much less likely to be documented. As do their counterparts in the more-developed parts of the world, researchers in developing countries face many challenges in their work. In both environments, success is tied to the availability of resources and the overall intellectual climate (Adair, 1995). Significant achievements as a scholar in a university or research institute require the ability to attract funding for research and to publish research findings, preferably in journals of high repute. Although the expectations from employers and the public might be the same, both activities are not always easy to execute by scholars in poor countries in which there are virtually no local resources for research.

When asked about the major problems encountered in their work, researchers and service providers affiliated with drug-demand–reduction organizations in Nigeria not surprisingly identified lack of funding as the leading challenge (Obot, 2004). Indeed, it is a rare country in Africa and other low-income parts of the world in which one can find consistent and near-sufficient outlay for scientific research on any topic, including addiction and other public health issues. This is especially the case for researchers in countries that constitute the "bottom billion" (Collier, 2007) or countries often described as least developed. In addition, competing for scarce resources with colleagues who are in resource-rich countries is often an impossible challenge. For the enterprising researcher, the response to this dearth of local funding opportunities is to conduct self-sponsored research (with all the limitations that this entails) or seek support from less-competitive external sources. This situation provides a good opportunity for organizations with ideological positions to propagate their interests and for others with economic interests to gain a foothold through financial support for research and training in these countries.

This is a potential source of danger for research in many developing countries and one that has not received sufficient attention. Although there has been active discussion about unfair distribution of benefits of international research, especially coming from concerns about the ethical dimensions of clinical trials in developing countries (e.g., Bhutta, 2002), the exploitation that is implicit in some sources of funding for research in developing countries deserves greater scrutiny. Exploitation is more likely to occur in situations in which there is little understanding of competing interests, low economic capacity, limited infrastructure, and lack of ethical oversight—all of which are conditions that characterize many low-income countries.

In the field of alcohol research, developing countries are experiencing a growing interest by representatives of the alcoholic-beverage industry masquerading as social-aspects organizations and seeking partnerships with researchers and policy makers. Usually the amount of money involved is a fraction of what would be spent for similar efforts in western countries, but it goes a long way for the scholar to whom such support is a lifeline, enabling research and the publication of a book with an international imprint. In Africa, for example, the International Center for Alcohol Policies (ICAP; Box 16.3 above) has provided support for data collection, write-up, and publication of work with the potential of influencing local alcohol policy (e.g., Haworth & Simpson, 2004). For the funding organization, association with (usually) a high-profile academic or policy expert in a developing country validates their professed selfless motives. This can be a particularly pernicious strategy, because the developing-country scholar who has been co-opted by the alcohol, tobacco, or pharmaceutical industry might be the same scholar on whom government depends for advice when needed.

It is not always lack of financial resources that drives the accommodation to untested imported theories and practices. Sometimes it is lack of knowledge, or even naïveté. A researcher in a developing country might find it difficult to suspect the motives of a funding agency that is acceptable to that country's government and one that is supported or led by internationally recognized academics or professionals. To guard against establishing or sustaining relationships with funding agencies that might lead to bad science or bad policy, it is important for researchers in developing countries to be more skeptical of easy money by questioning its source and the motives of its providers. That is easier to do today than it might have been 10 years ago, because most of the time all the information that is needed to decide whether to take the money can be found on the Internet.

Competing Interests: What are They, Why are They Important

As suggested by the examples reviewed above, funding sources can influence scientific integrity in a variety of ways, ranging from subtle bias in the way research findings are presented to outright distortion of the research agenda or the scientific literature. One way to approach the ethical implications of many of the issues raised in this chapter is through the concept of competing interests. Competing interests can be financial, personal, ideological, political, and academic. A competing interest does not in itself constitute wrongdoing; rather, it acknowledges that the researcher has an interest that may be put above the integrity of the research being conducted. It is only the failure to declare real or potential competing interests to an editor, one's co-authors, and the readers of an article that constitutes scientific misconduct. Potential competing interests are very important when it comes to the ability of the reader to assess the validity of any piece of scientific work. As noted above and in Chapter 14, competing interests may take many forms. For example, the issue of ideological bias has been raised as a possible competing interest in medical research. A series of articles and responses about prayer as medicine has raised substantial concerns about the interface between faith and science (Clarke, 2007; Jantos & Kiat, 2007). It has been suggested that "for the benefit of a secular readership, in articles concerning religion and medicine in the Journal, the Editor should require the authors' religious position to be stated under 'competing interests'" (Clarke, 2007, p. 422).

How to Avoid Competing Interests and Other Threats to Scientific Integrity and Academic Freedom

Just as there are many forms of competing interests, so too are there many different ways to avoid or reduce undue influence, although many commentators believe that none of the possible options is entirely satisfactory or risk free (Adams & Rossen, 2006). By far the most commonly proposed way to avoid or ameliorate competing interests is through communication with one's peers, particularly when done alongside ethics-awareness exercises (e.g., White & Popovits, 2001). Adams (2007) recommends that individuals, organizations, and others involved with interested parties engage in processes that raise ethical consciousness in conjunction with transparent regulatory frameworks that ensure accountability and independence from organizations and governmental and professional associations. This kind of communication and awareness raising has begun to occur at a number of levels.

Recently, the institutions responsible for the production and dissemination of research (i.e., journals, professional societies, and academic institutions) have taken some important initiatives. Academic journals have increasingly begun to enact competing interest strategies including (a) requiring author statements that declare funding source, which are then published with the article; (b) a positive statement that all authors had complete control over the research process; (c) reviewer and editor statements similar to those of authors; and (d) prior registration with an approved clinical-trials register as a prerequisite for publication. Journal editors have also begun to look at strategies for assessing publication bias within their journals and at a more general level. Some journals have used their editorial pages to name and shame parties that behave inappropriately (e.g., Edwards et al., 2005) and to educate the scientific community about the need for competing interest policies (Babor & Miller, 2014).

Professional associations have begun to draw up guidelines regarding the behavior of acceptable funding bodies, competing interests, and related issues. For example, the Federation of American Societies for Experimental Biology (2007) has issued a call to the scientific community to adopt more-consistent policies and practices for disclosing and managing financial relationships between academia and industry in biomedical research. The Federation of American Societies for Experimental Biology Toolkit (Federation of American Societies for Experimental Biology) consists of a set of model guidelines that speaks specifically to institutions that develop and enforce policies for their investigators, editors who develop disclosure policies for authors, and scientific and professional societies that have a role in promoting professional ethics. Similarly, the RESPECT Code of Practice (Dench et al., 2004) is a voluntary code of practice regarding the conduct of socioeconomic research. The proposed guidelines are a synthesis of several professional and ethical codes of practice designed to protect researchers from unprofessional or unethical demands. In one of the most thorough policy statements on the subject of competing interests, the International Network on Brief Interventions for Alcohol and Other Drugs issued a position statement that is summarized in Box 16.5.

---

(1) INEBRIA believes that the commercial activities of the alcohol industry pose a conflict of interest of such magnitude that any form of engagement with the alcohol industry may influence its independence, objectivity, integrity, and credibility internationally.

(2) All individuals wishing to present at an INEBRIA meeting will be required to complete a conflict-of-interest declaration for the work being presented.

(3) Members of the coordinating committee will sign a conflict-of-interest declaration and may not have worked with or received funding from the alcohol industry, directly or indirectly, in the five years before their election date or during their term of office.

Box 16.5: Summary of the International Network on Brief Interventions for Alcohol and Other Drugs (INEBRIA) Position Statement on the Alcohol Industry.

Institutions such as universities and research centers have developed policies regarding acceptable funding bodies, and some scrutinize research contracts for possible competing interests. A growing number of universities (e.g., Kings College London) have refused to accept funding from the tobacco industry, and some research centers have developed their own internal policies (Box 16.6). Deakin University (Australia) now prohibits the receipt of research funding from the tobacco and gambling industries, as well as social, health or epidemiological research funded by the alcohol industry. There is also scope for institutional ethics review boards to assess the appropriateness of funder–researcher relationships. Questions regarding such relationships are now incorporated in the Australian National Ethics Application Form (www.nhmrc.gov.au/health-ethics/human-research-ethics-committees-hrecs/hrec-forms/neaf-national-ethics-application-for). Such responses are designed to support individual researchers in the decision-making process and provide more-reliable and consistent approaches to this complex issue (Babor & McGovern, 2007; Miller et al., 2006).

However, resolving these issues remains in large part the responsibility of individual authors, many of whom have a limited ability to understand or act upon the complex ethical, political, clinical, and scientific issues surrounding the initiatives coming from a particular funding source. Fortunately, most addiction scientists have chosen to eliminate themselves from participation in activities with obvious competing interests, such as consulting arrangements with the tobacco and alcohol industries and restrictions from funding sources that prevent them from retaining ownership of data and the investigator's right to publish it (Babor & McGovern, 2007). Nevertheless, what is needed is a more-systemic set of procedures that allows individuals to conduct a risk analysis of different funding opportunities.

Decision-Making Approaches

Several approaches have been suggested to guide decision making by independent scientists when they consider collaboration with the alcoholic-beverage industry and other dangerous consumption industries (Babor, 2009; Babor & McGovern, 2007; Stenius & Babor, 2010). Decisions regarding collaboration with bodies that may seek to influence research can range from a "hands-off" position to full collaboration. Adopting a hands-off position, in which members of the scientific community and their organizational sponsors refuse to engage in communication or collaboration with industry representatives, is based on the assumption that commercial interests are incompatible with the values and aims of public health in general and with health-related scientific research in particular. Some have argued that the main effect of industry's recent cooperation with scientists and public health professionals has been to improve their corporate image with the public and with governmental policy

Dealing with Possible Competing Interests Related to the Financing of Our Research Projects

The proportion of industry research funding within the financial budget of the institute has been very low since the foundation of the IFT Institut für Therapieforschung in 1973. But caution is needed, because this part of research support is provided by organizations and companies that produce or distribute psychoactive substances (e.g., alcohol or pharmaceutical industry) or are active in the gambling business (including gambling companies licensed or owned by the German States) and because of the internationally known incidents of scientific misconduct.

The IFT does not reject funding of research by commercial institutions in principle but is aware of the particular responsibility in this area. In times of short or even declining public research funding and direct demands of the public to cooperate with industry and to expand commercial third-party funds for research, it is hardly possible to abandon such sources of funding in principle. The institute has in this context the following rules:

- Research requests to conduct a study on a given research question will be accepted only if (a) the research question is formulated globally and is undirected (e.g., the extent of drug abuse in the population) and not biased (e.g., the study is not expected to demonstrate that a certain medicine bears no risk for the population), (b) the research question is scientifically relevant, and (c) the free and unrestricted further design of the study is guaranteed.
- A further precondition for accepting funding by industry sources is the guaranteed independent formulation of the research objectives, hypotheses, and study methodology, and the unrestricted statistical analysis, interpretation, and publication of results. The funds have to be granted to the IFT as unrestricted educational grants or donations.
- We do not accept funding of research projects by the tobacco industry (reasons: evidence of long-lasting, one-sided, and unacceptable manipulation of scientists and scientific results).
- A single funding source must not contribute to more than 10% of the annual budget, and all industry funds should not exceed 20%. It is notable that these limits have never been reached: The average contribution is about 2%, and it has never exceeded 5% in the past.
- All results will be published.
- Lectures given in the context of industry organizations are accessible via the website of the IFT.

(Box continued on next page)

> ### Funding in the "Gray Area" between Public and Commercial Organizations
>
> Examples are charitable organizations, (nonprofit) health insurance companies, and industry associations. In most cases, these organizations are accountable to the public or the commercial sector. The IFT applies in each case the same rules as for commercial organizations.

Box 16.6: One research institution's guidelines on acceptable research funding. Source: Institut für Therapieforschung, München, Germany (www.ift.de).

makers, rather than to promote science (Babor & Robaina, 2013; Gmel et al., 2003; McCreanor et al., 2000; Munro, 2004).

The other end of the spectrum is to engage in dialogue with industry representatives, accept industry funding for research, and participate as "partners" in industry-funded scientific activities such as the publication of books (e.g., Stimson et al., 2006).

A third approach is based on the growing number of case studies, ethical reviews, and documentary information now available with respect to industries that have an important stake in products that affect public health (Brennan et al., 2006; Hirshhorn et al., 2001; Rampton & Stauber, 2002; Rundall, 1998). This approach avoids categorical recommendations to either allow or discourage relationships between science and industry in favor of a more-nuanced set of guidelines that outlines conditions of cooperation between science and industry (Adams, 2007).

PERIL

Adams' (2007) PERIL framework (purpose, extent, relevant harm, identifiers, link) provides a structured means of evaluating individual situations from an ethical perspective. Depending on circumstances, each of the five PERIL subcontinuums is influenced in varying ways by the different domains of risk.

Purpose refers to the degree to which purposes are divergent between funder and recipient. For example, if the primary purpose of the recipient is the advancement of public good, receiving funds from dangerous consumption industries such as tobacco, alcohol, and gambling companies will probably conflict with this purpose. Similarly, the risk is mitigated partially if the funder has a clear public-good role. For example, the provincial government of Ontario runs a state monopoly on liquor distribution, the profits from which they invest in a broad range of research (Adams, 2007).

Extent is the degree to which the recipient relies on this source of funding. As the proportion of income increases, it becomes more difficult to separate one's research from expectations associated with the source. For example, a young investigator may find an award from an industry-sponsored organization is the sole source of salary support, which could create pressure to obtain industry-favorable results to ensure the continuation of funding.

Relevant harm is the degree of harm associated with this form of consumption. The level of harm generated by different forms of consumption varies. Lower potency products, such as lottery tickets or low-alcohol beer, are on the whole less likely to lead to problems than more-potent products, such as electronic gambling machines or alcoholic energy drinks.

Funders are unlikely to contribute anonymously, because for them the point of the exercise is often to be *identified,* to form a visible association with public-good activities for the purposes of positive branding. This in turn can be used for political or commercial purposes. The extent of visible association can be reduced by moving away from high-profile advertisements (such as media releases of findings) to more-discrete acknowledgements on plaques or at the end of publications. Through reputational risk, this strategy indirectly discourages engaging in industry-supported research.

The more direct the *link* is between funder and researcher, the stronger the influence and the more visible the association are. For example, direct funding by a tobacco company involves more exposure than receiving the funding via an independent intermediary agency, such as a foundation or governmental funding body. As long as there are no major competing interests for the intermediary agency, the separation reduces the likelihood that recipients will feel obligations, even coercion, for their activities to comply with the interests of the donor. The overall extent of moral jeopardy ranges from very high levels, as indicated by high ratings on all five subcontinuums, to very low levels, as indicated by consistently low ratings. Decisions regarding future industry relationships are made accordingly. Boxes 16.7 and 16.8 provide two case studies to illustrate how a PERIL analysis can be applied to specific funding opportunities.

Is Industry Funding of Research the only Peril that Matters?

A new genre of policy analysis suggests that vested interests use research to achieve their ultimate goals of profit maximization (Babor & Robaina, 2013).

In their illuminating series of articles, Hawkins, McCambridge and colleagues highlight the way in which the alcohol industry uses both industry-funded research and their relationships with researchers to demonstrate their credibility and good intentions (Hawkins & Holden, 2014; Hawkins et al., 2012; McCambrige et al., 2013; Hawkins & Holden, 2014). These public-relations activities are commonly hidden in the rhetoric of corporate social responsibility, which is particularly important to recognize when considering the long-term relationships between the alcohol industry and most politicians and the way in which these relationships are formed. Although politicians might read a newspaper article about new alcohol trends, they are easily calmed when their likeable industry representative, who knows their kids' names and the schools they go to, assures them that there is no need to worry because the industry — often through one of its front bodies such as Drinkwise (Australia), Drinkaware (United Kingdom), EURAB (Europe), or the ABMRF/The Foundation for Alcohol Research (United States) — is working with a group of respected researchers to deal with the issue. Hawkins and Holden (2014) demonstrated convincingly just how effective this strategy is, especially when it is combined with the very long-term engagement approach that the alcohol industry adopts with politicians from all sides of the political fence. It is even more effective when they are able to suggest that the industry has actually funded the research into this important issue and that they have found it not to be so important or that the interventions they recommend are effective and much more palatable politically than "nanny state" interventions, such as raising taxes or restricting trading hours (Miller et al., 2011).

In the end, whether or not other elements of the PERIL analysis such as reputational risk or extent of funding are of concern, the overriding consideration in the strategic funding of research by the alcohol industry is their ability to use those relationships to gain a place at the discussion table regarding policy at the state, federal, and global levels.

A university-based school of medicine distributes an email announcing to all faculty and staff the availability of a new research funding opportunity. The announcement reads: "Please see the link below for an available funding opportunity from the Philip Morris External Research Foundation The website invited scientists to submit funding proposals to Philip Morris's independent, peer-reviewed, external research program, which is willing to support research on the disease mechanisms and health endpoints of tobacco smoking and smoke exposure. The program's scientific advisory board members are listed on one of the pages of the request for applications, an impressive-looking group of academics, including department chairs, distinguished professors, and even the President of the Hungarian Academy of Sciences. This announcement raises a number of questions about the moral hazards of industry sponsorship of scientific research.

Assume you are a tobacco researcher at a large academic medical center whose dissertation was recently completed on a topic related to the announcement. Should you apply for the funds? A PERIL analysis along the lines recommended in the Adams article would require some independent research and a review of the literature on tobacco-industry tactics.

PERIL Analysis

Is the *purpose* of your academic institution (e.g., "excellent medical care through research and education") consistent with the stated purpose of Phillip Morris (i.e., to sell cigarettes to adults, without taking any responsibility for the millions of adolescents who become addicted before they can legally purchase tobacco products)? If your institution is in any way devoted to health, the answer is that the purposes are incompatible. In addition, some have pointed to the anti-scientific record of Phillip Morris. The reason Phillip Morris's research foundation is now called "external" is that the company was ordered to disband a prior organization that was found by a U.S. court to be biased in the way it awarded grants to scientists.

What about the *extent* of the funding? Is it sufficient to compromise the independence of an academic medical center with a large portfolio of research grants and contracts? It probably is not, but for individual investigators it could create a dependence on tobacco money when other sources of funding become more scarce.

Is there *relevant harm* associated with Phillip Morris's continued marketing of tobacco products? The evidence is incontrovertible.

Will the recipient of the funds be *identified* with the funder so that Phillip Morris might benefit from its support of university-based scientists? And could funded scientists eventually be exposed to reputational risk if their names were associated with Phillip Morris? The answer is a possible yes to both questions.

Finally, is the nature of the *link* between recipient and donor direct or indirect? In this case it is indirect; therefore, it may not involve a major competing interest, and there are no limitations on publication imposed by the funder.

In summary, the analysis indicates that there are incompatible institutional interests, a potential for developing dependence on an industry funding source, relevant harms to the public if tobacco sales continue as more research is conducted, a potential for future reputational risk, and a possible political benefit for Phillip Morris.

Box 16.7: PERIL analysis of a funding opportunity from Phillip Morris.

A residential rehabilitation charity approaches you to collaborate in an application to fund doctoral research into the long-term effectiveness of its project. The charity reports that it has been involved in research previously and has found it beneficial. The methodology is discussed and agreed. The application is designed to go to a governmental funding body that provides matching funds for collaborations between community organizations and universities. The charity expresses concern about the confidentiality of its service users and requests that "We would, however, want the research findings to be kept confidential except in so far as they are needed to fulfill the requirements for the degree." Subsequent investigation shows that, although the charity refers to a strong research pedigree, findings have been published only in non-peer-reviewed trade magazines or internal reports.

PERIL Analysis

Is the *purpose* of your academic institution (e.g., excellent medical care through research and education) consistent with the stated purpose of the charity? At first glance it would appear that the charity has the laudable goal of assessing its effectiveness through independent research. However, its desire to control dissemination (presumably in case of unfavorable findings) and its previous track record of publishing only in non-peer-reviewed journals would suggest that its goal might not be excellence.

What about the *extent* of the funding? In this example, this is unlikely to be a major factor because the amount involved would be comparatively small.

Is there *relevant harm?* There is a chance of some harm in this case if the findings are unfavorable and the charity chooses not to disseminate the report. In this situation, the charity is clearly providing ineffective treatment and using resources that might be better used elsewhere. In addition, it may be skewing the knowledge base through omission of negative findings.

There is also a significant issue that the researchers and university will be *identified* with the evaluation. It is within the interest of the charity to point to the fact that the research was independently conducted.

Finally, is the nature of the *link* between recipient and donor direct or indirect? In this case it is indirect; therefore, it may not involve a major competing interest, and there are no limitations on publication imposed by the funder. In this case, it would be possible for the researchers or the university to insist that the charity remove its right to control release of the data. If that were done, the PERIL analysis would suggest that the funding is worth pursuing.

Box 16.8: PERIL analysis of a funding opportunity limited by conditions imposed by a collaborating organization.

Conclusion

Every individual, discipline, and funding organization brings its own agenda to the research process. The practical and ethical conundrums associated with research funding are becoming increasingly complex in a context in which research plays a greater role in the regulation and marketing of potentially addictive products. The examples reviewed in this chapter suggest that addiction scientists should be vigilant and critically reflective about the funding they accept from any source, particularly in relation to the ultimate purpose of such funding. This is even more so the case when there are restrictions on the design, interpretation, and publication of the resulting data. Thus, researchers should always be very wary about accepting research funding directly from dangerous consumption industries, their trade associations, and public-relations organizations. Consulting arrangements wherein scientists are paid by parties with a clear competing interest to critique the work of other scientists can constitute a serious financial competing interest that is unlikely to benefit either science or the investigator. Acceptance of fees for writing book chapters, preparing background reports, attending industry-organized conferences, and writing letters to the editor should be prefaced by careful consideration of the following questions:

(1) To what extent is the scientific activity designed to promote the commercial interests of a particular industry?
(2) Will the funding source be acknowledged?
(3) How could this research or my institution's relationship with this company be used to undermine the implementation of effective policy?

Addiction scientists also need to be careful that their objectivity and independence are not compromised by fraternizing with industry executives as well as paid travel to meeting sites and consulting fees (Wagner & Steinzor, 2007). Investigators in particular need to be attentive to the possibility that industry funding in many health areas is being contested on both ethical and scientific grounds (Foxcroft, 2005; King, 2006; Brennan et al., 2006). Finally, researchers should examine all funding sources using a framework such as the PERIL analysis, which allows the individual scientist and his or her institution to review relevant information about the motives of the funding source and the uses of the research that will be conducted.

Please visit the website of the International Society of Addiction Journal Editors (ISAJE) at www.isaje.net to access supplementary materials related to this chapter. Materials include additional reading, exercises, examples, PowerPoint presentations, videos, and e-learning lessons.

References

Adair, J. G. (1995). The research environment in developing countries: Contributing to the national development of discipline. *International Journal of Psychology, 30,* 643–662.

Adams, P. J. (2007). Assessing whether to receive funding support from tobacco, alcohol, gambling and other dangerous consumption industries. *Addiction, 102,* 1027–1033.

Adams, P. J., & Rossen, F. (2006). Reducing the moral jeopardy associated with receiving funds from the proceeds of gambling. *Journal of Gambling Issues, 17,* 1–21.

Ashcroft, R. E. (2006). Getting what you pay for? The ethics of selective publication. *International Journal of Drug Policy, 17,* 238–239.

Babor, T. F. (2006). Diageo, University College Dublin and the integrity of alcohol science: It's time to draw the line between public health and public relations. *Addiction, 101,* 1375–1377.

Babor, T. F. (2009). Alcohol research and the alcoholic beverage industry issues, concerns and conflicts of interest. *Addiction, 104,* 34–47.

Babor, T. F., Edwards, G., & Stockwell, T. (1996). Science and the drinks industry: Cause for concern. *Addiction, 91,* 5–9.

Babor, T. F., & McGovern, T. F. (2007). Minimizing moral jeopardy: Perils of the slippery slope. *Addiction, 102,* 1037–1038.

Babor, T., & Miller, P. (2014). McCarthyism, conflict of interest and *Addiction's* new transparency declaration procedures. *Addiction, 109,* 341–344.

Babor, T. F., & Robaina, K. (2013). Public health, academic medicine, and the alcohol industry's corporate social responsibility activities. *American Journal of Public Health, 103,* 206–214.

Babor, T. F., & Xuan, Z. (2004). Alcohol policy research and the grey literature. *Nordic Studies on Alcohol and Drugs,* 125–137.

Bhutta, A. (2002). Ethics in international health research: A perspective from the developing world. *Bulletin of the World Health Organization, 80,* 114–120.

Brennan, T. A., Rothman, D. J., Blank, L., Blumenthal, D., Chimonas, S. C., Cohen, J. J., . . ., Smelser, N. (2006). Health industry practices that create conflicts of interest: A policy proposal for academic medical centers. *JAMA, 295*, 429–433.

Caetano, R. (2008). About smoke and mirrors: The alcohol industry and the promotion of science. *Addiction, 103*, 175–178.

Carpenter, C. M., Wayne, G. F., & Connolly, G. N. (2005). Designing cigarettes for women: New findings from the tobacco industry documents. *Addiction, 100*, 837–851.

Clarke, J. (2007). Religion as a competing interest. *Medical Journal of Australia, 187*, 421–422.

Collier, P. (2007). *The bottom billion: Why the poorest countries are failing and what can be done about it.* New York, NY: Oxford University Press.

Dench, S., Iphofen, R., & Huws, U. (2004). *RESPECT code of practice for socio-economic research.* Brighton, UK: Institute for Employment Studies. Retrieved from http://www.respectproject.org/code/respect_code.pdf.

Drope, J., Bialous, S. A., & Glantz, S. A. (2004). Tobacco industry efforts to present ventilation as an alternative to smoke-free environments in North America. *Tobacco Control, 13*(Supplement 1), i41–i47.

Edwards, G., West, R., Babor, T. F., Hall, W., & Marsden, J. (2005). The integrity of the science base: A test case. *Addiction, 100*, 581–584.

Einstein, A. (1934). The world as I see it. *Forum and Century: Living Philosophies, 84*, 193–194.

Etter, J.-F., Burri, M., & Stapleton, J. (2007). The impact of pharmaceutical company funding on results of randomized trials of nicotine replacement therapy for smoking cessation: A meta-analysis. *Addiction, 102*, 815–822.

Federation of American Societies for Experimental Biology. Federation of American Societies for Experimental Biology Toolkit 2015 [20 Sep 2015]. Available from: http://www.webcitation.org/6bgp0m2Ap.

Fischer, B., Jones, W., Krahn, M., & Rehm, J. (2011). Differences and over-time changes in levels of prescription opioid analgesic dispensing from retail pharmacies in Canada, 2005–2010. *Pharmacoepidemiology and Drug Safety, 20*(12), 1269–1277.

Fischer, B., Keates, A., Bühringer, G., Reimer, J., & Rehm, J. (2013). Non-medical use of prescription opioids and prescription opioid-related harms: why so markedly higher in North America compared to the rest of the world? *Addiction, 109*(2), 177–181.

Foxcroft, D. (2005). International Center for Alcohol Policies (ICAP)'s latest report on alcohol education: A flawed peer review process. *Addiction, 100*, 1066–1068.

Friedman, L. S., & Richter, E. D. (2004). Relationship between conflicts of interest and research results. *Journal of General Internal Medicine, 19*, 51–56.

Gmel, G., Heeb, J. L., & Rehm, J. (2003). Research and the alcohol industry [Letter to the editor]. *Addiction, 98*, 1773–1774.

Gruning, T., Gilmore, A. B., & McKee, M. (2006). Tobacco industry influence on science and scientists in Germany. *American Journal of Public Health, 96*, 20–32.

Gupta, R., & Derevensky, J. L. (1998). Adolescent gambling behavior: A prevalence study and examination of the correlates associated with problem gambling. *Journal of Gambling Studies, 14*, 319–345.

Hall, W. (2006a). Ensuring that addiction science is deserving of public trust. *Addiction, 101*, 1223–1224.

Hall, W. (2006b). Minimising research censorship by government funders. *International Journal of Drug Policy, 17*, 240–241.

Hawkins, B., & Holden, C. (2014). 'Water dripping on stone'? Industry lobbying and UK alcohol policy. *Policy & Politics, 42*, 55–70.

Hawkins, B., Holden, C., & McCambridge, J. (2012). Alcohol industry influence on UK alcohol policy: A new research agenda for public health. *Critical Public Health, 22*, 297–305.

Haworth, A., & Simpson, R. (Eds.). (2004). *Moonshine markets: Issues in unrecorded alcohol beverage production and consumption.* Florence, KY: Brunner-Routledge.

Hirshhorn, N., Aguinaga-Bialous, S., & Shatenstein, S. (2001). Philip Morris' new scientific initiative: An analysis. *Tobacco Control, 10*, 247–252.

Hough, M., & Turnbull, P. (2006). Over-regulation or legitimate control? *International Journal of Drug Policy, 17*, 242–243.

Iida, K., & Proctor, R. N. (2004). Learning from Philip Morris: Japan Tobacco's strategies regarding evidence of tobacco health harms as revealed in internal documents from the American tobacco industry. *The Lancet, 363*, 1820–1824.

Jantos, M., & Kiat, H. (2007). Prayer as medicine: How much have we learned? *Medical Journal of Australia, 186*(Supplement), S51–S53.

Kassirer, J. P. (2005). *On the take: How medicine's complicity with big business can endanger your health.* New York, NY: Oxford University Press.

Khoshnood, K. (2006). The regulation of research by funding bodies: A wake-up call. *International Journal of Drug Policy, 17*, 246–247.

King, J. (2006). Accepting tobacco industry money for research: Has anything changed now that harm reduction is on the agenda? *Addiction, 101*, 1067–1069.

Kraus, L., Pabst, A., Piontek, D., & Gomes de Matos, E. (2013). Substanzkonsum und substanzbezogene Störungen: Trends in Deutschland 1980–2012 [Consumption of substances and substance-related disorders: Trends in Germany 1980–2012]. *SUCHT, 59*(6), 333–345.

Lenton, S., & Midford, R. (2006). Research regulation by omission and by publication. *International Journal of Drug Policy, 17*, 244–245.

McCambridge, J., Hawkins, B., & Holden, C. (2013). Industry use of evidence to influence alcohol policy: A case study of submissions to the 2008 Scottish government consultation. *PLoS Medicine, 10*(4), e1001431.

McCreanor, T., Caswell, S., & Hill, L. (2000). ICAP and the perils of partnership. *Addiction, 95,* 179–185.

Miller, P. (2013). Energy drinks and alcohol: Research supported by industry may be downplaying harms. *BMJ, 347,* f5345. DOI: https://doi.org/10.1136/bmj.f5345

Miller, P. G., Groot, F. d., McKenzie, S., & Droste, N. (2011). Alcohol industry use of social aspect public relations organisations against preventative health measures. *Addiction, 106,* 1560–1567.

Miller, P. G., Moore, D., & Strang, J. (2006). The regulation of research by funding bodies: An emerging ethical issue for the alcohol and other drug sector. *International Journal of Drug Policy, 17,* 12–16.

Muggli, M. E., LeGresley, E. M., & Hurt, R. D. (2004). Big tobacco is watching: British American Tobacco's surveillance and information concealment at the Guildford depository. *The Lancet, 363,* 1812–1819.

Munro, G. (2004). An addiction agency's collaboration with the drinks industry: MooJoose as a case study. *Addiction, 99,* 1370–1374.

Obot, I. S. (2004). Responding to drug problems in Nigeria: The role of civil society organizations. *Substance Use and Misuse, 39,* 1289–1301.

Ong, E. K., & Glantz, S. A. (2000). Tobacco industry efforts subverting International Agency for Research on Cancer's second-hand smoke study. *The Lancet, 355,* 1253–1259.

OpenSecrets.org. (2015). Global Alcohol Producers Group: OpenSecrets.org [20 Sep 2015]. Available from http://www.webcitation.org/6bgoUXaPE.

Parascandola, M. (2005). Science, industry, and tobacco harm reduction: A case study of tobacco industry scientists' involvement in the National Cancer Institute's Smoking and Health Program, 1964–1980. *Public Health Reports, 120,* 338–349.

Pollak, H. (2007, March). *How should harm reduction interventions be evaluated?* Paper presented at the International Society for the Study of Drug Policy, Oslo, Norway.

Rampton, S., & Stauber, J. (2002). Research funding, conflicts of interest, and the meta-methodology of public relations. *Public Health Reports, 117,* 331–339.

Room, R. (2005). Drinking patterns as an ideology. *Addiction, 100,* 1803–1804.

Rundall, P. (1998). Should industry sponsor research? How much research in infant feeding comes from unethical marketing? *BMJ, 317,* 338–339.

Savell, E., Gilmore, A. B., & Fooks, G. (2014). How does the tobacco industry attempt to influence marketing regulations? A systematic review. *PLoS ONE, 9*(2), e87389.

Singer, M. (2008). *Drugging the poor: Legal and illegal drugs and social inequality.* Long Grove, IL: Waveland Press.

Small, D., & Drucker, E. (2006). Policy makers ignoring science and scientists ignoring policy: The medical ethical challenges of heroin treatment. *Harm Reduction Journal, 3,* 16.

Stenius, K., & Babor, T. F. (2010). The alcohol industry and public interest science. *Addiction, 105,* 191–198.

Stimson, G., Grant, M., Choquet, M., & Garrison, P. (2006). *Drinking in context: Patterns, interventions, and partnerships.* Washington, DC: Routledge.

Wagner, W., & Steinzor, R. (Eds.). (2007). *Rescuing science from politics: Regulation and the distortion of scientific research.* New York, NY: Cambridge University Press.

Wander, N., & Malone, R. E. (2006). Making big tobacco give in: You lose, they win. *American Journal of Public Health, 96,* 2048–2054.

White, W. L., & Popovits, R. M. (2001). *Critical incidents: Ethical issues in the prevention and treatment of addiction* (2nd ed.). Bloomington, IL: Chestnut Health Systems.

Willis, E. (1989). *Medical dominance: The division of labour in Australian health care* (Rev. ed.). Sydney, Australia: Allen & Unwin.

Wilton, R., & DeVerteuil, G. (2006). Spaces of sobriety/sites of power: Examining social model alcohol recovery programs as therapeutic landscapes. *Social Science & Medicine, 63,* 649–661.

SECTION 5

Conclusion

CHAPTER 17

Addiction Publishing and the Meaning of [Scientific] Life

Thomas F. Babor, Kerstin Stenius and Jean O'Reilly

Introduction

The global scientific discipline of addiction studies that has developed during the past half century would be impossible without the infrastructure of the publishing enterprise. At the core of this infrastructure lie the peer-reviewed scientific article and the expanding network of journals that publish such articles. Throughout this book, we have focused on publishing scientific articles in peer-reviewed journals because this is a key part of the meaning of scientific life. Publishing allows the scientist to communicate findings, ideas, and opinions within a forum representing the scientific community. In this final chapter, we will explore this theme in relation to addiction science, which for many highly trained researchers throughout the world has become a career commitment that is not only personally rewarding but also beneficial to society.

In brief, our argument is as follows: Science is meaningless unless it is communicated. Publication communicates scientific findings, and it is also the hallmark of a productive scientific career. Scientific integrity is another core feature of a successful career, and it must be nurtured by individuals, groups, and institutions, including scientific journals. To the extent that science constitutes a universal language, there is a special need to foster addiction careers in low- and middle-income countries.

How to cite this book chapter:
Babor, T F, Stenius, K and O'Reilly, J. 2017. Addiction Publishing and the Meaning of [Scientific] Life. In: Babor, T F, Stenius, K, Pates, R, Miovský, M, O'Reilly, J and Candon, P. (eds.) *Publishing Addiction Science: A Guide for the Perplexed*, Pp. 355–364. London: Ubiquity Press. DOI: https://doi.org/10.5334/bbd.q. License: CC-BY 4.0.

The Meaning of Science

A seminal article by Ilkka Niiniluoto (2002), professor of philosophy at Helsinki University, traces the history of science through the various milestones in the search for knowledge from the time of the ancient Greeks to the present time.

The first milestone, which is the legacy of Aristotle, lies above all in the organized description of how we come to know the world and its generally accepted laws ("why" knowledge). A second phase in the history of science came with Galileo's search for regularities in how the world changes ("how" knowledge).

Compared with these steps, the third one is more complicated. A much later advance in the development of science began at the end of the 19th century when Charles Pierce introduced the notion of fallibility, which claimed that human beings constantly make mistakes in their search for knowledge and that all claims about the real world should be questioned. "This is true also of research, even if the scientific method of the research community, at least in the long run, is the most reliable way to produce and motivate conceptions of the world" (Niiniluoto, 2002, p. 32, authors' translation).

Niiniluoto talks about science as a self-correcting process. The modern scientific community has its own quality-assessment system (e.g., the peer review process), scientific claims are public, and all parties in the scientific community have the right to discuss, criticize, or refute those claims. According to Niiniluoto, contemporary science is characterized by objectivity (gaining as true a picture of the object studied as possible), a critical attitude (research should be public and open for critical discussion in the research community), autonomy (the scientific community operates independently of religious, political, economic, personal, and social influences), and progressivity (science creatively seeks new solutions and builds on old ones).

Arguing further that science is a social institution, Niiniluoto refers to Merton's (1973) four imperatives for the ethos of science: (a) universalism (the truth of claims shall be judged on impersonal grounds irrespective of the race, nationality, class, or personal characteristics of the person who presents them), (b) communism (scientific findings result from social cooperation and should be common property), (c) disinterestedness (scientists present and analyze scientific knowledge without considering the career or prestige of the researcher), and (d) organized skepticism (scientists assess scientific results on the bases of empirical and theoretical criteria).

According to Niiniluoto, Merton's principles have been criticized as deficient, insufficient, and inconsistent with the everyday life of research in the contemporary world. "Big science," increasing competition for personal repute, and the inequitable concentration of resources have eroded the ethos of science, as has the use of science in war and commercial production, which has produced a form of applied science that is businesslike and breaches the "communism" principle of common ownership of intellectual property. Niiniluoto argues, though, that this activity is not really *scientia* and should be viewed as something other than academic research.

In addiction research, the increasing competition for research positions and financial resources can foster the temptation to neglect ethical rules as well as the ethos of science. Career considerations can orient one's research to what is popular or fundable rather than toward what is interesting or important. The growth in private research funding may lead to secrecy instead of the open exchange of new ideas and research results, and may lead to new priorities that favor business interests rather than the public good.

If we accept Niiniluoto's assertions, we can understand why good publication practices, of the type described in this book, are crucial for science and the search for meaning in scientific life. Good publication practices represent the principles that should guide the quest for truth and, at the same time, demonstrate how to become a respected member of the scientific community. If science is to be used properly in the search for meaning as well as the basis for the betterment of humankind, there needs to be free access to the enormous reservoir of scientific knowledge in the world. That knowledge not only needs to be readily available, but it must also be recorded in a way that is understandable, useable, and certifiably scrutinized for error and bias. This is the role of journals and the responsibility of their authors. As noted by LaFollette (1992), a journal serves as the arbitrator of the authenticity and legitimacy of knowledge. It provides a historical record of a particular area of knowledge and confers implicit certification on authors for the originality of their work.

Careers in Addiction Science

Publishing with scientific integrity is for many the *sine qua non* of a productive scientific career in addiction science. The remarkable growth of addiction science worldwide (Babor, 1993, 2002; see also Chapter 2) coincides with the development of a variety of career options for those interested in basic, clinical, or social research. Research societies, subspecialties within professional organizations, and research centers have proliferated in many parts of the world, as has the availability of addiction specialty journals (see Chapter 3). There is growing evidence that a career in addiction science has become a viable and rewarding way to spend one's professional life (Edwards, 1991, 2002). As noted in Chapter 3, journals and the process of scientific publication serve the interests of career advancement and provide a vehicle for scholarly achievement. Indeed, the easiest way to understand a scientist's career is to review the publications proudly listed in his or her curriculum vitae. When one looks at the seminal thinkers and scientists in the field, published works constitute the main record of their professional lives. Boxes 17.1–17.2 provide examples of how productive and influential addiction researchers reflect on their research and scientific communications.

Born in Antwerp, Belgium, in 1931, Charles S. Lieber received his medical degree in 1955. Soon thereafter, he moved to the United States and obtained senior research appointments at Harvard Medical School, Cornell Medical College, and Mount Sinai Medical School. His research focused on alcohol abuse and its biological components, including the mechanisms underlying the development of alcoholic cirrhosis of the liver (Edwards, 2002). In the following, he describes one of the discoveries that changed the course of biological research on alcohol:

"There seemed to be an adaptive system which helps us survive in modern society because it is relatively non-specific and detoxifies foreign compounds even when the body has never been exposed to them before. When we observed a similar morphological response after alcohol, I postulated that alcohol may therefore also be a substrate for this system. This hypothesis led to the discovery of the microsomal ethanol oxidizing system (MEOS) as a new pathway of ethanol metabolism" (Edwards, 2002, p. 19).

Box 17.1: Charles S. Lieber, M.d. (1931–2009).

In 1972, Martha Sanchez-Craig took a position as director of a halfway house for homeless alcoholics at the Addiction Research Foundation in Toronto, Canada. Five years later, she became a senior scientist at the Clinical Institute of the Addiction Research Foundation. Here, her research centered on brief interventions for people with alcohol- and other drug-related problems (Edwards, 2002). Despite her extensive publication career, she cautions about the "publish or perish" mentality:

". . . one of the senior people, who was conducting experiments with small numbers of non-human subjects, said 'I don't have much regard for any scientist who doesn't publish at least six papers a year in peer-reviewed journals". I was very worried about that. I met colleagues who would get depressed or seriously worried if they couldn't publish a paper every month. I began to think that there are a lot of people here who like to do science that looks good, and only a few who like to do good science" (Edwards, 2002, p. 124).

Box 17.2: Martha Sanchez-Craig, Ph.d.

Individual Responsibility

Research can be a solitary endeavor, involving late nights spent in your study or laboratory, preparations to defend a thesis or to question someone else's dissertation, and standing alone on a podium to present a scientific article. In many cases, it is impossible, at least without considerable effort, for an outsider to know whether a researcher has conducted his research ethically. All researchers are thus responsible for guarding the integrity of the public trust in research.

But research is also a highly social enterprise, which introduces its own ethical concerns. Much scientific research is now conducted via teams of investigators and support staff that share responsibility for the completion of a project and the publication of a scientific report. In this context, individual responsibility sometimes becomes diluted and ambiguous in relation to ethical matters. The research world is also very hierarchical. Younger researchers are like apprentices being trained by their masters, economically dependent on them for positions and promotions. These differential power relations can further dilute ethical responsibility.

Despite these threats to research integrity, addiction scientists must adhere to the ideal of the *polis* of the ancient Greeks, whereby every free man (we will have to ignore the gender discrimination of the time) was an equal, with similar responsibilities to decide matters of importance and civil rights to support those responsibilities. Similarly, every researcher must accept his or her personal responsibility for creating a more transparent and ethical addiction research community, which includes young investigators and senior researchers alike, as well as editors of journals and peer reviewers. Everyone, for example, has a responsibility to use citations in a fair and informative way (Chapter 10), to ensure the proper assignment of authorship credits (Chapter 11), and to adhere to ethical rules (Chapters 14, 15 and 16). When all researchers view themselves as equals in the republic of science, they will create the best foundation for creative discussions, which in turn will lead to progress in research.

Creating Good Institutions

In many instances, exhortations to individual responsibility are not enough to guarantee scientific integrity. Good institutions must support creative research milieus with sound ethical principles. Informal structures, such as open communication within departments (not only about research but also about ethical problems), the reading and critiquing of each other's work, democratic decision making, and cooperation on multidisciplinary projects all emanate from participatory norms and strong leadership. In Boxes 17.3–17.4, two influential addiction researchers reflect on the social and institutional aspects of their research and scientific communications.

Mustapha I. Soueif was born in 1924. He completed his graduate studies in psychology at University of Cairo, Egypt. In addition to teaching psychology at the University of Cairo, he also worked for the World Health Organization (Edwards (1991). Here, he describes the challenges of publishing in different languages and the conflicts between having a national commitment and an international vision:

"It is a long time now that I have been living with this double identity; on the one hand I feel a world-citizen, on the other I belong to Egypt. This complex 'consciousness' or oscillating began in the late fifties when I was carrying out my first piece of clinical research in Egypt (at Abbassia Psychiatric Hospital) while keeping an eye on getting it published abroad. This was the paper on 'Testing for organicity in Egyptian psychiatric patients'. It was accepted for publication in Acta Psychologica *(in Amsterdam). That was the first step towards establishing my reference group, defined in this case as a group of international scientists who would judge the worth of my research on its objective merits. My international identity, however, was definitely promoted through my contact with the WHO in Geneva. In 1966 I was approached by the WHO people to prepare a paper for publication in the UN* Bulletin on Narcotics *reporting on our work on 'Hashish Consumption in Egypt' which has been under way since 1957. This I did, and the paper was published in 1967. In 1970 I was invited to participate in a 'Scientific group' meeting to be held at WHO headquarters. The recognition my work received there was deeply gratifying"* (Edwards, 1991, p. 436).

Box 17.3: Mustapha Soueif, Ph.d.

Also helpful to scientific integrity are more formal structures, such as policies for the ethical conduct of research (Chapter 15) and procedures for the determination of authorship credits (Chapter 11). In recent years addiction journals have emerged from their relatively obscure and modest origins to take a leadership role in the prevention of scientific misconduct. The ethical principles for authors included in this book represent the consensus of editors who are members of the International Society of Addiction Journal Editors. Integrity in scientific publishing can be enhanced only by education, vigilance, clear policies, and institutional norms that put science first.

Awareness of Global Inequality

Addiction is a global concern, and the concepts of universalism and autonomy suggest that knowledge gained from research should be shared

Kettil Edmund Bruun received his doctoral training in sociology from the University of Helsinki. He is perhaps best known for his influential book, *Alcohol Control Policies in Public Health Perspective,* published in 1975 under the auspices of the World Health Organization. Sometimes called the "purple book" (owing to its cover in the English-language version), the publication gained wide attention for its basic tenet: "*changes in the overall consumption of alcoholic beverages have a bearing on the health of the people in any society. Alcohol control measures can be used to limit consumption: thus, control of alcohol availability becomes a public health issue*" (Bruun et al., 1975, p. 90; see also Edwards, 1991, and Room, 1986). In the following, Bruun describes with characteristic modesty the process that gave rise to the book:

"*The background was that I had to rethink my ideas of alcohol control in the light of the Finnish experience in 1968/69 when controls had been suddenly relaxed with dramatic increase in consumption and harmful effects. My own liberal views on alcohol policies had received a blow. Then I was confronted in the European Office with international issues. I thought that I had to reconsider my position and that probably the best way to do it was to try to have a group which could develop a perspective beyond the specific situation in Finland. The situation was fortunate because many of the relevant questions had by then been focused for research. The group which emerged from my invitation did a marvelous job*" (Edwards, 1991, pp. 371–372).

Box 17.4: Kettil Edmund Bruun, Ph.d. (1924–1985).

throughout the world. Unfortunately, resources for both research and scientific communications are limited in many parts of the world, and research conducted in the more-resourced countries often follows parochial national interests. Moreover, the dominance of English as the *de facto* language of science comes at a price for the majority of the world, in which other languages predominate.

Addiction researchers in the English-speaking and the more-developed countries have a special obligation to conduct and present their research, whenever possible, in a way that benefits the rest of humankind. The peer-review process should be open to scientists from all languages and nationalities, as should the editorial boards of the journals serving as the gatekeepers for scientific truth. Language and culture should not limit publication in addiction science. Not only is this a question of fairness, but it also speaks to the cross-cultural generalizability of scientific findings and the need to discover universal truths.

Conclusion: The Meaning of Scientific Life

In Chapter 1, we referred to the medieval philosopher Maimonides and his *Guide for the Perplexed*. This was perhaps not a very modest analogy. We do not want to suggest that this book—or any book for that matter—can remove all confusion and provide a researcher with the guidance needed to have a successful career in addiction science. Rather, we hope the information in this book will lead its readers to the *agora* of science, a community square or common ground on which open and democratic discussions can take place among equals about the difficult problems all researchers, novices and career professionals alike, encounter in their everyday work.

One of those difficult problems is the meaning of scientific life itself. It is a question perhaps secondary to the broader question of life's meaning in general, but it is nevertheless worth asking if we want to make our own lives meaningful as addiction scientists. Various spiritual, religious, and philosophical traditions from the East, West, North, and South have contributed to this profound line of questioning.

Despite their important insights, biologist Edward O. Wilson (2014) believes philosophy is ill-equipped to tackle the meaning of existence. Wilson concludes that, by default, the task of explaining meaning necessarily falls to science itself. Among the disciplines that he favors in determining meaning are evolutionary biology and neuroscience. To those we would add the behavioral, social, and population sciences, which may help us understand how addiction is the antithesis of harmony with the natural world and how modern civilization seems designed to make that harmony difficult for many to achieve. And we should not defer entirely to science when meaning can surely be derived from religion, literature and other areas of knowledge.

In the most spiritual and reflective period of his life, Leo Tolstoy (1886) wrote a novella called *The Death of Ivan Ilyich*, which tells the story of the last days of a high-court judge in 19th-century Russia. It is at its core a philosophical commentary on the meaning of life as revealed in the interactions one has with family, work colleagues, and people encountered in day-to-day living at all social levels. What are the lessons for us, the living? One lesson is that if our lives are intimately invested in addiction science, this would be a good time to take inventory of what we have accomplished and what remains to be done. Have we avoided meaningless writing projects that lead to publications that nobody reads or values? Have we worked amicably with colleagues, supported their ideas, and given credit where it is due? Have we considered the plight of the alcoholic and the drug addict; the families who lose children to drunk drivers; and the evidence-based policies that could prevent drunk driving, fetal alcohol spectrum disorder, and underage drinking?

Beyond literature, science, and philosophy, perhaps the answer lies elsewhere. At the end of the Monty Python film, aptly called *The Meaning of Life*, the Lady

Presenter addresses the question this way: "Well, it's nothing very special. Try to be nice to people, avoid eating fat, read a good book every now and then, get some walking in, and try and live together in peace and harmony with people of all creeds and nations." Or, as comedian Groucho Marx observed, "If you're not having fun, you're doing something wrong."

Ultimately, the meaning of scientific life is a question you will have to answer yourself. Even if a single answer to the question may elude you, that elusiveness is no great tragedy. More important is the search itself and the insights you gain as you realize that addiction science is a wonderful way to add benefit to society and depth to your own understanding of human nature. And finally, it is a way to have fun.

Please visit the website of the International Society of Addiction Journal Editors (ISAJE) at www.isaje.net to access supplementary materials related to this chapter. Materials include additional reading, exercises, examples, PowerPoint presentations, videos, and e-learning lessons.

References

Babor, T. F. (2002). In their own words: Conversations about the evolution of a specialist field. In G. Edwards (Ed.). *Addiction: Evolution of a specialist field* (pp. 383–389). Oxford, England: Blackwell Science.

Babor, T. F. (1993). Megatrends and dead ends: Alcohol research in global perspective. *Alcohol Health and Research World, 17,* 177–186.

Bruun, K., Edwards, G., & Lumio, M., Mäkelä, K., Pan, L., Popham, R. E., . . ., Österberg, E. (1975). *Alcohol control policies in public health perspective.* Helsinki, Finland: Finnish Foundation for Alcohol Studies, Vol. 25.

Edwards, G. (Ed.). (1991). *Addictions: Personal influences and scientific movements.* New Brunswick, NJ: Transaction Publishers.

Edwards, G. (Ed.). (2002). *Addiction: Evolution of a specialist field.* Oxford, England: Blackwell Science.

LaFollette, M. C. (1992). *Stealing into print: Fraud, plagiarism, and misconduct in scientific publishing.* Berkeley, CA: University of California Press.

Merton, R. (1973). *The sociology of science: Theoretical and empirical investigations.* Chicago, IL: University of Chicago Press.

Niiniluoto, I. (2002). Tieteen Tunnuspiirteet [The characteristics of science]. In S. Karjalainen, V. Launis, R. Pelkonen, & J. Pietarinen (Eds.), *Tutkijan Eettiset Valinnat* [The ethical choices of the researcher], (pp. 30–41). Tampere, Finland: Gaudeamus.

Room, R. (1986). Kettil Bruun, 1924–1985: An appreciation. *The Drinking and Drug Practices Surveyor, 21,* 1, 42–49.

Tolstoy, L. (1886). *The Death of Ivan Ilyich.* Russia.

Wilson, E. O. (2014). *The meaning of human existence.* New York, NY: Liveright Publishing Corporation.

Index

study design   116, 138, 160, 175,
177, 180, 277, 310
study generalizability   184
style   80, 81, 96–97, 114, 127, 138,
142, 148, 153, 182, 231, 242
style guide   136–38, 140
submissions   49, 53, 58, 108–9,
136–37, 169, 208, 213–14,
230–31, 246–52, 255–58,
260–62, 287–88, 310, 316
concurrent   252
first   81, 247
initial   100, 109, 256
rejected   248
submitting   50, 54, 79, 96, 103, 127,
137–38, 145–46, 230, 260,
274, 276, 290
Substance Abuse Librarians &
Information Specialists
22, 70
survey findings   309
survey results   38
survey studies   269, 287
systematic review and meta-
analysis   6, 173, 175, 188, 205
systematic reviews   62, 64, 107,
139–40, 174–78, 181, 184–89,
193, 202–4, 296, 351

T

tenure   54, 91, 255, 325
terminology   10, 49, 110, 174, 177,
183, 196
right   81
thesis   67, 90, 92, 94–96, 98–100,
107–9, 113–14, 359
compilation   89
graduate   177
master's   112, 147
publishing graduate-level   90
thesis advisor   100
thesis authorship   112
thesis defense   100
thesis research   117–18

dissertation or   116
thesis topic   114
Thomson Reuters   52, 60–61, 63, 65,
69–70, 75
titles   24, 38, 57, 67, 69, 142, 144,
147, 149, 152, 164–65, 242,
274, 276
tobacco industry   184, 213, 278, 282,
321, 326–27, 332, 340–41, 351
trainees   90–91, 94, 104, 253, 257
postdoctoral   91
postgraduate   90
training   10, 16, 22, 25, 27, 107, 116,
243, 253, 292, 336, 361
ethical   292, 314
training programs   11, 25, 27, 29, 38
translation   144, 295, 320, 356
treatment research   18, 51, 76
triage   230–31
trials   122, 124, 128, 180, 226, 292,
296, 328, 332
cluster   139
independent   283, 332
industry-supported   283, 332
multicenter   226
multisite   18
nonrandomized controlled   139
pharmaceutical company   215
randomised   189, 242, 244
randomised controlled   123–24,
139, 143, 203

U

Ulrich's Periodicals Directory   54
Undeclared COIs   284
Undeserved authorships   208
unethical research   300–301
unfair authorship practices   280, 292
universalism   356, 360
Unpublished manuscript   34, 298

V

verbatim material   310
verbs, strong   148

CPSIA information can be obtained
at www.ICGtesting.com
Printed in the USA
BVHW052131180719
553750BV00018B/319/P

9 781911 529088